METHODS IN MOLECULAR BIOLOGY

Series Editor
John M. Walker
School of Life and Medical Sciences
University of Hertfordshire
Hatfield, Hertfordshire, AL10 9AB, UK

For further volumes:
http://www.springer.com/series/7651

Clinical Metabolomics

Methods and Protocols

Edited by

Martin Giera

Center for Proteomics and Metabolomics, Leiden University Medical Center
Leiden, The Netherlands

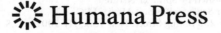

Editor
Martin Giera
Center for Proteomics and Metabolomics
Leiden University Medical Center
Leiden, The Netherlands

ISSN 1064-3745   ISSN 1940-6029   (electronic)
Methods in Molecular Biology
ISBN 978-1-4939-8530-2   ISBN 978-1-4939-7592-1   (eBook)
https://doi.org/10.1007/978-1-4939-7592-1

Cover illustration: Cover art design by Dr. Magnus Palmblad

Printed on acid-free paper

This Humana Press imprint is published by Springer Nature
The registered company is Springer Science+Business Media, LLC
The registered company address is: 233 Spring Street, New York, NY 10013, U.S.A.

Preface

In recent years, metabolomics has become an inevitable tool in several clinical research fields, helping to discover new diagnostic markers and molecules and furthering our understanding of pathophysiological processes. Unlike the field of clinical chemistry which today is integrated into many clinical processes, clinical metabolomics is a much more "juvenile" discipline still on its way to become fully integrated into modern health care. Nevertheless, metabolomics is at the core of several very promising initiatives evolving around personalized health care and precision medicine.

Ideally, clinical metabolomics should be seen as a complimentary discipline to clinical chemistry. The much more hypothesis-driven exploratory nature of clinical metabolomics allows it to fill the pipelines of clinical chemistry with novel disease markers and diagnostic patterns. Besides this, clinical metabolomics is very well suited to help clinicians and biologists understand pathophysiological processes in detail, hopefully allowing us to design novel treatment strategies and therapies. Its multidisciplinary nature covering (analytical) chemistry, biology, bioinformatics, and pathology necessitates that scientists from various fields understand each other. Hence, a common fundament for communication is a mandatory prerequisite for the successful embedding of clinical metabolomics into modern disease-related research. When communicating with colleagues from various disciplines, it is of utmost importance to the planning of joint studies that every partner understands the needs and limitations of one another. In multidisciplinary projects, this particularly applies to the fact that each partner should be aware of practical requirements and limitations of the different methods and technologies used. Therefore, exchanging experimental protocols and making colleagues aware of critical practical considerations is of vital importance for a successful study outcome.

With this book, we hope to present a comprehensive compendium of clinical metabolomics protocols covering LC-MS-, GC-MS-, CE-MS-, and NMR-based clinical metabolomics as well as bioinformatics and study design considerations. We hope that this book will serve as the basis for the successful (practical) communication between scientists from several fields, including chemists, biologists, bioinformaticians, and clinicians, ultimately leading to successful study design and completion.

Leiden, The Netherlands *Martin Giera*

Contents

Part VI   MALDI-Based Techniques and Mass Spectrometry
           Imaging of Clinical Samples

Part VII   Study Design, Data Analysis, and Bioinformatics

Contributors

BENJAMIN BALLUFF • *Maastricht MultiModal Molecular Imaging Institute (M4I), Maastricht University, Maastricht, The Netherlands*

ANNE BARDEN • *Medical School, Royal Perth Hospital Unit, University of Western Australia, Perth, WA, Australia*

SUSEN BECKER • *Institute of Laboratory Medicine, Clinical Chemistry and Molecular Diagnostics, University Hospital Leipzig, Leipzig, Germany; LIFE-Leipzig Research Center for Civilization Diseases, Leipzig University, Leipzig, Germany*

JUSTINE BERTRAND-MICHEL • *MetaToul-Lipidomic MetaboHUB Core Facility, INSERM U1048, Toulouse Cedex 4, France*

JULIEN BOCCARD • *School of Pharmaceutical Sciences, University of Geneva, University of Lausanne, Geneva, Switzerland; Swiss Centre for Applied Human Toxicology (SCAHT), Universities of Basel and Geneva, Basel, Switzerland*

BO BURLA • *Singapore Lipidomics Incubator (SLING), Life Sciences Institute, National University of Singapore, Singapore, Singapore*

UTA CEGLAREK • *Institute of Laboratory Medicine, Clinical Chemistry and Molecular Diagnostics, University Hospital Leipzig, Leipzig, Germany; LIFE-Leipzig Research Center for Civilization Diseases, Leipzig University, Leipzig, Germany*

ROMANAS CHALECKIS • *Division of Physiological Chemistry 2, Department of Medical Biochemistry and Biophysics, Karolinska Institutet, Stockholm, Sweden; Gunma University Initiative for Advanced Research (GIAR), Gunma University, Gunma, Japan*

YOUNG HAE CHOI • *Natural Product Laboratory, Institute of Biology, Leiden University, Leiden, The Netherlands*

ROMAIN A. COLAS • *Lipid Mediator Unit, Biochemical Pharmacology, William Harvey Research Institute, Barts and the London School of Medicine, Queen Mary University of London, London, UK*

JESMOND DALLI • *Lipid Mediator Unit, Biochemical Pharmacology, William Harvey Research Institute, Barts and the London School of Medicine, Queen Mary University of London, London, UK*

PATRICIA T.N. VAN DAM • *Department of Biotechnology, Faculty of Applied Sciences, Delft University of Technology, Delft, The Netherlands*

KO WILLEMS VAN DIJK • *Einthoven Laboratory for Experimental Vascular Medicine, Department of Human Genetics, Leiden University Medical Center (LUMC), Leiden, The Netherlands; Division of Endocrinology, Department of Medicine, Leiden University Medical Center (LUMC), Leiden, The Netherlands*

STÉPHANIE DURAND • *Université Clermont Auvergne, INRA, UNH, Plateforme d'Exploration du Métabolisme, MetaboHUB Clermont, Clermont-Ferrand, France*

LARS F. EGGERS • *Division of Bioanalytical Chemistry, Research Center Borstel, Borstel, Germany*

BART EVERTS • *Department of Parasitology, Leiden University Medical Center (LUMC), Leiden, The Netherlands*

MARTIN GIERA • *Center for Proteomics and Metabolomics, Leiden University Medical Center (LUMC), Leiden, The Netherlands*

BRUCE D. HAMMOCK • *Department of Entomology and Nematology, UC Davis Comprehensive Cancer Center, University of California, Davis, Davis, CA, USA*

WEI HAN • *Department of Chemistry, University of Alberta, Edmonton, AB, Canada*

VANESSA VAN HARMELEN • *Einthoven Laboratory for Experimental Vascular Medicine, Department of Human Genetics, Leiden University Medical Center (LUMC), Leiden, The Netherlands*

RIANNE HAUMANN • *Department of Biobanking, Leiden University Medical Center, Leiden, The Netherlands*

MARIEKE HEIJINK • *Center for Proteomics and Metabolomics, Leiden University Medical Center (LUMC), Leiden, The Netherlands*

PATRICK O. HELMER • *Institute of Inorganic and Analytical Chemistry, University of Münster, Münster, Germany*

MICHAL HOLČAPEK • *Department of Analytical Chemistry, Faculty of Chemical Technology, University of Pardubice, Pardubice, Czech Republic*

LISA R. HOVING • *Einthoven Laboratory for Experimental Vascular Medicine, Department of Human Genetics, Leiden University Medical Center (LUMC), Leiden, The Netherlands*

BORA INCEOGLU • *Department of Entomology and Nematology, UC Davis Comprehensive Cancer Center, University of California, Davis, Davis, CA, USA*

JULIJANA IVANISEVIC • *Metabolomics Platform, Faculty of Biology and Medicine, University of Lausanne, Lausanne, Switzerland*

HYE KYONG KIM • *Natural Product Laboratory, Institute of Biology, Leiden University, Leiden, The Netherlands*

SARANTOS KOSTIDIS • *Center for Proteomics and Metabolomics, Leiden University Medical Center (LUMC), Leiden, The Netherlands*

SABRINA KRAUTBAUER • *Institute of Clinical Chemistry and Laboratory Medicine, University of Regensburg, Regensburg, Germany*

JETTY CHUNG-YUNG LEE • *School of Biological Sciences, The University of Hong Kong, Hong Kong, SAR, China*

YIU YIU LEE • *School of Biological Sciences, The University of Hong Kong, Hong Kong, SAR, China*

LIANG LI • *Department of Chemistry, University of Alberta, Edmonton, AB, Canada*

MIROSLAV LÍSA • *Department of Analytical Chemistry, Faculty of Chemical Technology, University of Pardubice, Pardubice, Czech Republic*

GERHARD LIEBISCH • *Institute of Clinical Chemistry and Laboratory Medicine, University of Regensburg, Regensburg, Germany*

LIAM A. MCDONNELL • *Fondazione Pisana per la Scienza ONLUS, Pisa, Italy*

ISABEL MEISTER • *Division of Physiological Chemistry 2, Department of Medical Biochemistry and Biophysics, Karolinska Institutet, Stockholm, Sweden; Gunma University Initiative for Advanced Research (GIAR), Gunma University, Gunma, Japan*

CAROLE MIGNÉ • *Université Clermont Auvergne, INRA, UNH, Plateforme d'Exploration du Métabolisme, MetaboHUB Clermont, Clermont-Ferrand, France*

TREVOR A. MORI • *Medical School, Royal Perth Hospital Unit, University of Western Australia, Perth, WA, Australia*

CHRISTOPHE MORISSEAU • *Department of Entomology and Nematology, UC Davis Comprehensive Cancer Center, University of California, Davis, Davis, CA, USA*

SNEHA MURALIDHARAN • *Singapore Lipidomics Incubator (SLING), Department of Biological Sciences, National University of Singapore, Singapore, Singapore*

SHAMA NAZ • *Division of Physiological Chemistry 2, Department of Medical Biochemistry and Biophysics, Karolinska Institutet, Stockholm, Sweden*

JOHN W. NEWMAN • *Obesity and Metabolism Research Unit, United States Department of Agriculture (USDA), Agricultural Research Service (ARS), Western Human Nutrition Research Center, University of California, Davis, Davis, CA, USA; Department of Nutrition, University of California, Davis, Davis, CA, USA; NIH West Coast Metabolomics Center, University of California, Davis, Davis, CA, USA*

THERESA L. PEDERSEN • *Advanced Analytics, Woodland, CA, USA*

ANGELA TEN PIERICK • *BioAnalytical Chemistry, Department of Chemistry and Pharmaceutical Sciences, Vrije Universiteit Amsterdam, Amsterdam, The Netherlands*

ESTELLE PUJOS-GUILLOT • *Université Clermont Auvergne, INRA, UNH, Plateforme d'Exploration du Métabolisme, MetaboHUB Clermont, Clermont-Ferrand, France*

RAWI RAMAUTAR • *Division of Systems Biomedicine and Pharmacology, Leiden Academic Center for Drug Research, Leiden University, Leiden, The Netherlands*

AMY A. RAND • *Department of Entomology and Nematology, UC Davis Comprehensive Cancer Center, University of California, Davis, Davis, CA, USA*

MADLEN REINICKE • *Institute of Laboratory Medicine, Clinical Chemistry and Molecular Diagnostics, University Hospital Leipzig, Leipzig, Germany; LIFE-Leipzig Research Center for Civilization Diseases, Leipzig University, Leipzig, Germany*

FABIEN RIOLS • *MetaToul-Lipidomic MetaboHUB Core Facility, INSERM U1048, Toulouse Cedex 4, France*

KARINA TREVISAN RODRIGUES • *Institute of Chemistry, University of Sao Paulo (USP), Sao Paulo, SP, Brazil; Department of Pharmaceutical and Pharmacological Sciences, Pharmaceutical Analysis, KU Leuven-University of Leuven, Leuven, Belgium*

HUUB H. VAN ROSSUM • *Laboratory of Clinical Chemistry and Hematology, The Netherlands Cancer Institute, Amsterdam, The Netherlands*

SERGE RUDAZ • *School of Pharmaceutical Sciences, University of Geneva, University of Lausanne, Geneva, Switzerland; Swiss Centre for Applied Human Toxicology (SCAHT), Universities of Basel and Geneva, Basel, Switzerland*

ANN VAN SCHEPDAEL • *Department of Pharmaceutical and Pharmacological Sciences, Pharmaceutical Analysis, KU Leuven-University of Leuven, Leuven, Belgium*

DOMINIK SCHWUDKE • *Division of Bioanalytical Chemistry, Research Center Borstel, Borstel, Germany*

REZA MALEKI SEIFAR • *Department of Biotechnology, Faculty of Applied Sciences, Delft University of Technology, Delft, The Netherlands; DSM Food Specialties B.V., Delft, The Netherlands*

CHARLES N. SERHAN • *Department of Anesthesiology, Perioperative and Pain Medicine, Center for Experimental Therapeutics and Reperfusion Injury, Brigham and Women's Hospital, Harvard Medical School, Boston, MA, USA*

MARINA FRANCO MAGGI TAVARES • *Institute of Chemistry, University of Sao Paulo (USP), Sao Paulo, SP, Brazil*

OLAF VAN TELLINGEN • *Laboratory of Clinical Chemistry and Hematology, The Netherlands Cancer Institute, Amsterdam, The Netherlands*

AURELIEN THOMAS • *Unit of Toxicology, CURML, CHUV Lausanne University Hospital, HUG Geneva University Hospitals, Lausanne, Switzerland; Faculty of Biology and Medicine, University of Lausanne, Lausanne, Switzerland*

FEDERICO TORTA • *Singapore Lipidomics Incubator (SLING), Department of Biochemistry, YLL School of Medicine, National University of Singapore, Singapore, Singapore*

BALJIT K. UBHI • *SCIEX, Redwood City, CA, USA*

HEIN W. VERSPAGET • *Department of Biobanking, Leiden University Medical Center, Leiden, The Netherlands; Parelsnoer Institute, Utrecht, The Netherlands*

MARY E. WALKER • *Lipid Mediator Unit, Biochemical Pharmacology, William Harvey Research Institute, Barts and the London School of Medicine, Queen Mary University of London, London, UK*

MARKUS R. WENK • *Singapore Lipidomics Incubator (SLING), Department of Biochemistry, YLL School of Medicine, National University of Singapore, Singapore, Singapore*

CRAIG E. WHEELOCK • *Division of Physiological Chemistry 2, Department of Medical Biochemistry and Biophysics, Karolinska Institutet, Stockholm, Sweden; Gunma University Initiative for Advanced Research (GIAR), Gunma University, Gunma, Japan*

LENNART J. VAN WINDEN • *Laboratory of Clinical Chemistry and Hematology, The Netherlands Cancer Institute, Amsterdam, The Netherlands*

Part I

Clinical Metabolomics and Lipidomics

Chapter 1

Metabolomics as a Tool to Understand Pathophysiological Processes

Julijana Ivanisevic and Aurelien Thomas

Abstract

Multiple diseases have a strong metabolic component, and metabolomics as a powerful phenotyping technology, in combination with orthogonal biological and clinical approaches, will undoubtedly play a determinant role in accelerating the understanding of mechanisms that underlie these complex diseases determined by a set of genetic, lifestyle, and environmental exposure factors. Here, we provide several examples of valuable findings from metabolomics-led studies in diabetes and obesity metabolism, neurodegenerative disorders, and cancer metabolism and offer a longer term vision toward personalized approach to medicine, from population-based studies to pharmacometabolomics.

Key words Metabolomics, Obesity metabolism, Diabetes metabolism, Neurodegenerative diseases, Cancer metabolism, Personalized medicine, Pharmacometabolomics, Population studies

1 Metabolome as a Complement to Other 'Omes

Metabolomics is making a major impact in a wide variety of scientific areas such as functional genomics, fundamental biochemistry, toxicology, environmental sciences, food safety, drug dosage and discovery, new-born screening, disease diagnostics, and biomedical and clinical research in general [1, 2].

To get a complete picture about the potential of metabolomics and the advantages of studying the metabolome in this post-genomic era of biology, we should first take a look at the central dogma of molecular biology (Fig. 1). Using high-throughput 'omics technologies, we can measure the biological variability at four main biochemical or functional levels of a system's organization, each one providing complementary information about the phenotype. The **genome** or complete information encoded in genes, represented by both the coding and noncoding sequences of DNA, provides us with data on how an organism should be built and maintained or *what may happen*. Compared to other 'omes, the genome is relatively static although subject to regulatory processes

Martin Giera (ed.), *Clinical Metabolomics: Methods and Protocols*, Methods in Molecular Biology, vol. 1730,
https://doi.org/10.1007/978-1-4939-7592-1_1, © Springer Science+Business Media, LLC 2018

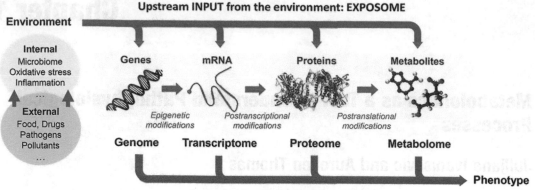

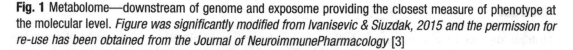

Fig. 1 Metabolome—downstream of genome and exposome providing the closest measure of phenotype at the molecular level. *Figure was significantly modified from Ivanisevic & Siuzdak, 2015 and the permission for re-use has been obtained from the Journal of NeuroimmunePharmacology* [3]

such as epigenetic modifications that influence the gene expression and transcription by switching it "on" or "off." The **transcriptome** or the full qualitative set of messenger RNA molecules transcribed from the "expressed" part of the genome and **proteome** or the complete set of proteins translated from messenger RNA are more dynamic, thus providing the information on *what is happening* and *what makes it happen*, respectively. Both are subject to posttranscriptional and posttranslational modifications, respectively. As the final piece of the 'omic puzzle, the **metabolome** is highly dynamic providing the snapshots of information about *what has happened* [4]. More explicitly in the biological context, while the information in genes is inherited and genotypes describe the potential of the system or *what may happen*, metabolites tell us *what has really happened* and describe the current functional (physiological or developmental) status of the system. Metabolites are the end products of genome, transcriptome, and proteome activity, thus representing the direct signatures of cellular activity and the reason why the metabolome is addressed as the **closest link to phenotype**, providing the direct and sensitive measure of the dynamic changes at the molecular level [5, 6]. In addition to mediating biochemical processes as substrates and products of energy conversion reactions and signaling molecules (or secondary messengers), metabolites also have far-reaching biochemical actions in the regulation of epigenetic modifications and protein activity [7, 8].

A further important advantage of studying the metabolome lies in the fact that in addition of being the downstream output of the genome, the metabolome also directly reflects the upstream input from the environment, allowing us to explore the interaction between the genome and the environment [1]. Metabolite levels are significantly influenced by synergistic effects of many different environmental risk factors, including the internal microbiome, oxidative stress and inflammation, and the external climate, pollutants, pathogens, food, drugs, and social interactions, we are exposed to. These exposures we are subjected to throughout our life also define our phenotype [9]. As the end product of the genome under the direct influence of the environmental stressors, the metabolome allows us to measure the integrated effect of both, providing the assessment of the **exposome** (Fig. 1). Several metabolomics-led studies have recently revealed the highly relevant influence that the environment, and more particularly our microbiome (gastrointestinal, nasal, oral, urogenital, skin population, etc.), has on disease development, in cardiovascular disease, obesity, cancer, etc., [10, 11]. It is also assumed that the microbiome's metabolic activity actively influences our mind [12]!

Overall, the discovery of a dominant metabolic component in many diseases (rather than genetic as widely accepted), such as in cancer, atherosclerosis, diabetes, and neurodegenerative disorders, has stimulated the rethinking toward a systems biology approach [1]. By integration of metabolomics into systems biology, we will go beyond the biomarker discovery and toward mechanistic understanding of pathophysiological processes [2, 13]. The advancements in technology, computing power, and back-end bioinformatics solutions [14–17] made it possible not only to rapidly measure thousands of metabolites from only minimal amounts of biological sample (biofluids, cells, and tissues) but also to go from metabolites to pathways (Fig. 2) and to link these altered identified pathways back to the phenotype in combination with orthogonal molecular approaches (interfering with gene expression, enzyme activity, cell signaling, etc.) and technologies (epigenomics, transcriptomics, proteomics) [2].

As described in several case studies in the following section, disease mechanisms can be better understood, and significantly more variance explained, when viewed from the metabolite and not only gene perspective [1]. In these lines, metabolomics is also bringing new insights into medical approaches, from a reactive, one-size-fits-all model of care toward a predictive, preventive, and patient-centric model of care. As a powerful phenotyping technology, metabolomics will play a key role in further development of this **personalized approach to medicine** [19]. The advantage of metabolomics is high throughput and scalability, thus offering the possibility to acquire specific metabolic profiles for each individual in the emerging population-based metabolomics studies [20]. The great potential of population-based studies is the evaluation of the effects of environmental exposures (the "exposome"), including pharmaceutical and nutritional, on disease onset, diagnosis, and progression.

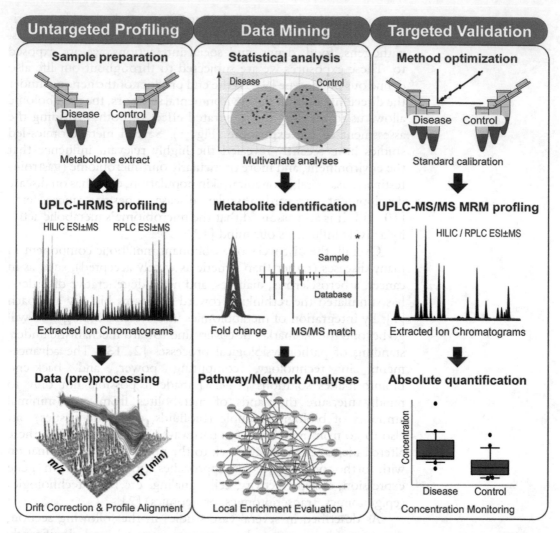

Fig. 2 Discovery-based clinical metabolomic workflow, combining untargeted and targeted metabolomic approach. From comprehensive untargeted profiling, through data processing, statistical analyses, metabolite identification, and pathway analyses, to targeted tandem MS quantification. *Figure was significantly modified from Ivanisevic et al. Scientific Reports 2015 and the permission for re-use is specified under Creative Commons License* [18]

2   Metabolomic Technology and Technical Approaches

The human metabolome is the final level of cellular regulatory processes, constituted of more than 30,000 low-molecular-weight molecules (<1500 Da) with a great chemical diversity and a wide concentration range [21]. Due to its complexity and size, comprehensive coverage of the metabolome is challenging but necessary when studying phenotypes of complex biological systems [22]. Integration of metabolomic strategies in the bioanalytical research pipeline has created great opportunities to bring new biological and clinical insights, out of which some will be developed in this chapter [23, 24].

Large-scale coverage of the metabolome requires the combination of multiple orthogonal and complementary analytical platforms, non-exhaustively including liquid chromatography-mass spectrometry (LC-MS), nuclear magnetic resonance (NMR), gas chromatography-MS (GC-MS), direct injection MS (DIMS), and capillary electrophoresis-MS (CE-MS) [25, 26]. Several metabolomic workflows involving various combinations of these platforms have been successfully demonstrated [27, 28]. For instance, GC- and LC-MS analyses based on full-scan acquisition mode are highly complementary, as each gives access to metabolites with different physicochemical properties and usually a limited number of detected species by both platforms [29]. GC provides efficient separation of relatively polar metabolites with a molecular weight range below 350 Da, primarily including amino and organic acids, fatty acids, and carbohydrates, whereas LC-MS non-exhaustively allows the detection of higher-molecular-weight compounds [26].

Despite this analytical diversity, emergence of ultrahigh pressure LC (UHPLC) hyphenated to electrospray ionization source high-resolution MS (ESI-HRMS) has deeply extended our capacity of deciphering the metabolome [30]. The recent advancements and innovations of HRMS in terms of sensitivity, resolution, and throughput have driven the development of untargeted metabolomics by overcoming technical limitations associated with the complexity of this field [24]. The latest generation of instrumentation, notably the Fourier transform MS, routinely reaches mass resolution above 200,000 (m/Δm, FWHM) at 1 ppm of mass accuracy [31]. Associated with continual improvement of scan rate frequencies and dynamic range, these platforms provide unprecedented insights into the ultimate response of biological systems to genetic or environmental changes [1, 32].

In combination to the high spectral capacities of HRMS, the use of orthogonal UHPLC separation modes that involve different physicochemical interactions is suitable since the chromatographic peak capacity (i.e., the number of peaks that can be theoretically separated with a chromatographic resolution of 1) increases drastically for complex biological extracts [22, 33]. In this way, HILIC provides a good alternative to RP separation for polar and ionic compounds and is attractive because both techniques use the same type of eluent: a mixture of organic solvent (typically acetonitrile) and water with a volatile buffer [34, 35]. Combination of both modes enables the observation of metabolites associated to lipid and central carbon metabolism from a single biological extract with the highest number of unique detected features by RP and HILIC in ESI positive and negative modes, respectively (Fig. 2) [36].

In order to better understand pathophysiological processes, a main advantage of applying untargeted metabolomics is the amount of data that can be generated in a single run. This advantage also represents one of the main challenges in the field considering our capability to maximize the computational mining of this *big data*. Gigabytes of data per run can be easily generated raising the question of the amount of information which can indeed be currently explored. To overcome this issue, different aspects have to be considered and optimized ranging from data pre- and post-processing, statistical analysis to metabolite assignment, database curation, and biological interpretation. From a bioinformatic point of view, computational resources have greatly been improved in the last few years helping the scientific community with this fastidious task. Notably, different Web-based software, such as XCMS Online [37] and MetaboAnalyst [38], have reached a state of maturity for accelerating and automating the computational workflow, providing user-friendly tools for both novice and expert bioinformaticians [2].

Unlike other 'omics approaches, annotation, identification, and validation of metabolite features is still considered as a bottleneck due to the wide physicochemical diversity of the metabolome. Basically, metabolites cannot be "sequenced" per se leading to challenging in silico prediction of a specific chemical structure. However, UHPLC-HRMS provides high mass and MS/MS spectral accuracies, which represent a key step for generating putative structure attribution to a given peak of interest when used with an adapted heuristic filtering procedure [39]. This information can be even complemented by the use of authentic standard when available in order to obtain the reference MS/MS spectrum and chromatographic retention time specific to each dedicated instrumentation and method [2]. Despite the current progress observed in the database curation like for HMDB [40] and METLIN [16], only a very small subset (roughly below 10%) of all the metabolite features detected in an untargeted approach can be undoubtedly identified. However, this matching score is perpetually improved with the experimental characterization of the metabolome growing number of records in databases and mass spectral libraries.

Facing these metabolite assignment issues in untargeted experiments, a paradigm shift has been observed toward the emergence of large-scale targeted metabolomics. Because of its high sensitivity, selectivity, robustness, and throughput, triple quadrupole instruments (QqQ) working in the multiple reaction monitoring (MRM) mode provide an excellent approach in metabolomic profiling allowing simultaneous targeted analysis of hundreds of metabolites involved in multiple major metabolic pathways (Fig. 2) [24, 41]. Considerable efforts have been made to increase the number of measured metabolites, making this approach more popular within the clinical community [42]. The possibility of monitoring a broad range of identified metabolites is speeding up the use and application of metabolomics in clinical studies, bringing new insights into personalized approaches to medicine.

In addition to the conventional metabolomic strategies discussed in this section, imaging MS is one of the most recent and innovative approaches for direct analysis of tissues or individual cells. Indeed, imaging MS will allow the mapping of hundreds of metabolites in tissue sections while maintaining a high correlation between the histological features and the obtained molecular images [43–45]. Enriched by a decade of development, the technology has reached a stage of maturity bringing promising opportunities as a tool to understand pathophysiological processes [28]. Many clinical applications of the technology have been demonstrated, notably in cancer research, cardiovascular diseases, and neurosciences, and some of them will be discussed in more details further on in this chapter [46–49]. In the MALDI Biotyper® era with the successful implementation in clinical microbiology laboratories, imaging MS may find concrete applications in clinical pathology by fostering the discovery of molecular signatures associated to pathophysiological conditions.

3   Novel Biological Insights into Disease Pathology Led by Metabolomics

Multiple diseases, including diabetes, obesity, cancer, cardiovascular disease, and neurodegenerative disorders, have a dominant metabolic component or even cause in terms of metabolic switch or reprogramming that likely occurs prior to disease onset. Applying metabolomics, as a powerful phenotyping technology, in combination with orthogonal biological and clinical approaches, could significantly accelerate the understanding of mechanisms that underlie these complex diseases determined by a set of genetic, lifestyle, and environmental risk factors [50, 51]. Here, we provide several examples of valuable findings from metabolomics-led studies in diabetes and obesity metabolism, neurodegenerative disorders, and cancer metabolism.

3.1   Obesity Metabolism

Incidence of obesity has doubled since 1980, making it a major public health concern worldwide. In 2014, more than 1.9 billion adults were overweight and 600 million were obese, numbers that are expected to increase in the future [52]. Obesity is a complex chronic disorder with a multifactorial etiology, involving increased intake of high caloric food and a decrease of physical activity that arise from a combination of genetic and environmental factors. Recently, genome-wide association studies have identified common single-nucleotide polymorphisms (SNPs) associated with obesity [53, 54]. While the catalogue of susceptibility loci in obesity keeps growing, only few specific genes may currently support the biological processes of the identified genetic association. For the vast majority of loci, the causal relationship between genes and pathways remains unknown [54]. Metabolomics has emerged as a promising approach to better characterize molecular mechanisms involved in pathophysiology of obesity [55, 56]. Studies in both human and animal models have revealed strong associations between obesity and different metabolic species, including amino acids and acylcarnitines [57, 58]. For instance, Newgard et al. found that the dysregulation of branched-chain amino acid (BCAA) metabolism in humans contributes to the development of obesity-related insulin resistance, ultimately leading to type 2 diabetes mellitus (T2DM) [59]. They validated their finding by showing that rats fed on a high-fat diet supplemented by BCAA developed insulin resistance despite having reduced food intake compared to the high-fat diet group [59].

In the last decade, the potential role of adipose tissue (AT) in the onset of obesity and associated metabolic disorders has gained attention [60, 61]. Initially viewed as a simple storage tissue for accumulation of xenobiotics, AT is becoming more recognized as an active metabolic and endocrine organ, essential for regulation of appetite, energy metabolism, and inflammatory response [60, 62, 63]. However, the response of different AT depots to diet-induced obesity is variable and increasingly studied. While visceral AT is more prone to inflammation compared to subcutaneous AT, less is known about the molecular-signature-induced phenotype that differs between these two AT depots. Metabolomics studies have demonstrated significant difference in the molecular compositions between AT depots, opening the route for further research to decipher the pathophysiological phenotype of visceral AT [55, 64–66].

In addition, metabolomics has shown promising results in clinical research to assess and predict the response to weight loss intervention [57]. For instance, alterations in saturated fatty acid, monounsaturated fatty acid, and BCAA levels were observed in serum samples of overweight and obese elder subjects demonstrating a reduction of weight of 7% after 8 weeks of energy-restricted diets [67]. In another study, a metabolomic signature of BCAAs and biologically associated metabolites predicted improvement in insulin resistance, thus helping with the identification of patients who will most likely benefit from a moderate weight lost [68]. For severely obese patients, the Roux-en-Y gastric bypass (RYGB) is considered as the most common and efficient bariatric surgery to reduce body weight but also to solve comorbid conditions, such as remission of T2DM [69, 70]. However, one third of postoperative patients will still remain "nonrespondent" without significant weight loss following the surgery [71, 72]. Due to this, bariatric surgery is another field of interest in which metabolomics could bring new opportunities to identify predictive biomarkers of "nonrespondent" individuals and thus help physicians in the selection of candidate patients prior to surgery [57, 73, 74].

3.2   Diabetes Metabolism

Type 2 diabetes mellitus (T2DM) is a major public health concern worldwide. According to estimations roughly half of 400 million individuals suffering of T2DM are undiagnosed [75, 76]. Since the disease remains asymptomatic for years, its clinical assessment is difficult. Due to this, patients are often identified when complications, such as chronic kidney diseases and cardiovascular diseases, among others, are already present. In order to prevent or postpone the onset of T2DM, there is a need to develop predictive biomarkers to identify the individuals at risk for developing the disease. Currently approved clinical and biological markers were not sufficient to efficiently address the complexity of T2DM pathogenesis [77, 78]. The addition of different panels of T2DM-associated genetic markers allowed for a limited improvement in the predictive capacity of usual clinical scores [79–81].

Looking at the complexity of T2DM pathophysiology involving numerous organs such as the pancreas, liver, blood vessels, kidneys, and eyes, this multifaceted disease requires the development and application of systems medicine approach including metabolomics to increase translational potential toward clinical applications [57, 82]. Metabolomics studies have been applied to population-based cohorts to identify individuals at risk to develop diabetes [83]. These studies emphasized the role of amino acid and lipid metabolism in the underlying mechanism of the disease [57]. For instance, Wang-Sattler et al. quantified 140 metabolites in 4297 serum samples in the population-based cooperative health research of Augsburg (KORA) [84]. Their study revealed glycine and lysophosphatidylcholine (18:2) as strong predictors of glucose tolerance, even 7 years before disease onset. In another study, a panel of 196 metabolites enabled the identification of a metabolic signature associated with the early manifestations of T2DM, characterized by the significant increase of glyoxylate [85]. In a case-control study, Wang et al. performed a metabolomic profiling of 180 metabolites in 189 individuals who developed new-onset diabetes during a 12-year follow-up period and 189 propensity-matched control subjects from the same baseline examination [86]. They identified a metabolic signature marked by five BCAA (leucine, isoleucine, and valine) and aromatic amino acids (phenylalanine and tyrosine) significantly correlated with diabetes onset. By integrating three of these metabolites (isoleucine, tyrosine, and phenylalanine), individuals in the top quartile exhibited fivefold higher risk to become diabetes converters, thus demonstrating the predictive value of this signature.

These studies are in agreement with the potential key role of BCAA metabolism in adipose tissue accompanied with the increase of circulating BCAA observed in obese and prediabetic patients [87, 88]. This is emphasized by the high correlation between insulin sensitivity and expression of BCAA metabolism genes observed in adipose tissue of diabetic patients [57, 89]. These examples demonstrate the great potential of metabolomics in the study of metabolic diseases not only for determining new clinical biomarkers but also to decipher potential biological pathways associated with the onset and development of such diseases. However, most of the population-based studies discussed in this section have been carried out by targeted metabolomics focusing largely on amino acid and lipid metabolism. Considering the methodological advances in the field of metabolomics, we foresee that the application of untargeted strategies could potentially bring additional clues about the complexity of T2DM by expanding the metabolome coverage.

3.3 Aging and Age-Related Neurodegenerative Diseases

In this last decade, the great emphasis has been placed on the understanding of biological mechanisms of aging representing the greatest risk factor for the onset of multiple age-related declines in organ function and thus the whole spectra of diseases (including cancer, cardiovascular disease, neurological disorders, diabetes, etc.). There is an obvious added value in targeting the processes of aging itself to improve healthy longevity by delaying the onset and progression of many different diseases [90]. It is assumed that diabetes and obesity as consequences of excessive energy intake accelerate brain aging and increase the risk for neurodegenerative diseases (NDDs) and stroke. In today's aging society, the NDDs, including Alzheimer disease (AD), Parkinson disease (PD), and amyotrophic lateral sclerosis (ALS), present one of the major health care concerns [91]. According to the National Institute of Aging in the USA, one in four individuals over 60 years of age are affected by AD as the most common cause of dementia. Although clinical studies have already revealed a great deal of information about NDDs, the changes at the molecular level that cause these diseases are poorly understood, and it is still unknown why some diseases, like AD, are tenfold more frequent than the others (PD or ALS) and why cognitive decline occurs more rapidly in some individuals than in others [91]. Currently approved therapeutic strategies that target the abnormal metabolism of amyloid and tau protein in AD (followed by extracellular deposition of amyloid oligomer plaques and intracellular formation of neurofibrillary tangles, respectively) have so far been unsuccessful treating the symptoms or consequences of disease progression rather than the causes of disease itself. In this surge to understand the cause of NDDs, few recent studies suggest the importance of exploring energy metabolism and the role it plays in brain function [91–94]. It is well known that the human brain is metabolically expensive using up to 20% of the total body energy to sustain its function [95]. This is supported by the evolutionary findings of massive increase of genes involved in energy production in human cortex compared to nonhuman primates. Elevated expression of genes involved in cellular energy metabolism and neuronal function, as highest energy-demanding cells, has led to specific metabolic profiles, highly sensitive to changes in energy homeostasis and oxidative stress [91]. Energy homeostasis is maintained by small-molecule metabolites (among other *actors* at the molecular level) that mediate biochemical reactions in the interconnected central carbon and signaling pathways. Neurodegenerative disorders are thought to be associated with and perhaps even caused by alterations in central carbon energy metabolism and derived oxidative stress [93, 96, 97]. However, these metabolic alterations, beyond the reduced glucose metabolism that was demonstrated in functional neuroimaging studies, remain largely unexplored. In this context, the utilization of metabolomics to comprehensively investigate metabolic alterations associated with the pathophysiological process of NDDs is achieving considerable interest [94]. Multiple metabolomics studies have explored the metabolic traits of AD, PD, and ALS in several transgenic animal models and small human cohorts, using different metabolomic platforms, from NMR and GC-MS to CE-MS, DIMS, and LC-MS. Various sample types, from biofluids such as plasma, serum, urine, saliva, and CSF to brain tissue and several peripheral organs, have been analyzed at different stages of disease severity [93, 98–102]. Although mainly putative, the output of these studies implies the cellular traits of accelerated aging related to the intrinsic neuronal vulnerability to oxidative stress. Alterations at the metabolite and pathway level highlight the perturbed energy production in mitochondria accompanied by excessive reactive oxygen species (ROS) formation leading to oxidative damage to DNA, electron transport chain proteins, and plasma membrane redox system (due to the loss of function and structural integrity of amino acids, nucleic acids, and lipids as a consequence of oxidation) [50, 51]. These changes are thought to trigger the production and appearance of disease-specific factors, amyloid-β and p-tau proteins in AD, α-synuclein in PD, huntingtin in Huntington's disease, and Cu/Zn-superoxide dismutase (SOD) in ALS, further inducing a neurodegenerative cascade of perturbed nutrient transport, neurotransmission, and cell-to-cell communication across the brain, leading to synapse dysfunction and loss [51, 103, 104]. Beyond these findings related to free radical theory of aging, many new additional facets of mitochondrial dysfunction associated with aging and NDDs are constantly being revealed [51, 92, 105, 106]. These include a marked deregulation of mitochondrial aspartate metabolism [96] as well as stage- and region-dependent deregulation of purine metabolism in postmortem frontal cortex tissue of confirmed AD-affected brain [102]. Metabolic profiling of biofluids from AD-confirmed patients has revealed the affected metabolism of specific amino acids with implications on carnitine synthesis and fatty-acid oxidation in mitochondria and the potential influence on the neurotransmitter synthesis [93]. Two interesting studies, although based on small number of samples

(<20 subjects in each group), have yielded compelling AD prediction models applying the chemometrics on the acquired CE-MS and LC-MS profiles of CSF and postmortem brain tissue [107, 108]. The latest comparative analysis across multiple cohorts within the Alzheimer Disease Metabolomics Consortium has shown the alterations in sphingomyelins and ether-containing phosphatidylcholines in preclinical biomarker-defined AD stages, whereas the later symptomatic stages were characterized by altered levels of acylcarnitines and several amines [94]. Overall, there is a need to validate the acquired data in the independent cohorts of clinically well-characterized patients at different stages of disease and correlate the data with clinical manifestations.

By comprehensive evaluation of changes embedded in metabolic networks, metabolomics has the potential to advance the understanding of the mechanistic association between the energy intake and brain function in relation to the aging process, oxidative stress, and neurodegeneration. This will be essential for unraveling new pathways and small-molecule targets for diagnostic and therapeutic purposes to prevent the pathogenesis before irreparable cell death. Beyond the identification of small-molecule markers to diagnose the disease, the strength of the metabolomic approaches will be in the large-scale cohort studies of clinically well-characterized patients and healthy aged controls (without cognitive impairment). These population studies will allow us to correlate the revealed metabolic alterations with differentially expressed proteins and genes and clinical meta-data (including the patient's gender, body mass index, blood-brain barrier permeability, structural and functional neuroimaging results, assessment of cognitive decline, etc.) to provide the reliable clues about the mechanism of disease onset and progression for further investigation using in vivo modeling for design of efficient therapeutic strategies to slow down or prevent the disease.

3.4 Mechanistic Discoveries in Cancer Metabolism Research

Despite population-wide efforts in whole genome sequencing, transcript profiling, and single-nucleotide polymorphism (SNP) characterization, there is a lack of targetable disease genes, and it has become evident that many diseases can be better understood when viewed from the metabolic in addition to the genetic perspective [1, 109]. Cancer is a great example of a fairly complex disease whose understanding has evolved significantly since its metabolic component has been under tenacious investigation. Today, the metabolic reprogramming is considered as the core hallmark of cancer [110]. The first report of deregulated metabolism in cancer dates back to the early 1930s when high glucose consumption and lactate production, even in the presence of oxygen, were discovered by Otto Warburg [111]. After decades of intensive genetic exploration, the recent advancements in metabolomic technology have contributed to rethinking of cancer as a metabolic disorder via (re) discovery of many additional *oncometabolites*. Instead of challenging exploration of millions of possible mutations, recent metabolomics-assisted studies imply that the majority of the cancer-associated genes code for well-known metabolic enzymes and thus impact only several pathways, such as aerobic glycolysis, glutaminolysis, and one-carbon metabolism [112]. These *oncopathways* serve in the production of a number of *oncometabolites* whose upregulation in tumor tissue is essential for rapid tumor growth and propagation. Most oncometabolites play a key role in tumorigenesis by altering signaling pathways and cell division processes. For example, the accumulation of organic acids, succinate, fumarate, and 2-hydroxyglutarate leads to activation of cellular hypoxia pathways in the presence of normal oxygen concentrations (termed as *pseudo-hypoxia*), as well as far-reaching DNA methylation and histone modifications [113]. Latter epigenetic changes affect gene expression on a large scale, thus promoting the cancer development. Reliance or even "addiction" of different types of cancer cells on several nutrient oncometabolites is investigated for selective targeting of cancer proliferation [112]. These metabolic vulnerabilities include the dependence of most of the cancer types on glucose, lactate, serine, and polyamines and more specifically the reliance of Myc-activated tumors on glutamine, breast cancer on glycine synthesis, acute lymphoblastic leukemia on asparagine, and brain and prostate cancer on choline. The discovery of these metabolic vulnerabilities in cancer was made possible primarily due to the advancements in HRMS-based metabolomics, MS-based imaging, and magnetic resonance imaging (MRI) that allow us to detect and measure a broad range of small-molecule metabolites in tumor tissues. As a follow-up to comprehensive untargeted and targeted approaches that highlight the potentially affected pathways, the stable isotope-assisted metabolomic profiling and dynamic flux analyses are gaining significant interest in cancer research to provide further insights into direct pathway activity by tracing the nutrient utilization in cancer. The most recent study published in *Nature Neuroscience* demonstrates the application of nitrogen-labeling-assisted tracing experiment to decipher how brain tumor-initiating cells (BTICs) in glioma activate de novo purine synthesis to maintain self-renewal, proliferation, and tumor-forming capacity [114]. The cancer and immunology field is progressing at the moment due to application of isotopic profiling. Many efforts are invested toward understanding how to regulate the nutrient uptake and utilization in T cells that have an essential role in immunotherapy against cancer [115]. Overall, the combined usage of different metabolomic and genomic approaches is paving the way for manipulation of cellular metabolism and design of efficient anticancer therapies by targeting the well-known metabolic enzymes.

4 Toward Personalized Approach to Medicine: From Population-Based Studies to Pharmacometabolomics

The understanding of molecular mechanisms associated with the interindividual variability in response to drug treatment, including the acute or chronic toxicities, and the discovery of associated biomarkers represent key elements in the development of precision or personalized medicine. Combination of "omics" technologies, to analyze gene polymorphisms, epigenetic modifications, gene and protein expressions, and metabolite concentrations and interactions, is required to identify key players at the molecular level with the greatest detectable effect in a pathophysiological state. The increased reproducibility and reduced cost of "omics" approaches, together with development of computational tools, have allowed us to go toward multi-scale, network-embedded integration of different types of molecular data. The ultimate goal of multi-scale data integration is to link the changes at the metabolite level within affected biochemical pathways to the responsible enzymes and the underlying genetic alterations (Fig. 3).

Application of multi-omic approach in longitudinal population-based studies is one of the most promising strategies to better characterize and understand complex pathophysiological state at the system's level. Although genocentric in their beginnings, the population-based initiatives have recognized the value of comprehensive and high-throughput nature of metabolomics, as a powerful phenotyping technology that allows the cost-effective measurement of a massive panel of metabolites in diverse biological matrices from large population cohorts [1, 117]. In addition, the data generated by metabolomics is remarkably consistent with the data provided by genomic analysis. Integration of genome-wide associations studies (GWAS) and metabolomic profiling has thus revealed unprecedented insights into how genetic polymorphisms influence human metabolome, providing strong association between metabolic traits and loci of biomedical and pharmaceutical interest [116, 118]. For example, combining GWAS and metabolomics, Shin et al. reported a comprehensive exploration of genetic loci influencing human metabolism based on 7824 participants from two European population cohorts (Fig. 3) [116]. Their study provided information on hundreds of single genotype-metabolite associations, greatly expanding the assignment of gene functions and disease-associated gene-metabolite links. Among 37 genetic loci associated with blood metabolite levels, Suhre et al. identified a strong association between the rs662138 SNP and isobutyrylcarnitine [118]. Among genetically determined metabotypes (GDMs), they reported the polymorphism in the SLC22A1 loci associated with the pharmacokinetics of the antidiabetic drug metformin that can be used as a typical example of how

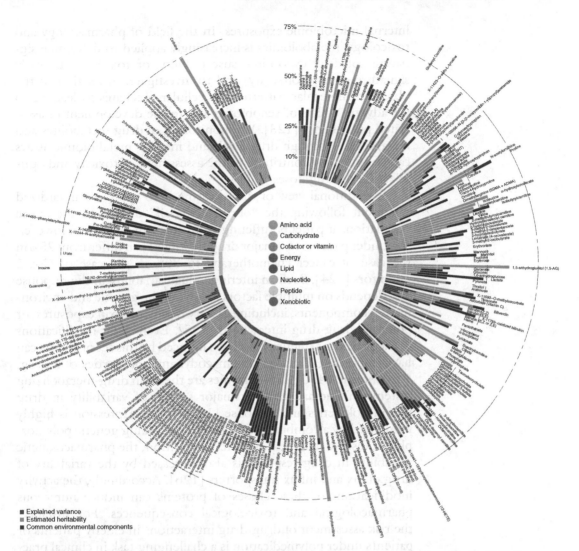

Fig. 3 Map of associations between the single-nucleotide polymorphism (SNP) and metabolite levels in blood. The contribution of genetic vs. environmental factors to metabolite level variance was estimated on the basis of GWAS study results. The proportion of heritability explained by all SNPs associated with a given metabolite at the genome-wide level is shown in red. *Figure was reproduced from Shin et al. Nature genetics 2014 with the permission for re-use from Nature publishing group* [116]

the association between genetic variants and specific product metabolites allows for the early determination of adverse pharmacological effects [118].

In the era of pharmacogenomics and personalized medicine, the term "pharmacometabolomics" has emerged to define the use of metabolomics in measurement of drug or disease biomarker signatures to predict and assess therapeutic response [119–122]. The measured changes in endogenous and exogenous metabolite levels reflect the end response of biological systems as a combined effect of genetic determinants and environmental and internal microbiome exposures. In the field of pharmacology and toxicology, metabolomics is increasingly applied to determine signatures of drug exposition, susceptibility, or toxicity (e.g., early signature of hepatotoxicity) and to investigate, in relation to the observed molecular patterns, the cellular mechanisms leading to the adverse effects of xenobiotics and to the development of associated pathologies [123]. Via monitoring of drug metabolites and metabolism (through drug levels and intake), metabolomic assays can considerably contribute to the assessment of toxicity and optimization of drug doses.

A traditional view of the medicine is to apply a standardized treatment following the "one size fits all" dogma to the entire population of patients suffering from the same disease. However, responder patients to a major drug treatment can differ from 25% in the case of cancer chemotherapy to 80% in the case of Cox-2 inhibitors [124]. Such an interindividual variability in the response rate depends on multiple factors involving the genetic and environmental components, including dietary regimen, toxin exposures, or potential drug-drug interactions [125]. Even slight modifications of the pharmacokinetic parameters induced by this set of factors can have important effect on pharmacodynamic properties of a drug. Cytochrome P450 (CYP) enzymes are the main drug-metabolizing systems in humans and the major source of variability in drug pharmacokinetics and response [125]. Their expression is highly influenced by a combination of factors, including genetic polymorphism, disease state, sex, age, etc. In addition, the pharmacokinetic variability in drug response is also increased by the variability of drug efflux and influx transporters [126]. Accordingly, the activity modifications of these classes of proteins can induce numerous pharmacological and toxicological consequences. For instance, the risk assessment of drug-drug interactions in elderly patients or patients under polymedication is a challenging task in clinical practice [127]. The in vivo assessment of the activities of enzymes and transporters to adapt the therapeutic strategy, known as phenotyping, is already used in clinical setting. Phenotyping consists of the controlled administration of a cocktail containing multiple probes, each of them targeting one specific enzyme or transporter [128, 129]. Unlike genotyping, phenotyping presents the advantage of measuring the combined effect of genetic, environmental, and endogenous factors on the activity of proteins involved in the drug pharmacokinetic variability [130, 131]. To overcome the potential issues with a drug cocktail administration in patients at risk, the possibility of developing metabolomics-based endogenous biomarkers is a promising alternative to current phenotyping assays. For example, change in blood levels of an unidentified metabolite was recently shown to be significantly associated with polymorphism in genomic region corresponding to the CYP3A5 gene locus [132]. Another recent untargeted metabolomics study allowed the identification of a urinary metabolite associated with CYP2D6 activity [133]. This association was validated in an adult cohort with the observation of a strong decrease of the metabolite abundance following the administration of fluoxetine as an inhibitor of CYP2D6.

In a systems biology approach, the integration of genomics, epigenomics, transcriptomics, proteomics, and metabolomics to investigate the response of an organism to xenobiotics represents a unique opportunity to accelerate the understanding of alterations in biological pathways associated with pathophysiological traits and the discovery of more classified biomarkers toward personalized approach to the medicine [134]. The integration of metabolomics to systems biology approach is rapidly evolving to provide a more complete picture of the dynamics of molecular systems [135]. Due to the fact that many diseases have a strong metabolic basis (that is heavily influenced by external and internal environmental exposures), in this post-genomic era, metabolomics will undoubtedly play a determinant role, offering powerful phenotyping capacity for more cost-effective models in drug discovery, efficient monitoring of therapeutic response, and customization of drug dosing.

References

1. Wishart DS (2016) Emerging applications of metabolomics in drug discovery and precision medicine. Nat Rev Drug Discov 15 (7):473–484. https://doi.org/10.1038/nrd.2016.32

2. Johnson CH, Ivanisevic J, Siuzdak G (2016) Metabolomics: beyond biomarkers and towards mechanisms. Nat Rev Mol Cell Biol 17(7):451–459. https://doi.org/10.1038/nrm.2016.25

3. Ivanisevic J, Siuzdak G (2015) The role of metabolomics in brain metabolism research. J Neuroimmune Pharmacol 10:391–395

4. Schmidt CW (2004) Metabolomics: what's happening downstream of DNA. Environ Health Perspect 112(7):A410–A415

5. Zamboni N, Saghatelian A, Patti GJ (2015) Defining the metabolome: size, flux, and regulation. Mol Cell 58(4):699–706. https://doi.org/10.1016/j.molcel.2015.04.021

6. Patti GJ, Yanes O, Siuzdak G (2012) Innovation: metabolomics: the apogee of the omics trilogy. Nat Rev Mol Cell Biol 13 (4):263–269. https://doi.org/10.1038/nrm3314

7. Sperber H, Mathieu J, Wang Y, Ferreccio A, Hesson J, Xu Z, Fischer KA, Devi A, Detraux D, Gu H, Battle SL, Showalter M, Valensisi C, Bielas JH, Ericson NG, Margaretha L, Robitaille AM, Margineantu D, Fiehn O, Hockenbery D, Blau CA, Raftery D, Margolin AA, Hawkins RD, Moon RT, Ware CB, Ruohola-Baker H (2015) The metabolome regulates the epigenetic landscape during naive-to-primed human embryonic stem cell transition. Nat Cell Biol 17(12):1523–1535. https://doi.org/10.1038/ncb3264

8. Sabari BR, Zhang D, Allis CD, Zhao Y (2017) Metabolic regulation of gene expression through histone acylations. Nat Rev Mol Cell Biol 18(2):90–101. https://doi.org/10.1038/nrm.2016.140

9. Siroux V, Agier L, Slama R (2016) The exposome concept: a challenge and a potential driver for environmental health research. Eur Respir Rev 25(140):124–129. https://doi.org/10.1183/16000617.0034-2016

10. Pedersen HK, Gudmundsdottir V, Nielsen HB, Hyotylainen T, Nielsen T, Jensen BAH, Forslund K, Hildebrand F, Prifti E, Falony G, Le Chatelier E, Levenez F, Doré J, Mattila I, Plichta DR, Pöhö P, Hellgren LI, Arumugam M, Sunagawa S, Vieira-Silva S, Jørgensen T, Holm JB, Trošt K, MetaHIT Consortium, Kristiansen K, Brix S, Raes J, Wang J, Hansen T, Bork P, Brunak S, Oresic M, Ehrlich SD, Pedersen O (2016) Human gut microbes impact host serum

metabolome and insulin sensitivity. Nature 535(7612):376–381

11. Bucci M (2016) Gut microbiome: branching into metabolic disease. Nat Chem Biol 12 (9):657–657. https://doi.org/10.1038/nchembio.2164

12. Mayer EA, Knight R, Mazmanian SK, Cryan JF, Tillisch K (2014) Gut microbes and the brain: paradigm shift in neuroscience. J Neurosci 34(46):15490–15496. https://doi.org/10.1523/jneurosci.3299-14.2014

13. Mamas M, Dunn WB, Neyses L, Goodacre R (2011) The role of metabolites and metabolomics in clinically applicable biomarkers of disease. Arch Toxicol 85(1):5–17. https://doi.org/10.1007/s00204-010-0609-6

14. Weber RJM, Lawson TN, Salek RM, Ebbels TMD, Glen RC, Goodacre R, Griffin JL, Haug K, Koulman A, Moreno P, Ralser M, Steinbeck C, Dunn WB, Viant MR (2017) Computational tools and workflows in metabolomics: an international survey highlights the opportunity for harmonisation through galaxy. Metabolomics 13(2):12. https://doi.org/10.1007/s11306-016-1147-x

15. Benton HP, Ivanisevic J, Mahieu NG, Kurczy ME, Johnson CH, Franco L, Rinehart D, Valentine E, Gowda H, Ubhi BK, Tautenhahn R, Gieschen A, Fields MW, Patti GJ, Siuzdak G (2015) Autonomous metabolomics for rapid metabolite identification in global profiling. Anal Chem 87(2):884–891. https://doi.org/10.1021/ac5025649

16. Tautenhahn R, Cho K, Uritboonthai W, Zhu Z, Patti GJ, Siuzdak G (2012) An accelerated workflow for untargeted metabolomics using the METLIN database. Nat Biotechnol 30(9):826–828. https://doi.org/10.1038/nbt.2348

17. Johnson CH, Ivanisevic J, Benton HP, Siuzdak G (2015) Bioinformatics: the next frontier of metabolomics. Anal Chem 87 (1):147–156. https://doi.org/10.1021/ac5040693

18. Ivanisevic J, Elias D, Deguchi H, Averell PM, Kurczy M, Johnson CH, Tautenhahn R, Zhu Z, Watrous J, Jain M (2015) Arteriovenous blood metabolomics: a readout of intratissue metabostasis. Sci Rep 5:12757

19. Beger RD, Dunn W, Schmidt MA, Gross SS, Kirwan JA, Cascante M, Brennan L, Wishart DS, Oresic M, Hankemeier T, Broadhurst DI, Lane AN, Suhre K, Kastenmüller G, Sumner SJ, Thiele I, Fiehn O, Kaddurah-Daouk R, for "Precision M, Pharmacometabolomics Task Group"-Metabolomics Society I (2016) Metabolomics enables precision medicine: "a white paper, community perspective".

Metabolomics 12(9):149. https://doi.org/10.1007/s11306-016-1094-6

20. Su LJ, Fiehn O, Maruvada P, Moore SC, O'Keefe SJ, Wishart DS, Zanetti KA (2014) The use of metabolomics in population-based research. Adv Nutr 5(6):785–788. https://doi.org/10.3945/an.114.006494

21. Psychogios N, Hau DD, Peng J, Guo AC, Mandal R, Bouatra S, Sinelnikov I, Krishnamurthy R, Eisner R, Gautam B, Young N, Xia J, Knox C, Dong E, Huang P, Hollander Z, Pedersen TL, Smith SR, Bamforth F, Greiner R, McManus B, Newman JW, Goodfriend T, Wishart DS (2011) The human serum metabolome. PLoS One 6 (2):e16957. https://doi.org/10.1371/journal.pone.0016957

22. Yanes O, Tautenhahn R, Patti GJ, Siuzdak G (2011) Expanding coverage of the metabolome for global metabolite profiling. Anal Chem 83(6):2152–2161. https://doi.org/10.1021/ac102981k

23. Zhang A, Sun H, Yan G, Wang P, Han Y, Wang X (2014) Metabolomics in diagnosis and biomarker discovery of colorectal cancer. Cancer Lett 345(1):17–20. https://doi.org/10.1016/j.canlet.2013.11.011

24. Thomas A, Lenglet S, Chaurand P, Deglon J, Mangin P, Mach F, Steffens S, Wolfender JL, Staub C (2011) Mass spectrometry for the evaluation of cardiovascular diseases based on proteomics and lipidomics. Thromb Haemost 106(1):20–33. https://doi.org/10.1160/TH10-12-0812

25. Dunn WB, Bailey NJ, Johnson HE (2005) Measuring the metabolome: current analytical technologies. Analyst 130(5):606–625. https://doi.org/10.1039/b418288j

26. Dunn WB, Broadhurst D, Begley P, Zelena E, Francis-McIntyre S, Anderson N, Brown M, Knowles JD, Halsall A, Haselden JN, Nicholls AW, Wilson ID, Kell DB, Goodacre R, Human Serum Metabolome Consortium (2011) Procedures for large-scale metabolic profiling of serum and plasma using gas chromatography and liquid chromatography coupled to mass spectrometry. Nat Protoc 6 (7):1060–1083. https://doi.org/10.1038/nprot.2011.335

27. Schweiger R, Baier MC, Persicke M, Muller C (2014) High specificity in plant leaf metabolic responses to arbuscular mycorrhiza. Nat Commun 5:3886. https://doi.org/10.1038/ncomms4886

28. Tugizimana F, Steenkamp PA, Piater LA, Dubery IA (2014) Multi-platform metabolomic analyses of ergosterol-induced dynamic changes in Nicotiana tabacum cells. PLoS

One 9(1):e87846. https://doi.org/10.1371/journal.pone.0087846

29. Bouatra S, Aziat F, Mandal R, Guo AC, Wilson MR, Knox C, Bjorndahl TC, Krishnamurthy R, Saleem F, Liu P, Dame ZT, Poelzer J, Huynh J, Yallou FS, Psychogios N, Dong E, Bogumil R, Roehring C, Wishart DS (2013) The human urine metabolome. PLoS One 8(9):e73076. https://doi.org/10.1371/journal.pone.0073076

30. Want EJ, Masson P, Michopoulos F, Wilson ID, Theodoridis G, Plumb RS, Shockcor J, Loftus N, Holmes E, Nicholson JK (2013) Global metabolic profiling of animal and human tissues via UPLC-MS. Nat Protoc 8(1):17–32. https://doi.org/10.1038/nprot.2012.135

31. Junot C, Fenaille F, Colsch B, Becher F (2013) High resolution mass spectrometry based techniques at the crossroads of metabolic pathways. Mass Spectrom Rev 33(6):471–500. https://doi.org/10.1002/mas.21401

32. Fiehn O (2002) Metabolomics – the link between genotypes and phenotypes. Plant Mol Biol 48(1–2):155–171

33. Wolfender JL, Marti G, Thomas A, Bertrand S (2014) Current approaches and challenges for the metabolite profiling of complex natural extracts. J Chromatogr A 1382:136–164. https://doi.org/10.1016/j.chroma.2014.10.091

34. Thomas A, Deglon J, Steimer T, Mangin P, Daali Y, Staub C (2010) On-line desorption of dried blood spots coupled to hydrophilic interaction/reversed-phase LC/MS/MS system for the simultaneous analysis of drugs and their polar metabolites. J Sep Sci 33(6–7):873–879. https://doi.org/10.1002/jssc.200900593

35. McCalley DV (2007) Is hydrophilic interaction chromatography with silica columns a viable alternative to reversed-phase liquid chromatography for the analysis of ionisable compounds? J Chromatogr A 1171(1–2):46–55. https://doi.org/10.1016/j.chroma.2007.09.047

36. Ivanisevic J, Zhu ZJ, Plate L, Tautenhahn R, Chen S, O'Brien PJ, Johnson CH, Marletta MA, Patti GJ, Siuzdak G (2013) Toward 'omic scale metabolite profiling: a dual separation-mass spectrometry approach for coverage of lipid and central carbon metabolism. Anal Chem 85(14):6876–6884. https://doi.org/10.1021/ac401140h

37. Tautenhahn R, Patti GJ, Rinehart D, Siuzdak G (2012) XCMS Online: a web-based platform to process untargeted metabolomic data. Anal Chem 84(11):5035–5039. https://doi.org/10.1021/ac300698c

38. Xia J, Wishart DS (2011) Web-based inference of biological patterns, functions and pathways from metabolomic data using MetaboAnalyst. Nat Protoc 6(6):743–760. https://doi.org/10.1038/nprot.2011.319

39. Eugster PJ, Glauser G, Wolfender JL (2013) Strategies in biomarker discovery. Peak annotation by MS and targeted LC-MS micro-fractionation for de novo structure identification by micro-NMR. Methods Mol Biol 1055:267–289. https://doi.org/10.1007/978-1-62703-577-4_19

40. Wishart DS, Jewison T, Guo AC, Wilson M, Knox C, Liu Y, Djoumbou Y, Mandal R, Aziat F, Dong E, Bouatra S, Sinelnikov I, Arndt D, Xia J, Liu P, Yallou F, Bjorndahl T, Perez-Pineiro R, Eisner R, Allen F, Neveu V, Greiner R, Scalbert A (2013) HMDB 3.0— The Human Metabolome Database in 2013. Nucleic Acids Res 41(Database issue): D801–D807. https://doi.org/10.1093/nar/gks1065

41. Yuan M, Breitkopf SB, Yang X, Asara JM (2012) A positive/negative ion-switching, targeted mass spectrometry-based metabolomics platform for bodily fluids, cells, and fresh and fixed tissue. Nat Protoc 7(5):872–881. https://doi.org/10.1038/nprot.2012.024

42. Cai Y, Weng K, Guo Y, Peng J, Zhu Z-J (2015) An integrated targeted metabolomic platform for high-throughput metabolite profiling and automated data processing. Metabolomics 11(6):1575–1586. https://doi.org/10.1007/s11306-015-0809-4

43. Norris JL, Caprioli RM (2013) Analysis of tissue specimens by matrix-assisted laser desorption/ionization imaging mass spectrometry in biological and clinical research. Chem Rev 113(4):2309–2342. https://doi.org/10.1021/cr3004295

44. Sun N, Ly A, Meding S, Witting M, Hauck SM, Ueffing M, Schmitt-Kopplin P, Aichler M, Walch A (2014) High-resolution metabolite imaging of light and dark treated retina using MALDI-FTICR mass spectrometry. Proteomics 14(7-8):913–923. https://doi.org/10.1002/pmic.201300407

45. Thomas A, Charbonneau JL, Fournaise E, Chaurand P (2012) Sublimation of new matrix candidates for high spatial resolution imaging mass spectrometry of lipids: Enhanced information in both positive and negative polarities after 1,5-diaminonapthalene deposition. Anal

Chem 84(4):2048–2054. https://doi.org/10.1021/ac2033547

46. Ly A, Buck A, Balluff B, Sun N, Gorzolka K, Feuchtinger A, Janssen KP, Kuppen PJ, van de Velde CJ, Weirich G, Erlmeier F, Langer R, Aubele M, Zitzelsberger H, McDonnell L, Aichler M, Walch A (2016) High-mass-resolution MALDI mass spectrometry imaging of metabolites from formalin-fixed paraffin-embedded tissue. Nat Protoc 11 (8):1428–1443. https://doi.org/10.1038/nprot.2016.081

47. Patterson NH, Alabdulkarim B, Lazaris A, Thomas A, Marcinkiewicz MM, Gao ZH, Vermeulen PB, Chaurand P, Metrakos P (2016) Assessment of pathological response to therapy using lipid mass spectrometry imaging. Sci Rep 6:36814. https://doi.org/10.1038/srep36814

48. Aichler M, Walch A (2015) MALDI Imaging mass spectrometry: current frontiers and perspectives in pathology research and practice. Lab Invest 95(4):422–431. https://doi.org/10.1038/labinvest.2014.156

49. Hamilton LK, Dufresne M, Joppe SE, Petryszyn S, Aumont A, Calon F, Barnabe-Heider F, Furtos A, Parent M, Chaurand P, Fernandes KJ (2015) Aberrant lipid metabolism in the forebrain niche suppresses adult neural stem cell proliferation in an animal model of Alzheimer's disease. Cell Stem Cell 17(4):397–411. https://doi.org/10.1016/j.stem.2015.08.001

50. Tonnies E, Trushina E (2017) Oxidative stress, synaptic dysfunction, and Alzheimer's disease. J Alzheimer's Dis 57(4):1105–1121. https://doi.org/10.3233/jad-161088

51. Stranahan AM, Mattson MP (2012) Recruiting adaptive cellular stress responses for successful brain ageing. Nat Rev Neurosci 13 (3):209–216. https://doi.org/10.1038/nrn3151

52. http://www.who.int/mediacentre/factsheets/fs311/en/

53. Jiao H, Arner P, Hoffstedt J, Brodin D, Dubern B, Czernichow S, van't Hooft F, Axelsson T, Pedersen O, Hansen T, Sorensen TI, Hebebrand J, Kere J, Dahlman-Wright K, Hamsten A, Clement K, Dahlman I (2011) Genome wide association study identifies KCNMA1 contributing to human obesity. BMC Med Genomics 4:51. https://doi.org/10.1186/1755-8794-4-51

54. Locke AE, Kahali B, Berndt SI, Justice AE, Pers TH, Day FR, Powell C, Vedantam S, Buchkovich ML, Yang J, Croteau-Chonka DC, Esko T, Fall T, Ferreira T, Gustafsson S, Kutalik Z, Luan J, Mägi R, Randall JC, Winkler TW, Wood AR, Workalemahu T, Faul JD, Smith JA, Zhao JH, Zhao W, Chen J, Fehrmann R, Hedman ÅK, Karjalainen J, Schmidt EM, Absher D, Amin N, Anderson D, Beekman M, Bolton JL, Bragg-Gresham JL, Buyske S, Demirkan A, Deng G, Ehret GB, Feenstra B, Feitosa MF, Fischer K, Goel A, Gong J, Jackson AU, Kanoni S, Kleber ME, Kristiansson K, Lim U, Lotay V, Mangino M, Leach IM, Medina-Gomez C, Medland SE, Nalls MA, Palmer CD, Pasko D, Pechlivanis S, Peters MJ, Prokopenko I, Shungin D, Stančáková A, Strawbridge RJ, Sung YJ, Tanaka T, Teumer A, Trompet S, van der Laan SW, van Setten J, Van Vliet-Ostaptchouk JV, Wang Z, Yengo L, Zhang W, Isaacs A, Albrecht E, Ärnlöv J, Arscott GM, Attwood AP, Bandinelli S, Barrett A, Bas IN, Bellis C, Bennett AJ, Berne C, Blagieva R, Blüher M, Böhringer S, Bonnycastle LL, Böttcher Y, Boyd HA, Bruinenberg M, Caspersen IH, Chen YI, Clarke R, Daw EW, de Craen AJM, Delgado G, Dimitriou M, Doney ASF, Eklund N, Estrada K, Eury E, Folkersen L, Fraser RM, Garcia ME, Geller F, Giedraitis V, Gigante B, Go AS, Golay A, Goodall AH, Gordon SD, Gorski M, Grabe HJ, Grallert H, Grammer TB, Gräßler J, Grönberg H, Groves CJ, Gusto G, Haessler J, Hall P, Haller T, Hallmans G, Hartman CA, Hassinen M, Hayward C, Heard-Costa NL, Helmer Q, Hengstenberg C, Holmen O, Hottenga JJ, James AL, Jeff JM, Johansson Å, Jolley J, Juliusdottir T, Kinnunen L, Koenig W, Koskenvuo M, Kratzer W, Laitinen J, Lamina C, Leander K, Lee NR, Lichtner P, Lind L, Lindström J, Lo KS, Lobbens S, Lorbeer R, Lu Y, Mach F, Magnusson PKE, Mahajan A, McArdle WL, McLachlan S, Menni C, Merger S, Mihailov E, Milani L, Moayyeri A, Monda KL, Morken MA, Mulas A, Müller G, Müller-Nurasyid M, Musk AW, Nagaraja R, Nöthen MM, Nolte IM, Pilz S, Rayner NW, Renstrom F, Rettig R, Ried JS, Ripke S, Robertson NR, Rose LM, Sanna S, Scharnagl H, Scholtens S, Schumacher FR, Scott WR, Seufferlein T, Shi J, Smith AV, Smolonska J, Stanton AV, Steinthorsdottir V, Stirrups K, Stringham HM, Sundström J, Swertz MA, Swift AJ, Syvänen AC, Tan ST, Tayo BO, Thorand B, Thorleifsson G, Tyrer JP, Uh HW, Vandenput L, Verhulst FC, Vermeulen SH, Verweij N, Vonk JM, Waite LL, Warren HR, Waterworth D, Weedon MN, Wilkens LR, Willenborg C, Wilsgaard T, Wojczynski MK,

Wong A, Wright AF, Zhang Q, LifeLines Cohort Study, Brennan EP, Choi M, Dastani Z, Drong AW, Eriksson P, Franco-Cereceda A, Gådin JR, Gharavi AG, Goddard ME, Handsaker RE, Huang J, Karpe F, Kathiresan S, Keildson S, Kiryluk K, Kubo M, Lee JY, Liang L, Lifton RP, Ma B, McCarroll SA, McKnight AJ, Min JL, Moffatt MF, Montgomery GW, Murabito JM, Nicholson G, Nyholt DR, Okada Y, JRB P, Dorajoo R, Reinmaa E, Salem RM, Sandholm N, Scott RA, Stolk L, Takahashi A, Tanaka T, van't Hooft FM, AAE V, Westra HJ, Zheng W, Zondervan KT, ADIPOGen Consortium, AGEN-BMI Working Group, CARDIOGRAMplusC4D Consortium, CKDGen Consortium, GLGC, ICBP, MAGIC Investigators, MuTHER Consortium, MIGen Consortium, PAGE Consortium, ReproGen Consortium, GENIE Consortium, International Endogene Consortium, Heath AC, Arveiler D, SJL B, Beilby J, Bergman RN, Blangero J, Bovet P, Campbell H, Caulfield MJ, Cesana G, Chakravarti A, Chasman DI, Chines PS, Collins FS, Crawford DC, Cupples LA, Cusi D, Danesh J, de Faire U, den Ruijter HM, Dominiczak AF, Erbel R, Erdmann J, Eriksson JG, Farrall M, Felix SB, Ferrannini E, Ferrières J, Ford I, Forouhi NG, Forrester T, Franco OH, Gansevoort RT, Gejman PV, Gieger C, Gottesman O, Gudnason V, Gyllensten U, Hall AS, Harris TB, Hattersley AT, Hicks AA, Hindorff LA, Hingorani AD, Hofman A, Homuth G, Hovingh GK, Humphries SE, Hunt SC, Hyppönen E, Illig T, Jacobs KB, Jarvelin MR, Jöckel KH, Johansen B, Jousilahti P, Jukema JW, Jula AM, Kaprio J, Kastelein JJP, Keinanen-Kiukaanniemi SM, Kiemeney LA, Knekt P, Kooner JS, Kooperberg C, Kovacs P, Kraja AT, Kumari M, Kuusisto J, Lakka TA, Langenberg C, Marchand LL, Lehtimäki T, Lyssenko V, Männistö S, Marette A, Matise TC, McKenzie CA, McKnight B, Moll FL, Morris AD, Morris AP, Murray JC, Nelis M, Ohlsson C, Oldehinkel AJ, Ong KK, PAF M, Pasterkamp G, Peden JF, Peters A, Postma DS, Pramstaller PP, Price JF, Qi L, Raitakari OT, Rankinen T, Rao DC, Rice TK, Ridker PM, Rioux JD, Ritchie MD, Rudan I, Salomaa V, Samani NJ, Saramies J, Sarzynski MA, Schunkert H, Schwarz PEH, Sever P, Shuldiner AR, Sinisalo J, Stolk RP, Strauch K, Tönjes A, Trégouët DA, Tremblay A, Tremoli E, Virtamo J, Vohl MC, Völker U, Waeber G, Willemsen G, Witteman JC, Zillikens MC, Adair LS, Amouyel P, Asselbergs FW, Assimes TL, Bochud M, Boehm BO, Boerwinkle E, Bornstein SR, Bottinger EP, Bouchard C, Cauchi S, Chambers JC, Chanock SJ, Cooper RS, de Bakker PIW, Dedoussis G, Ferrucci L, Franks PW, Froguel P, Groop LC, Haiman CA, Hamsten A, Hui J, Hunter DJ, Hveem K, Kaplan RC, Kivimaki M, Kuh D, Laakso M, Liu Y, Martin NG, März W, Melbye M, Metspalu A, Moebus S, Munroe PB, Njølstad I, Oostra BA, Palmer CNA, Pedersen NL, Perola M, Pérusse L, Peters U, Power C, Quertermous T, Rauramaa R, Rivadeneira F, Saaristo TE, Saleheen D, Sattar N, Schadt EE, Schlessinger D, Slagboom PE, Snieder H, Spector TD, Thorsteinsdottir U, Stumvoll M, Tuomilehto J, Uitterlinden AG, Uusitupa M, van der Harst P, Walker M, Wallaschofski H, Wareham NJ, Watkins H, Weir DR, Wichmann HE, Wilson JF, Zanen P, Borecki IB, Deloukas P, Fox CS, Heid IM, O'Connell JR, Strachan DP, Stefansson K, van Duijn CM, Abecasis GR, Franke L, Frayling TM, McCarthy MI, Visscher PM, Scherag A, Willer CJ, Boehnke M, Mohlke KL, Lindgren CM, Beckmann JS, Barroso I, North KE, Ingelsson E, Hirschhorn JN, Loos RJF, Speliotes EK (2015) Genetic studies of body mass index yield new insights for obesity biology. Nature 518(7538):197–206. https://doi.org/10.1038/nature14177

55. Moreno-Navarrete JM, Jove M, Ortega F, Xifra G, Ricart W, Obis E, Pamplona R, Portero-Otin M, Fernandez-Real JM (2016) Metabolomics uncovers the role of adipose tissue PDXK in adipogenesis and systemic insulin sensitivity. Diabetologia 59(4):822–832. https://doi.org/10.1007/s00125-016-3863-1

56. Abu Bakar MH, Sarmidi MR, Cheng KK, Ali Khan A, Suan CL, Zaman Huri H, Yaakob H (2015) Metabolomics – the complementary field in systems biology: a review on obesity and type 2 diabetes. Mol Biosyst 11(7):1742–1774. https://doi.org/10.1039/c5mb00158g

57. Roberts LD, Koulman A, Griffin JL (2014) Towards metabolic biomarkers of insulin resistance and type 2 diabetes: progress from the metabolome. Lancet Diabetes Endocrinol 2(1):65–75. https://doi.org/10.1016/S2213-8587(13)70143-8

58. Morris C, O'Grada C, Ryan M, Roche HM, Gibney MJ, Gibney ER, Brennan L (2012) The relationship between BMI and metabolomic profiles: a focus on amino acids. Proc Nutr Soc 71(4):634–638. https://doi.org/10.1017/S0029665112000699

59. Newgard CB, An J, Bain JR, Muehlbauer MJ, Stevens RD, Lien LF, Haqq AM, Shah SH, Arlotto M, Slentz CA, Rochon J, Gallup D, Ilkayeva O, Wenner BR, Yancy WS Jr, Eisenson H, Musante G, Surwit RS, Millington DS, Butler MD, Svetkey LP (2009) A branched-chain amino acid-related metabolic signature that differentiates obese and lean humans and contributes to insulin resistance. Cell Metab 9(4):311–326. https://doi.org/10.1016/j.cmet.2009.02.002

60. Makki K, Froguel P, Wolowczuk I (2013) Adipose tissue in obesity-related inflammation and insulin resistance: cells, cytokines, and chemokines. ISRN Inflamm 2013:139239. https://doi.org/10.1155/2013/139239

61. Sun K, Kusminski CM, Scherer PE (2011) Adipose tissue remodeling and obesity. J Clin Invest 121(6):2094–2101. https://doi.org/10.1172/JCI45887

62. La Merrill M, Emond C, Kim MJ, Antignac JP, Le Bizec B, Clement K, Birnbaum LS, Barouki R (2013) Toxicological function of adipose tissue: focus on persistent organic pollutants. Environ Health Perspect 121 (2):162–169. https://doi.org/10.1289/ehp.1205485

63. Frayn KN, Karpe F, Fielding BA, Macdonald IA, Coppack SW (2003) Integrative physiology of human adipose tissue. Int J Obes Relat Metab Disord 27(8):875–888. https://doi.org/10.1038/sj.ijo.0802326

64. Hanzu FA, Vinaixa M, Papageorgiou A, Parrizas M, Correig X, Delgado S, Carmona F, Samino S, Vidal J, Gomis R (2014) Obesity rather than regional fat depots marks the metabolomic pattern of adipose tissue: an untargeted metabolomic approach. Obesity (Silver Spring) 22 (3):698–704. https://doi.org/10.1002/oby.20541

65. Cao H, Gerhold K, Mayers JR, Wiest MM, Watkins SM, Hotamisligil GS (2008) Identification of a lipokine, a lipid hormone linking adipose tissue to systemic metabolism. Cell 134(6):933–944. https://doi.org/10.1016/j.cell.2008.07.048

66. Liesenfeld DB, Grapov D, Fahrmann JF, Salou M, Scherer D, Toth R, Habermann N, Bohm J, Schrotz-King P, Gigic B, Schneider M, Ulrich A, Herpel E, Schirmacher P, Fiehn O, Lampe JW, Ulrich CM (2015) Metabolomics and transcriptomics identify pathway differences between visceral and subcutaneous adipose tissue in colorectal cancer patients: the ColoCare study. Am J Clin Nutr 102(2):433–443. https://doi.org/10.3945/ajcn.114.103804

67. Perez-Cornago A, Brennan L, Ibero-Baraibar I, Hermsdorff HH, O'Gorman A, Zulet MA, Martinez JA (2014) Metabolomics identifies changes in fatty acid and amino acid profiles in serum of overweight older adults following a weight loss intervention. J Physiol Biochem 70(2):593–602. https://doi.org/10.1007/s13105-013-0311-2

68. Shah SH, Crosslin DR, Haynes CS, Nelson S, Turer CB, Stevens RD, Muehlbauer MJ, Wenner BR, Bain JR, Laferrere B, Gorroochurn P, Teixeira J, Brantley PJ, Stevens VJ, Hollis JF, Appel LJ, Lien LF, Batch B, Newgard CB, Svetkey LP (2012) Branched-chain amino acid levels are associated with improvement in insulin resistance with weight loss. Diabetologia 55 (2):321–330. https://doi.org/10.1007/s00125-011-2356-5

69. Sjostrom L (2013) Review of the key results from the Swedish Obese Subjects (SOS) trial – a prospective controlled intervention study of bariatric surgery. J Intern Med 273 (3):219–234. https://doi.org/10.1111/joim.12012

70. Buchwald H, Avidor Y, Braunwald E, Jensen MD, Pories W, Fahrbach K, Schoelles K (2004) Bariatric surgery: a systematic review and meta-analysis. JAMA 292 (14):1724–1737. https://doi.org/10.1001/jama.292.14.1724

71. Liu SY, Wong SK, Lam CC, Yung MY, Kong AP, Ng EK (2015) Long-term results on weight loss and diabetes remission after laparoscopic sleeve gastrectomy for a morbidly obese Chinese population. Obes Surg 25(10):1901–1908. https://doi.org/10.1007/s11695-015-1628-4

72. Parikh M, Pomp A, Gagner M (2007) Laparoscopic conversion of failed gastric bypass to duodenal switch: technical considerations and preliminary outcomes. Surg Obes Relat Dis 3(6):611–618. doi:S1550-7289(07)00569-2 [pii]. https://doi.org/10.1016/j.soard.2007.07.010

73. Mutch DM, Fuhrmann JC, Rein D, Wiemer JC, Bouillot JL, Poitou C, Clement K (2009) Metabolite profiling identifies candidate markers reflecting the clinical adaptations associated with Roux-en-Y gastric bypass surgery. PLoS One 4(11):e7905. https://doi.org/10.1371/journal.pone.0007905

74. Laferrere B, Reilly D, Arias S, Swerdlow N, Gorroochurn P, Bawa B, Bose M, Teixeira J, Stevens RD, Wenner BR, Bain JR, Muehl-

bauer MJ, Haqq A, Lien L, Shah SH, Svetkey LP, Newgard CB (2011) Differential metabolic impact of gastric bypass surgery versus dietary intervention in obese diabetic subjects despite identical weight loss. Sci Transl Med 3 (80):80re82.   https://doi.org/10.1126/scitranslmed.3002043

75. Whiting DR, Guariguata L, Weil C, Shaw J (2011) IDF diabetes atlas: global estimates of the prevalence of diabetes for 2011 and 2030. Diabetes Res Clin Pract 94(3):311–321. https://doi.org/10.1016/j.diabres.2011.10.029

76. Huber CA, Schwenkglenks M, Rapold R, Reich O (2014) Epidemiology and costs of diabetes mellitus in Switzerland: an analysis of health care claims data, 2006 and 2011. BMC Endocr Disord 14:44. https://doi.org/10.1186/1472-6823-14-44

77. Schmid R, Vollenweider P, Waeber G, Marques-Vidal P (2011) Estimating the risk of developing type 2 diabetes: a comparison of several risk scores: the Cohorte Lausannoise study. Diabetes Care 34(8):1863–1868. https://doi.org/10.2337/dc11-0206

78. Marques-Vidal P, Schmid R, Bochud M, Bastardot F, von Kanel R, Paccaud F, Glaus J, Preisig M, Waeber G, Vollenweider P (2012) Adipocytokines, hepatic and inflammatory biomarkers and incidence of type 2 diabetes. the CoLaus study. PLoS One 7 (12):e51768. https://doi.org/10.1371/journal.pone.0051768

79. Vaxillaire M, Yengo L, Lobbens S, Rocheleau G, Eury E, Lantieri O, Marre M, Balkau B, Bonnefond A, Froguel P (2014) Type 2 diabetes-related genetic risk scores associated with variations in fasting plasma glucose and development of impaired glucose homeostasis in the prospective DESIR study. Diabetologia 57(8):1601–1610. https://doi.org/10.1007/s00125-014-3277-x

80. Lin X, Song K, Lim N, Yuan X, Johnson T, Abderrahmani A, Vollenweider P, Stirnadel H, Sundseth SS, Lai E, Burns DK, Middleton LT, Roses AD, Matthews PM, Waeber G, Cardon L, Waterworth DM, Mooser V (2009) Risk prediction of prevalent diabetes in a Swiss population using a weighted genetic score – the CoLaus Study. Diabetologia 52(4):600–608. https://doi.org/10.1007/s00125-008-1254-y

81. Meigs JB, Shrader P, Sullivan LM, McAteer JB, Fox CS, Dupuis J, Manning AK, Florez JC, Wilson PW, D'Agostino RB Sr, Cupples LA (2008) Genotype score in addition to common risk factors for prediction of type

2 diabetes. N Engl J Med 359 (21):2208–2219. https://doi.org/10.1056/NEJMoa0804742

82. Kussmann M, Morine MJ, Hager J, Sonderegger B, Kaput J (2013) Perspective: a systems approach to diabetes research. Front Genet 4:205. https://doi.org/10.3389/fgene.2013.00205

83. Klein MS, Shearer J (2016) Metabolomics and type 2 diabetes: translating basic research into clinical application. J Diabetes Res 2016:3898502. https://doi.org/10.1155/2016/3898502

84. Wang-Sattler R, Yu Z, Herder C, Messias AC, Floegel A, He Y, Heim K, Campillos M, Holzapfel C, Thorand B, Grallert H, Xu T, Bader E, Huth C, Mittelstrass K, Doring A, Meisinger C, Gieger C, Prehn C, Roemisch-Margl W, Carstensen M, Xie L, Yamanaka-Okumura H, Xing G, Ceglarek U, Thiery J, Giani G, Lickert H, Lin X, Li Y, Boeing H, Joost HG, de Angelis MH, Rathmann W, Suhre K, Prokisch H, Peters A, Meitinger T, Roden M, Wichmann HE, Pischon T, Adamski J, Illig T (2012) Novel biomarkers for pre-diabetes identified by metabolomics. Mol Syst Biol 8:615. https://doi.org/10.1038/msb.2012.43

85. Padberg I, Peter E, Gonzalez-Maldonado S, Witt H, Mueller M, Weis T, Bethan B, Liebenberg V, Wiemer J, Katus HA, Rein D, Schatz P (2014) A new metabolomic signature in type-2 diabetes mellitus and its pathophysiology. PLoS One 9(1):e85082. https://doi.org/10.1371/journal.pone.0085082

86. Wang TJ, Larson MG, Vasan RS, Cheng S, Rhee EP, McCabe E, Lewis GD, Fox CS, Jacques PF, Fernandez C, O'Donnell CJ, Carr SA, Mootha VK, Florez JC, Souza A, Melander O, Clish CB, Gerszten RE (2011) Metabolite profiles and the risk of developing diabetes. Nat Med 17(4):448–453. https://doi.org/10.1038/nm.2307

87. Newgard CB (2012) Interplay between lipids and branched-chain amino acids in development of insulin resistance. Cell Metab 15 (5):606–614. https://doi.org/10.1016/j.cmet.2012.01.024

88. Herman MA, She P, Peroni OD, Lynch CJ, Kahn BB (2010) Adipose tissue branched chain amino acid (BCAA) metabolism modulates circulating BCAA levels. J Biol Chem 285(15):11348–11356. https://doi.org/10.1074/jbc.M109.075184

89. Sears DD, Hsiao G, Hsiao A, Yu JG, Courtney CH, Ofrecio JM, Chapman J, Subramaniam S (2009) Mechanisms of human insulin

resistance and thiazolidinedione-mediated insulin sensitization. Proc Natl Acad Sci U S A 106(44):18745–18750. https://doi.org/10.1073/pnas.0903032106

90. Kaeberlein M, Rabinovitch PS, Martin GM (2015) Healthy aging: the ultimate preventative medicine. Science 350 (6265):1191–1193. https://doi.org/10.1126/science.aad3267

91. Jove M, Portero-Otin M, Naudi A, Ferrer I, Pamplona R (2014) Metabolomics of human brain aging and age-related neurodegenerative diseases. J Neuropathol Exp Neurol 73 (7):640–657. https://doi.org/10.1097/nen.0000000000000091

92. Magistretti PJ, Allaman I (2015) A cellular perspective on brain energy metabolism and functional imaging. Neuron 86(4):883–901. https://doi.org/10.1016/j.neuron.2015.03.035

93. Trushina E, Dutta T, Persson XM, Mielke MM, Petersen RC (2013) Identification of altered metabolic pathways in plasma and CSF in mild cognitive impairment and Alzheimer's disease using metabolomics. PLoS One 8(5):e63644. https://doi.org/10.1371/journal.pone.0063644

94. Toledo JB, Arnold M, Kastenmüller G, Chang R, Baillie RA, Han X, Thambisetty M, Tenenbaum JD, Suhre K, Thompson JW, John-Williams LS, MahmoudianDehkordi S, Rotroff DM, Jack JR, Motsinger-Reif A, Risacher SL, Blach C, Lucas JE, Massaro T, Louie G, Zhu H, Dallmann G, Klavins K, Koal T, Kim S, Nho K, Shen L, Casanova R, Varma S, Legido-Quigley C, Moseley MA, Zhu K, Henrion MYR, van der Lee SJ, Harms AC, Demirkan A, Hankemeier T, van Duijn CM, Trojanowski JQ, Shaw LM, Saykin AJ, Weiner MW, Doraiswamy PM, Kaddurah-Daouk R (2017) Metabolic network failures in Alzheimer's disease – a biochemical road map. Alzheimers Dement 13(9):965–984. https://doi.org/10.1016/j.jalz.2017.01.020

95. Mink JW, Blumenschine RJ, Adams DB (1981) Ratio of central nervous system to body metabolism in vertebrates: its constancy and functional basis. Am J Physiol 241(3):R203–R212

96. Paglia G, Stocchero M, Cacciatore S, Lai S, Angel P, Alam MT, Keller M, Ralser M, Astarita G (2016) Unbiased metabolomic investigation of Alzheimer's disease brain points to dysregulation of mitochondrial aspartate metabolism. J Proteome Res 15 (2):608–618. https://doi.org/10.1021/acs.jproteome.5b01020

97. Kapogiannis D, Mattson MP (2011) Disrupted energy metabolism and neuronal circuit dysfunction in cognitive impairment and Alzheimer's disease. Lancet Neurol 10 (2):187–198. https://doi.org/10.1016/s1474-4422(10)70277-5

98. Trushina E, Nemutlu E, Zhang S, Christensen T, Camp J, Mesa J, Siddiqui A, Tamura Y, Sesaki H, Wengenack TM, Dzeja PP, Poduslo JF (2012) Defects in mitochondrial dynamics and metabolomic signatures of evolving energetic stress in mouse models of familial Alzheimer's disease. PLoS One 7(2):e32737. https://doi.org/10.1371/journal.pone.0032737

99. Han X, Rozen S, Boyle SH, Hellegers C, Cheng H, Burke JR, Welsh-Bohmer KA, Doraiswamy PM, Kaddurah-Daouk R (2011) Metabolomics in early Alzheimer's disease: identification of altered plasma sphingolipidome using shotgun lipidomics. PLoS One 6 (7):e21643. https://doi.org/10.1371/journal.pone.0021643

100. Mapstone M, Cheema AK, Fiandaca MS, Zhong X, Mhyre TR, MacArthur LH, Hall WJ, Fisher SG, Peterson DR, Haley JM, Nazar MD, Rich SA, Berlau DJ, Peltz CB, Tan MT, Kawas CH, Federoff HJ (2014) Plasma phospholipids identify antecedent memory impairment in older adults. Nat Med 20(4):415–418. https://doi.org/10.1038/nm.3466

101. Kang J, Lu J, Zhang X (2015) Metabolomics-based promising candidate biomarkers and pathways in Alzheimer's disease. Pharmazie 70(5):277–282

102. Ansoleaga B, Jove M, Schluter A, Garcia-Esparcia P, Moreno J, Pujol A, Pamplona R, Portero-Otin M, Ferrer I (2015) Deregulation of purine metabolism in Alzheimer's disease. Neurobiol Aging 36(1):68–80. https://doi.org/10.1016/j.neurobiolaging.2014.08.004

103. Mattson MP (1998) Modification of ion homeostasis by lipid peroxidation: roles in neuronal degeneration and adaptive plasticity. Trends Neurosci 21(2):53–57

104. Mattson MP, Gleichmann M, Cheng A (2008) Mitochondria in neuroplasticity and neurological disorders. Neuron 60 (5):748–766. https://doi.org/10.1016/j.neuron.2008.10.010

105. Payne BAI, Chinnery PF (2015) Mitochondrial dysfunction in aging: Much progress but many unresolved questions. Biochimica et Biophysica Acta 1847(11):1347–1353

106. Mouchiroud L, Houtkooper RH, Moullan N, Katsyuba E, Ryu D, Cantó C, Mottis A, Jo

Y-S, Viswanathan M, Schoonjans K, Guarente L, Auwerx J (2013) The NAD(+)/sirtuin pathway modulates longevity through activation of mitochondrial UPR and FOXO signaling. Cell 154(2):430–441. https://doi.org/10.1016/j.cell.2013.06.016

107. Ibanez C, Simo C, Martin-Alvarez PJ, Kivipelto M, Winblad B, Cedazo-Minguez A, Cifuentes A (2012) Toward a predictive model of Alzheimer's disease progression using capillary electrophoresis-mass spectrometry metabolomics. Anal Chem 84 (20):8532–8540. https://doi.org/10.1021/ac301243k

108. Graham SF, Chevallier OP, Roberts D, Holscher C, Elliott CT, Green BD (2013) Investigation of the human brain metabolome to identify potential markers for early diagnosis and therapeutic targets of Alzheimer's disease. Anal Chem 85(3):1803–1811. https://doi.org/10.1021/ac303163f

109. Wishart DS (2015) Is cancer a genetic disease or a metabolic disease? EBioMedicine 2 (6):478–479. https://doi.org/10.1016/j.ebiom.2015.05.022

110. Hanahan D, Weinberg RA Hallmarks of cancer: the next generation. Cell 144 (5):646–674. https://doi.org/10.1016/j.cell.2011.02.013

111. Otto AM (2016) Warburg effect(s)—a biographical sketch of Otto Warburg and his impacts on tumor metabolism. Cancer Metab 4:5. https://doi.org/10.1186/s40170-016-0145-9

112. Yang M, Soga T, Pollard PJ (2013) Oncometabolites: linking altered metabolism with cancer. J Clin Invest 123(9):3652–3658. https://doi.org/10.1172/JCI67228

113. Morin A, Letouze E, Gimenez-Roqueplo AP, Favier J (2014) Oncometabolites-driven tumorigenesis: From genetics to targeted therapy. Int J Cancer 135(10):2237–2248. https://doi.org/10.1002/ijc.29080

114. Wang X, Yang K, Xie Q, Wu Q, Mack SC, Shi Y, Kim LJ, Prager BC, Flavahan WA, Liu X, Singer M, Hubert CG, Miller TE, Zhou W, Huang Z, Fang X, Regev A, Suva ML, Hwang TH, Locasale JW, Bao S, Rich JN (2017) Purine synthesis promotes maintenance of brain tumor initiating cells in glioma. Nat Neurosci 20(5):661–673. https://doi.org/10.1038/nn.4537

115. Chang CH, Pearce EL (2016) Emerging concepts of T cell metabolism as a target of immunotherapy. Nat Immunol 17 (4):364–368. https://doi.org/10.1038/ni.3415

116. Shin SY, Fauman EB, Petersen AK, Krumsiek J, Santos R, Huang J, Arnold M, Erte I, Forgetta V, Yang TP, Walter K, Menni C, Chen L, Vasquez L, Valdes AM, Hyde CL, Wang V, Ziemek D, Roberts P, Xi L, Grundberg E, Multiple Tissue Human Expression Resource Consortium, Waldenberger M, Richards JB, Mohney RP, Milburn MV, John SL, Trimmer J, Theis FJ, Overington JP, Suhre K, Brosnan MJ, Gieger C, Kastenmuller G, Spector TD, Soranzo N (2014) An atlas of genetic influences on human blood metabolites. Nat Genet 46 (6):543–550. https://doi.org/10.1038/ng.2982

117. Draisma HH, Pool R, Kobl M, Jansen R, Petersen AK, Vaarhorst AA, Yet I, Haller T, Demirkan A, Esko T, Zhu G, Bohringer S, Beekman M, van Klinken JB, Romisch-Margl W, Prehn C, Adamski J, de Craen AJ, van Leeuwen EM, Amin N, Dharuri H, Westra HJ, Franke L, de Geus EJ, Hottenga JJ, Willemsen G, Henders AK, Montgomery GW, Nyholt DR, Whitfield JB, Penninx BW, Spector TD, Metspalu A, Eline Slagboom P, van Dijk KW, t Hoen PA, Strauch K, Martin NG, van Ommen GJ, Illig T, Bell JT, Mangino M, Suhre K, McCarthy MI, Gieger C, Isaacs A, van Duijn CM, Boomsma DI (2015) Genome-wide association study identifies novel genetic variants contributing to variation in blood metabolite levels. Nat Commun 6:7208. https://doi.org/10.1038/ncomms8208

118. Suhre K, Shin SY, Petersen AK, Mohney RP, Meredith D, Wagele B, Altmaier E, CARDIoGRAM, Deloukas P, Erdmann J, Grundberg E, Hammond CJ, de Angelis MH, Kastenmuller G, Kottgen A, Kronenberg F, Mangino M, Meisinger C, Meitinger T, Mewes HW, Milburn MV, Prehn C, Raffler J, Ried JS, Romisch-Margl W, Samani NJ, Small KS, Wichmann HE, Zhai G, Illig T, Spector TD, Adamski J, Soranzo N, Gieger C (2011) Human metabolic individuality in biomedical and pharmaceutical research. Nature 477(7362):54–60. https://doi.org/10.1038/nature10354

119. Kaddurah-Daouk R, Weinshilboum R, Pharmacometabolomics Research Network (2015) Metabolomic signatures for drug response phenotypes: pharmacometabolomics enables precision medicine. Clin Pharmacol Ther 98(1):71–75. https://doi.org/10.1002/cpt.134

120. Everett JR (2016) From metabonomics to pharmacometabonomics: the role of metabolic profiling in personalized medicine.

Front Pharmacol 7:297. https://doi.org/10.3389/fphar.2016.00297

121. Lewis JP, Yerges-Armstrong LM, Ellero-Simatos S, Georgiades A, Kaddurah-Daouk R, Hankemeier T (2013) Integration of pharmacometabolomic and pharmacogenomic approaches reveals novel insights into antiplatelet therapy. Clin Pharmacol Ther 94(5):570–573. https://doi.org/10.1038/clpt.2013.153

122. Neavin D, Kaddurah-Daouk R, Weinshilboum R (2016) Pharmacometabolomics informs Pharmacogenomics. Metabolomics 12(7). https://doi.org/10.1007/s11306-016-1066-x

123. Nicholson JK, Connelly J, Lindon JC, Holmes E (2002) Metabonomics: a platform for studying drug toxicity and gene function. Nat Rev Drug Discov 1(2):153–161. https://doi.org/10.1038/nrd728

124. Spear BB, Heath-Chiozzi M, Huff J (2001) Clinical application of pharmacogenetics. Trends Mol Med 7(5):201–204

125. Bosilkovska M, Samer CF, Deglon J, Rebsamen M, Staub C, Dayer P, Walder B, Desmeules JA, Daali Y (2014) Geneva cocktail for cytochrome p450 and P-glycoprotein activity assessment using dried blood spots. Clin Pharmacol Ther 96(3):349–359. https://doi.org/10.1038/clpt.2014.83

126. Konig J, Muller F, Fromm MF (2013) Transporters and drug-drug interactions: important determinants of drug disposition and effects. Pharmacol Rev 65(3):944–966. https://doi.org/10.1124/pr.113.007518

127. Kohler GI, Bode-Boger SM, Busse R, Hoopmann M, Welte T, Boger RH (2000) Drug-drug interactions in medical patients: effects of in-hospital treatment and relation to multiple drug use. Int J Clin Pharmacol Ther 38(11):504–513

128. Chainuvati S, Nafziger AN, Leeder JS, Gaedigk A, Kearns GL, Sellers E, Zhang Y, Kashuba AD, Rowland E, Bertino JS Jr (2003) Combined phenotypic assessment of cytochrome p450 1A2, 2C9, 2C19, 2D6, and 3A, N-acetyltransferase-2, and xanthine oxidase activities with the "Cooperstown 5+1 cocktail". Clin Pharmacol Ther 74(5):437–447. https://doi.org/10.1016/S0009-9236(03)00229-7

129. Dumond JB, Vourvahis M, Rezk NL, Patterson KB, Tien HC, White N, Jennings SH, Choi SO, Li J, Wagner MJ, La-Beck NM, Drulak M, Sabo JP, Castles MA, Macgregor TR, Kashuba AD (2010) A phenotype-genotype approach to predicting CYP450 and P-glycoprotein drug interactions with the mixed inhibitor/inducer tipranavir/ritonavir. Clin Pharmacol Ther 87(6):735–742. https://doi.org/10.1038/clpt.2009.253

130. Daali Y, Samer C, Deglon J, Thomas A, Chabert J, Rebsamen M, Staub C, Dayer P, Desmeules J (2012) Oral flurbiprofen metabolic ratio assessment using a single-point dried blood spot. Clin Pharmacol Ther. https://doi.org/10.1038/clpt.2011.247

131. Bosilkovska M, Samer C, Deglon J, Thomas A, Walder B, Desmeules J, Daali Y (2016) Evaluation of mutual drug-drug interaction within Geneva cocktail for cytochrome P450 phenotyping using innovative dried blood sampling method. Basic Clin Pharmacol Toxicol 119(3):284–290. https://doi.org/10.1111/bcpt.12586

132. Krumsiek J, Suhre K, Evans AM, Mitchell MW, Mohney RP, Milburn MV, Wagele B, Romisch-Margl W, Illig T, Adamski J, Gieger C, Theis FJ, Kastenmuller G (2012) Mining the unknown: a systems approach to metabolite identification combining genetic and metabolic information. PLoS Genet 8(10):e1003005. https://doi.org/10.1371/journal.pgen.1003005

133. Tay-Sontheimer J, Shireman LM, Beyer RP, Senn T, Witten D, Pearce RE, Gaedigk A, Gana Fomban CL, Lutz JD, Isoherranen N, Thummel KE, Fiehn O, Leeder JS, Lin YS (2014) Detection of an endogenous urinary biomarker associated with CYP2D6 activity using global metabolomics. Pharmacogenomics 15(16):1947–1962. https://doi.org/10.2217/pgs.14.155

134. Waters MD, Fostel JM (2004) Toxicogenomics and systems toxicology: aims and prospects. Nat Rev Genet 5(12):936–948. https://doi.org/10.1038/nrg1493

135. Bersanelli M, Mosca E, Remondini D, Giampieri E, Sala C, Castellani G, Milanesi L (2016) Methods for the integration of multi-omics data: mathematical aspects. BMC Bioinformatics 17(Suppl 2):15. https://doi.org/10.1186/s12859-015-0857-9

Chapter 2

Metabolomics in Immunology Research

Bart Everts

Abstract

There is a growing appreciation that metabolic processes and individual metabolites can shape the function of immune cells and thereby play important roles in the outcome of immune responses. In this respect, the use of MS- and NMR spectroscopy-based platforms to characterize and quantify metabolites in biological samples has recently yielded important novel insights into how our immune system functions and has contributed to the identification of biomarkers for immune-mediated diseases. Here, these recent immunological studies in which metabolomics has been used and made significant contributions to these fields will be discussed. In particular the role of metabolomics to the rapidly advancing field of cellular immunometabolism will be highlighted as well as the future prospects of such metabolomic tools in immunology.

Key words Mass spectrometry, NMR, Metabolite tracing, Immune cells, Immunometabolism, Cellular metabolism

1 Introduction

Over the past three decades, it has become increasingly clear that our immune system play a key role not only in the protection against invading pathogens but also in many physiological processes to maintain whole-body homeostasis as well as in the aetiology of a large number of noninfectious diseases. Therefore, there have been tremendous efforts to understand in great detail how our immune system functions, to be able to identify targets and design approaches to manipulate immune cells or immunological pathways for therapeutic purposes. In addition, there is a major interest in developing approaches to reliably track and predict certain immune response parameters that could be used as diagnostic and/or prognostic biomarkers for susceptibility, progression, and outcome of immune-mediated diseases.

In recent years, the development and use of a range of analytical platforms, including gas (GC) or liquid chromatography (LC) coupled to mass spectrometry (MS) or nuclear magnetic resonance (NMR) spectroscopy, to detect, characterize, and quantify intra- or

Martin Giera (ed.), *Clinical Metabolomics: Methods and Protocols*, Methods in Molecular Biology, vol. 1730,
https://doi.org/10.1007/978-1-4939-7592-1_2, © Springer Science+Business Media, LLC 2018

extracellular metabolites and related metabolic pathways, have made important contributions to our understanding of how our immune system functions as well as to the identification of biomarkers of immunological diseases. I will here highlight some of the recent immunological studies in which these metabolomic approaches have provided new insights in these fields, with particular emphasis on the field of cellular immunometabolism. In addition, the pros and cons as well as the future prospects of such metabolomic tools in immunology research will be discussed.

2 Metabolics as a Tool to Study Cellular Immunometabolism

It has long been appreciated, especially in the cancer field, that changes in cellular activation and proliferation coincide with, and are underpinned by, alterations in cellular metabolic state [1, 2]. During the last couple of years, immunologists have come to realize that also immune cell activation, proliferation, fate, and function are closely linked to and dependent on activation of specific intracellular metabolic pathways. This has resulted in the emergence of the currently burgeoning field of immunometabolism [3]. Thus far, most studies have focused on addressing the role of core metabolic pathways like glycolysis, lipid metabolism, mitochondrial fatty acid oxidation (FAO), and oxidative phosphorylation (OXPHOS) in the regulation of immune cell biology. However, apart from the classical role of metabolism in bioenergetics and biosynthesis, recent studies have also highlighted that metabolites or enzymes within these pathways can additionally act as signaling molecules or have metabolism-independent "moonlighting" functions, respectively. This further illustrates the intricate and multilayered regulation of immune cell biology by cellular metabolic pathways. Importantly, several seminal studies with T cells and macrophages have shown that targeting these metabolic pathways can greatly influence their functional properties. This has led to the emerging concept that manipulation of immune cell metabolism could be a powerful approach to direct immune responses [4].

Several methods are available to assess cellular metabolism of immune cells and range from fluorescently labeled mitochondrial dyes or glucose and fatty acids that can be analyzed in individual cells by flow cytometry to tracing of uptake and of enzymatic conversion of radioactively labeled metabolites by scintillation counting and to real-time analysis of oxygen consumption and extracellular acidification as measures of mitochondrial OXPHOS and glycolysis, respectively, using the Seahorse XF flux analyzer [5, 3]. However, a major limitation of all of these techniques is that they are targeted approaches that enable one to assess only the activity of predefined metabolic pathways, whereas in many cases unbiased metabolic analyses are needed to identify novel metabolic signatures and study their role in immune cell biology. This is where untargeted metabolomics has proven to be a valuable and essential tool in the field of cellular immunometabolism. In the next sections, the recent studies will be discussed in which metabolomic platforms have provided novel insights in the metabolic characteristics and regulation of different types of immune cells. These studies illustrate that metabolomics has become an important and integral part of immune cell metabolism research.

2.1 Metabolomics as a Tool to Study Macrophage Metabolism

The role of metabolism in immune cell function has been most well-characterized in macrophages and T cells. Already more than a decade ago, it was observed that pro-inflammatory classically activated ("M1") macrophages and anti-inflammatory alternatively activated ("M2") macrophages have very distinct metabolic properties on which they depend for their polarization [6]. M1 macrophages are characterized by enhanced aerobic glycolysis and reduced mitochondrial OXPHOS. The use of GC/MS-based untargeted metabolic profiling has been instrumental in providing new insights into how M1 macrophage metabolism shapes their biology. MS-based metabolomics revealed a dramatic buildup of both succinate, a Krebs cycle intermediate, and itaconate, a product of citrate, in M1-polarized macrophages [7–9]. This led to the characterization of succinate as being crucial for IL-1β expression by macrophages [7], a process that was later identified to be antagonized by itaconate [9]. Importantly, inhibition of succinate or itaconate synthesis reduced or enhanced the pro-inflammatory nature of the targeted macrophages, respectively, and thereby outcome of models of macrophage-driven diseases in vivo [7–9]. In addition, targeted MS-based platforms have been used to specifically study the lipid profile of macrophages in response to M1-polarizing stimuli. This revealed immediate responses in fatty acid metabolism represented by increases in eicosanoid synthesis and delayed responses characterized by sphingolipid and sterol biosynthesis [10, 11].

In contrast to M1 macrophages, M2 macrophages are metabolically characterized by enhanced mitochondrial OXPHOS fueled by FAO. Using GC/MS-based global metabolic profiling, it was shown that M2 macrophages have increased intracellular levels of various monoacylglycerides, which was suggestive of enhanced lipolysis in these cells [12]. This observation resulted in the identification of a cell-intrinsic lysosomal lipolysis pathway that is essential for alternative activation of macrophages presumably by means of generating free fatty acids (FAs) for FAO [12]. Another example of how MS-based metabolomics has contributed to our understanding of how M2 polarization and cellular metabolism are linked comes from a study in which an integrated network analysis was performed of both transcriptomic and metabolomic data [13]. This integrated approach resulted in the identification of

M2-specific metabolic nodes/pathways that only based on transcriptomic or metabolomic data analysis would have been missed. Specifically, M2 polarization was found to activate glutamine catabolism and UDP-GlcNAc-associated modules. Correspondingly, glutamine deprivation or inhibition of N-glycosylation decreased M2 polarization [13].

2.2 Metabolomics as a Tool to Study T-Cell Metabolism

The metabolic requirements for T-cell activation, proliferation, effector cytokine production, and memory formation have been extensively studied in recent years. In particular the interplay between metabolism and these processes has been well-characterized in $CD8^+$ T cells [14]. One of the early studies linking T-cell activation and proliferation to metabolic reprogramming used MS-based metabolomics to characterize the metabolic changes that occur in $CD8^+$ T cells upon activation [15]. Preceding cellular division, sharp increases in intracellular levels of many metabolites were observed, such as amino acids, lipids, and nucleotides, indicative of increased biosynthesis required for cell growth and proliferation. This was dependent on transcription factor c-Myc. Interestingly, using MS-based untargeted metabolomics, upon the first cell division of $CD8^+$ T cells, c-Myc, mTORC1 (a central regulator of anabolic metabolism), and several amino acids were found to be unequally split between daughter cells. This asymmetric division in metabolic regulators and metabolites resulted in stronger glutaminolysis and glycolysis in the cells containing higher levels of c-Myc and mTORC1 and subsequently in superior proliferative capacity and effector function compared to c-Myc^{low} daughter cells [16]. In further support for an important role for the need of c-Myc in effector T-cell function, another study demonstrated, through the use of LC/MS-based targeted quantification of UDP-GlcNAc, that c-Myc regulates intracellular protein O-GlcNAcylation at key stages of T-cell development and differentiation [17]. As a consequence, loss of O-GlcNAc transferase blocked T-cell progenitor renewal and T-cell clonal expansion.

The formation of memory cells requires distinct sets of metabolic processes and is known to be favored by mitochondrial-centered catabolic metabolism in which FAO plays an important role [14]. Also here, metabolomics has provided important new insights into the metabolic processes underpinning $CD8^+$ T-cell memory formation. For instance, targeted lipidomics has revealed that analogous to M2 macrophages, memory $CD8^+$ T cells use cell-intrinsic lipolysis to support the metabolic programming necessary for development [18]. Moreover, autophagy-deficient $CD8^+$ T cells ($atg7^{-/-}$) failed to efficiently develop into memory T cells. This was found to be the result of defects in FAO in $atg7^{-/-}$ cells as determined by LC/MS metabolomics [19].

The metabolic programs required for $CD4^+$ T-cell effector function and memory formation are less well defined. However, untargeted metabolome analysis revealed that also in early activated CD4[+] T cells, intracellular pools of metabolites, in particular lipids and FAs, are elevated, presumably acting as a pool to generate cell membranes that allows for cellular proliferation [20]. Consistent with a role for mTORC1 in CD8[+] T-cell activation, this increase in lipids was found to be dependent on mTORC1 activity [20]. CD4[+] T cells can differentiate into different T-helper (Th) cell subsets including T follicular helper (Tfh) cells, important for aiding B cells in producing antibodies, and regulatory T cells (Tregs) that are known for their ability to suppress immune responses. Pathway analyses of altered metabolites based on LC/MS-based untargeted metabolomics in Tfh cells showed that amino acid metabolism, amino sugar metabolism, and glycolytic intermediates trended higher compared to Th1 or Th17 cells, while Tfh cells had lower amount of TCA metabolites and intermediates in cysteine and glycerophospholipid metabolism, suggesting that different Th cell subsets have unique metabolic profiles [21]. Early studies have suggested that Tregs are mostly oxidative and rely on FAO for their differentiation and function [22]. More recently, it was found that Tregs are in fact metabolically flexible and engage glycolysis and anabolic metabolism in order to proliferate but dampen these metabolic pathways to acquire an immunosuppressive phenotype [23]. Mechanistically, MS-based metabolome analysis revealed that this is a direct result of the ability of Foxp3, the master regulator of Treg differentiation, to suppress expression of genes involved in glycolysis and anabolic metabolism [23]. Finally, by using LC/MS, a recent study found that in activated CD4[+] T cells, arginine is selectively depleted [24]. Restoring arginine levels in these cells promoted their ability to form memory cells, which was associated with enhanced mitochondrial metabolism. This might suggest that analogous to CD8[+] T cells, memory formation by CD4[+] T cells is promoted by increased oxidative metabolism.

2.3 Metabolomics as a Tool to Study Metabolism of Other Immune Cells

Thus far there have only been a handful of studies that have started to explore the metabolic properties through the use of metabolomics of other immune cells and include innate lymphoid cells (ILCs), dendritic cells (DCs), and monocytes. For instance, type 2 ILCs (ILC2s), which contribute to the initiation and maintenance of type 2 immune responses, were found to express high levels of arginase 1 (Arg1), an enzyme that converts arginine into ornithine [25]. Arg1 expression was shown to be essential for ILC2 function, as inhibition of Arg1 blocked ILC2 proliferation and effector function. LC/MS-based metabolomics revealed that Arg1 enzymatic activity was important for polyamine biosynthesis and promoting glycolysis [25], thereby linking metabolic reprogramming to cellular function. Furthermore, metabolomic platforms have been used to study metabolism of DCs, the cells that are crucial for priming of T-cell responses. It was found that activation of pro-inflammatory DCs by toll-like receptor (TLR) ligands resulted in rapid metabolic changes in these cells, characterized by increased glycolysis and TCA cycle activity as determined by untargeted MS-based metabolic profiling [26]. The authors went on to show that this glycolytic reprogramming was essential for TLR-induced activation by means of supporting de novo FA synthesis required for ER and Golgi membrane expansion and thereby efficient expression of activation markers and cytokines. More recently, it was demonstrated through the use of MS-based metabolomics that also in monocytes TLR ligands induce glycolysis [27]. Interestingly, in the same analysis, it was observed that LPS and Pam3Cys, ligands for TLR4 and TLR2, oppositely affected OXPHOS, suggesting that different TLR ligands can promote distinct metabolic programs, possibly tailored to the type of inflammatory response these different TLR ligands induce.

2.4 Isotope Labeling as a Tool to Study Immunometabolism

Global metabolomics can quantify relative or absolute differences in abundances of metabolites between different conditions, which may indicate altered activity of the associated enzymatic, nonenzymatic, or transport reactions. However, since an increase in metabolite concentration can be indicative of both increased activity of metabolite-generating enzymes as well as decreased activity of metabolite-consuming enzymes, concentration changes do not allow one to draw conclusions on metabolic rates and direction of fluxes. This requires dedicated tracing studies with isotope-labeled metabolites. Radioactive isotopes such as ^3H incorporated into glucose or palmitic acid are still being used to trace uptake and incorporation into lipids or oxidation of these metabolites in the mitochondria of immune cells. A downside of this technique is not only the fact that one has to work with radioactive material but also that these radiolabels do not allow for tracing of metabolite-derived atoms at high resolution into individual metabolites. This has spurred the utilization of nonradioactive heavy isotope (most frequently ^{13}C or ^{15}N)-labeled nutrients to examine intracellular fluxes. These heavy isotopes can be followed and quantified into individual intracellular metabolites using MS- or NMR-based techniques [28]. Tracing/flux studies are a crucial tool for interpreting differences in static metabolite profiles between different cells or conditions that are generated by untargeted metabolomic approaches. As such this technique has been applied in several immunometabolic studies, such as those that have defined metabolic fluxes in macrophages (tracing ^{13}C-glucose [29] and ^{13}C- and ^{15}N-glutamine [13]), T cells (tracing ^{13}C-acetate [30], serine [31], glutamine, or glucose [18, 32]), ILC2s (tracing ^{13}C-arginine [25]), and DCs (tracing ^{13}C-glucose [26]).

3   Metabolomics as a Tool to Study Effects of Extracellular Metabolites on Immune Cells

Apart from an important role for intracellular metabolites and associated metabolic pathways in regulating the biology of immune cells, it is also well appreciated that extracellular metabolites present in the microenvironment of immune cells can greatly influence their functional properties. In this respect extracellular metabolites can affect immune cell function by acting as direct nutrients for these cells or as signaling molecules. Some of these insights have been gained from studies using metabolomics.

3.1   Metabolomics as a Tool to Study Effects of Extracellular Nutrients on Immune Cell Function

The nutrients available to immune cells in their microenvironment are likely to dictate their functional properties. This can be particularly relevant in situations where local nutrient availability has become limiting. For instance, a recent study elegantly showed that in a tumor microenvironment, tumor-infiltrating CD8[+] T cells are impaired in their function, due to insufficient availability of glucose that has been depleted by highly glycolytic tumor cells [33]. Moreover, lactic acid buildup as a result of high glycolytic rates and subsequent acidosis in tumor microenvironments has also been described to impair the function of pro-inflammatory immune cells [34] as well as to foster Treg activity [35]. These studies focused only on effects of single metabolites. It is to be expected that unbiased extracellular metabolomics will help identifying additional metabolites that are depleted or accumulating in tumor microenvironments that may further explain the poor effector function of many immune cells present in tumors. Also in other conditions, such as in highly inflamed lymph nodes or poorly vascularized sites, it is likely that local nutrient levels can significantly drop to the extent that it could have a negative functional impact on immune cells. Conversely, exposure of immune cells to very nutrient- or lipid-rich environments such as those found in adipose tissues in obese individuals or atherosclerotic plaques can also have a profound impact on immune cell function [36–38]. However, an in-depth understanding of this metabolic interaction between immune cells and their environment is still missing, and extracellular metabolome analyses will become an important tool in the coming years to fully uncover the complex interplay between nutrient/metabolite availability and immune cell function.

3.2   Metabolomics as a Tool to Study Effects of Extracellular Signaling Metabolites on Immune Cell Function

The prototypical soluble mediators that mediate communication between cells are a family of proteins called cytokines. However, in addition to these proteins, various metabolites can also serve as signaling molecules to immune cells. In this respect, the most well-known family of metabolites that by interacting with G-protein-coupled receptors (GPCRs) act as signaling molecules in paracrine or autocrine manner are the eicosanoids which are lipids made by the enzymatic or nonenzymatic oxidation of arachidonic acid or other C20-polyunsaturated fatty acids. Classic examples include the prostaglandins, thromboxanes, leukotrienes, and resolvins that depending on the species involved promote pro- or antiinflammatory responses by immune cells. Major producers of these lipid mediators are macrophages. In several studies MS-based metabolomics has proven to be instrumental in generating a comprehensive picture of the eicosanoid profile produced by macrophages under different inflammatory conditions both in vitro and in vivo (reviewed by [39]).

Apart from host cell-derived metabolites, it is becoming increasingly clear that our microbiota, especially a variety of intestinal bacteria, produce metabolites that regulate various physiological processes including the immune system of the host. In this respect, one of the most well-studied groups of metabolites that acts as signaling molecules on immune cells and has been shown to play a crucial role in the proper development of a well-balanced immune system is the short-chain fatty acids (SCFAs). SCFAs are the products of digested fibers by certain microbial species in the gut and have been shown to primarily promote anti-inflammatory responses in immune cells, such as Treg responses, and as such have been implicated in protection against various inflammatory diseases [40, 41]. In two seminal studies, NMR-based [42] or MS-based [43] comparative metabolomics were used to identify butyrate as the main SCFA that could promote Tregs, illustrating how the use of metabolomics has contributed to this field. Moreover, several human studies have used NMR- and MS-based metabolome analyses to link butyrate concentrations in feces and other body fluids, with changes in microbial composition, host immune responses, and outcome of various diseases such as atopic dermatitis [44], cancer [45], ulcerative colitis [46], and gout [47]. Similar metabolomic approaches have been used to identify and track concentrations of other immune modulatory microbiota-derived metabolites and are recently reviewed in great detail elsewhere [48, 41].

Most signaling metabolites are recognized by immune cells via metabolite-specific GPCRs. Interestingly, many putative GPCRs are currently still orphan receptors, indicating that there are likely several more still unidentified host- and/or commensal microbiota-derived metabolites that could serve as signaling mediators and thereby potentially direct immune cell function [40]. For these kinds of studies, unbiased comparative "exo"-metabolomics will be an important tool to identify some of these new signaling metabolites [48].

4   Metabolomics as a Tool for Identification of Biomarkers of Immune Responses

As most diseases are immune-driven or have at least an immuno-logical component, there is currently great interest in identifying reliable biomarkers that allow one to track the immune status of diseased individuals. Such biomarkers can then be used as diagnostic and/or prognostic markers for susceptibility, progression, and outcome of immune-mediated diseases, which will provide essential pieces of information for effective management of these diseases. In this respect, metabolomic profiling using NMR spectroscopy of biofluids is proven to be a highly promising approach for biomarker identification as it is becoming clear that metabolite composition or individual metabolites within biofluids can serve as reliable biomarkers for certain immune parameters in both infectious [49, 50] and immune-mediated noninfectious diseases [51]. Moreover, a practical advantage is that it can be performed reliably on readily obtainable biofluids like urine, blood, and saliva, without the need for acquisition of samples from diseased tissues that often requires invasive techniques. Although the use of metabolomics as a tool for biomarker identification of immune responses and immune-mediated diseases is a recent development, it is likely to make important contributions to this rapidly advancing field.

5   Future Perspectives

The recent appreciation that metabolism and immune responses are tightly linked has spurred the use of MS- and NMR-based platforms to quantitatively measure metabolic profiles in immunological studies and is rapidly becoming part of the standard toolbox for immunologists to study the immune system. Until now, most studies in which metabolomics has been used to assess cellular immunometabolism have employed MS-based platforms as these currently have a higher sensitivity compared to NMR-based techniques, making them better suited for studying immune cell metabolism of primary immune cells that are often only available in low numbers. Conversely, NMR spectroscopy has been the primary platform for analysis of biological samples used for biomarker identification and profiling, mostly because of the simplicity of sample preparation that does not rely on prior separation of the analytes. Moreover, the lower sensitivity of NMR spectrometry is less of an issue for biofluids as these samples can often be obtained in larger quantities. However, now with the development of NMR spectrometers with stronger magnets that result in higher resolution and sensitivity of this platform, it is to be expected that this technique will also be more implemented in cellular immunometabolomic studies. An important consideration is that due to the complexity of the metabolome and the diverse properties of metabolites, no single analytical platform can be applied to detect all metabolites in a biological sample. Hence, integrated complementary platforms will need to be used to reliably cover the full range of metabolites present in biological samples [52].

With improved sensitivity of individual metabolomic platforms as well as increasing integration of data from different metabolic platforms, there is a strong need for well-defined computational modeling pipelines to identify and correlate individual metabolites or metabolic pathways to immunological parameters of interest. Moreover, within the field of immunology as well as immunometabolism, there is a growing trend to use systems biology, which applies various unbiased "omics" approaches (such as proteomics, transcriptomics, metabolomics), to address immunological and immunometabolic questions. This makes the development of bioinformatic tools that enable one to interpret as well as integrate these kinds of complex data sets even more important. Recently, for instance, a web-based tool for integrated analysis of steady-state metabolomic and transcriptional data has been developed that focuses on identification of the most changing metabolic networks between two conditions of interest [53]. This has already proven to be a valuable bioinformatic tool for studying macrophage metabolism [13]. Bioinformatic pipelines such as this will be key to reveal the full depth of metabolic information that is captured by these different omics approaches. As such it is to be expected that in the coming years, NMR- and MS-based metabolomic platforms in conjunction with novel bioinformatic pipelines will further establish themselves as critical tools for unraveling the role of metabolism in immune cell biology and identifying metabolic biomarkers for immune responses and immune-related diseases.

Acknowledgments

This work was in part funded by an LUMC and Marie Curie fellowship (#631585). I declare to have no competing interests.

References

1. Vander Heiden MG (2011) Targeting cancer metabolism: a therapeutic window opens. Nat Rev Drug Discov 10(9):671–684. https://doi.org/10.1038/nrd3504

2. Vander Heiden MG, Cantley LC, Thompson CB (2009) Understanding the Warburg effect: the metabolic requirements of cell proliferation. Science 324(5930):1029–1033. https://doi.org/10.1126/science.1160809

3. O'Neill LA, Kishton RJ, Rathmell J (2016) A guide to immunometabolism for immunologists. Nat Rev Immunol 16(9):553–565. https://doi.org/10.1038/nri.2016.70

4. O'Sullivan D, Pearce EL (2015) Targeting T cell metabolism for therapy. Trends Immunol 36(2):71–80. https://doi.org/10.1016/j.it.2014.12.004

5. Pelgrom LR, van der Ham AJ, Everts B (2016) Analysis of TLR-induced metabolic changes in dendritic cells using the seahorse XF(e)96 extracellular flux analyzer. Methods Mol Biol 1390:273–285. https://doi.org/10.1007/978-1-4939-3335-8_17

6. Van den Bossche J, O'Neill LA, Menon D (2017) Macrophage immunometabolism: where are we (Going)? Trends Immunol. https://doi.org/10.1016/j.it.2017.03.001

7. Tannahill GM, Curtis AM, Adamik J, Palsson-McDermott EM, McGettrick AF, Goel G, Frezza C, Bernard NJ, Kelly B, Foley NH, Zheng L, Gardet A, Tong Z, Jany SS, Corr SC, Haneklaus M, Caffrey BE, Pierce K, Walmsley S, Beasley FC, Cummins E, Nizet V, Whyte M, Taylor CT, Lin H, Masters SL, Gottlieb E, Kelly VP, Clish C, Auron PE, Xavier RJ, O'Neill LA (2013) Succinate is an inflammatory signal that induces IL-1beta through HIF-1alpha. Nature 496 (7444):238–242. https://doi.org/10.1038/nature11986

8. Mills EL, Kelly B, Logan A, Costa AS, Varma M, Bryant CE, Tourlomousis P, Dabritz JH, Gottlieb E, Latorre I, Corr SC, McManus G, Ryan D, Jacobs HT, Szibor M, Xavier RJ, Braun T, Frezza C, Murphy MP, O'Neill LA (2016) Succinate dehydrogenase supports metabolic repurposing of mitochondria to drive inflammatory macrophages. Cell 167(2):457–470.e413. https://doi.org/10.1016/j.cell.2016.08.064

9. Lampropoulou V, Sergushichev A, Bambouskova M, Nair S, Vincent EE, Loginicheva E, Cervantes-Barragan L, Ma X, Huang SC, Griss T, Weinheimer CJ, Khader S, Randolph GJ, Pearce EJ, Jones RG, Diwan A, Diamond MS, Artyomov MN (2016) Itaconate links inhibition of succinate dehydrogenase with macrophage metabolic remodeling and regulation of inflammation. Cell Metab 24(1):158–166. https://doi.org/10.1016/j.cmet.2016.06.004

10. Dennis EA, Deems RA, Harkewicz R, Quehenberger O, Brown HA, Milne SB, Myers DS, Glass CK, Hardiman G, Reichart D, Merrill AH Jr, Sullards MC, Wang E, Murphy RC, Raetz CR, Garrett TA, Guan Z, Ryan AC, Russell DW, McDonald JG, Thompson BM, Shaw WA, Sud M, Zhao Y, Gupta S, Maurya MR, Fahy E, Subramaniam S (2010) A mouse macrophage lipidome. J Biol Chem 285(51):39976–39985. https://doi.org/10.1074/jbc.M110.182915

11. Lee JW, Mok HJ, Lee DY, Park SC, Kim GS, Lee SE, Lee YS, Kim KP, Kim HD (2017) UPLC-QqQ/MS-based lipidomics approach to characterize lipid alterations in inflammatory macrophages. J Proteome Res 16(4): 1460–1469. https://doi.org/10.1021/acs.jproteome.6b00848

12. Huang SC, Everts B, Ivanova Y, O'Sullivan D, Nascimento M, Smith AM, Beatty W, Love-Gregory L, Lam WY, O'Neill CM, Yan C, Du H, Abumrad NA, Urban JF Jr, Artyomov MN, Pearce EL, Pearce EJ (2014) Cell-intrinsic lysosomal lipolysis is essential for alternative activation of macrophages. Nat Immunol 15(9):846–855. https://doi.org/10.1038/ni.2956

13. Jha AK, Huang SC, Sergushichev A, Lampropoulou V, Ivanova Y, Loginicheva E, Chmielewski K, Stewart KM, Ashall J, Everts B, Pearce EJ, Driggers EM, Artyomov MN (2015) Network integration of parallel metabolic and transcriptional data reveals metabolic modules that regulate macrophage polarization. Immunity 42(3):419–430. https://doi.org/10.1016/j.immuni.2015.02.005

14. van der Windt GJ, Pearce EL (2012) Metabolic switching and fuel choice during T-cell differentiation and memory development. Immunol Rev 249(1):27–42. https://doi.org/10.1111/j.1600-065X.2012.01150.x

15. Wang R, Dillon CP, Shi LZ, Milasta S, Carter R, Finkelstein D, McCormick LL, Fitzgerald P, Chi H, Munger J, Green DR (2011) The transcription factor Myc controls metabolic reprogramming upon T lymphocyte activation. Immunity 35(6):871–882. https://doi.org/10.1016/j.immuni.2011.09.021

16. Verbist KC, Guy CS, Milasta S, Liedmann S, Kaminski MM, Wang R, Green DR (2016) Metabolic maintenance of cell asymmetry following division in activated T lymphocytes. Nature 532(7599):389–393. https://doi.org/10.1038/nature17442

17. Swamy M, Pathak S, Grzes KM, Damerow S, Sinclair LV, van Aalten DM, Cantrell DA (2016) Glucose and glutamine fuel protein O-GlcNAcylation to control T cell self-renewal and malignancy. Nat Immunol 17 (6):712–720. https://doi.org/10.1038/ni.3439

18. O'Sullivan D, van der Windt GJ, Huang SC, Curtis JD, Chang CH, Buck MD, Qiu J, Smith AM, Lam WY, DiPlato LM, Hsu FF, Birnbaum MJ, Pearce EJ, Pearce EL (2014) Memory CD8(+) T cells use cell-intrinsic lipolysis to support the metabolic programming necessary for development. Immunity 41(1):75–88. https://doi.org/10.1016/j.immuni.2014.06.005

19. Xu X, Araki K, Li S, Han JH, Ye L, Tan WG, Konieczny BT, Bruinsma MW, Martinez J, Pearce EL, Green DR, Jones DP, Virgin HW, Ahmed R (2014) Autophagy is essential for effector CD8(+) T cell survival and memory formation. Nat Immunol 15(12):1152–1161. https://doi.org/10.1038/ni.3025

20. Angela M, Endo Y, Asou HK, Yamamoto T, Tumes DJ, Tokuyama H, Yokote K, Nakayama T (2016) Fatty acid metabolic reprogramming via mTOR-mediated inductions of PPAR-gamma directs early activation of T cells. Nat Commun 7:13683. https://doi.org/10.1038/ncomms13683

21. Zeng H, Cohen S, Guy C, Shrestha S, Neale G, Brown SA, Cloer C, Kishton RJ, Gao X, Youngblood B, Do M, Li MO, Locasale JW, Rathmell JC, Chi H (2016) mTORC1 and mTORC2 kinase signaling and glucose metabolism drive follicular helper T cell differentiation. Immunity 45(3):540–554. https://doi.org/10.1016/j.immuni.2016.08.017

22. Gerriets VA, Rathmell JC (2012) Metabolic pathways in T cell fate and function. Trends Immunol 33(4):168–173. https://doi.org/10.1016/j.it.2012.01.010

23. Gerriets VA, Kishton RJ, Johnson MO, Cohen S, Siska PJ, Nichols AG, Warmoes MO, de Cubas AA, MacIver NJ, Locasale JW, Turka LA, Wells AD, Rathmell JC (2016) Foxp3 and Toll-like receptor signaling balance Treg cell anabolic metabolism for suppression. Nat Immunol 17(12):1459–1466. https://doi.org/10.1038/ni.3577

24. Geiger R, Rieckmann JC, Wolf T, Basso C, Feng Y, Fuhrer T, Kogadeeva M, Picotti P, Meissner F, Mann M, Zamboni N, Sallusto F, Lanzavecchia A (2016) L-arginine modulates T cell metabolism and enhances survival and antitumor activity. Cell 167(3):829–842.e813. https://doi.org/10.1016/j.cell.2016.09.031

25. Monticelli LA, Buck MD, Flamar AL, Saenz SA, Tait Wojno ED, Yudanin NA, Osborne LC, Hepworth MR, Tran SV, Rodewald HR, Shah H, Cross JR, Diamond JM, Cantu E, Christie JD, Pearce EL, Artis D (2016) Arginase 1 is an innate lymphoid-cell-intrinsic metabolic checkpoint controlling type 2 inflammation. Nat Immunol 17(6):656–665. https://doi.org/10.1038/ni.3421

26. Everts B, Amiel E, Huang SC, Smith AM, Chang CH, Lam WY, Redmann V, Freitas TC, Blagih J, van der Windt GJ, Artyomov MN, Jones RG, Pearce EL, Pearce EJ (2014) TLR-driven early glycolytic reprogramming via the kinases TBK1-IKKvarepsilon supports the anabolic demands of dendritic cell activation.

Nat Immunol 15(4):323–332. https://doi.org/10.1038/ni.2833

27. Lachmandas E, Boutens L, Ratter JM, Hijmans A, Hooiveld GJ, Joosten LA, Rodenburg RJ, Fransen JA, Houtkooper RH, van Crevel R, Netea MG, Stienstra R (2016) Microbial stimulation of different Toll-like receptor signalling pathways induces diverse metabolic programmes in human monocytes. Nat Microbiol 2:16246. https://doi.org/10.1038/nmicrobiol.2016.246

28. Buescher JM, Antoniewicz MR, Boros LG, Burgess SC, Brunengraber H, Clish CB, DeBerardinis RJ, Feron O, Frezza C, Ghesquiere B, Gottlieb E, Hiller K, Jones RG, Kamphorst JJ, Kibbey RG, Kimmelman AC, Locasale JW, Lunt SY, Maddocks OD, Malloy C, Metallo CM, Meuillet EJ, Munger J, Noh K, Rabinowitz JD, Ralser M, Sauer U, Stephanopoulos G, St-Pierre J, Tennant DA, Wittmann C, Vander Heiden MG, Vazquez A, Vousden K, Young JD, Zamboni N, Fendt SM (2015) A roadmap for interpreting (13)C metabolite labeling patterns from cells. Curr Opin Biotechnol 34:189–201. https://doi.org/10.1016/j.copbio.2015.02.003

29. Rodriguez-Prados JC, Traves PG, Cuenca J, Rico D, Aragones J, Martin-Sanz P, Cascante M, Bosca L (2010) Substrate fate in activated macrophages: a comparison between innate, classic, and alternative activation. J Immunol 185(1):605–614. https://doi.org/10.4049/jimmunol.0901698

30. Balmer ML, Ma EH, Bantug GR, Grahlert J, Pfister S, Glatter T, Jauch A, Dimeloe S, Slack E, Dehio P, Krzyzaniak MA, King CG, Burgener AV, Fischer M, Develioglu L, Belle R, Recher M, Bonilla WV, Macpherson AJ, Hapfelmeier S, Jones RG, Hess C (2016) Memory CD8(+) T cells require increased concentrations of acetate induced by stress for optimal function. Immunity 44(6):1312–1324. https://doi.org/10.1016/j.immuni.2016.03.016

31. Ma EH, Bantug G, Griss T, Condotta S, Johnson RM, Samborska B, Mainolfi N, Suri V, Guak H, Balmer ML, Verway MJ, Raissi TC, Tsui H, Boukhaled G, Henriques da Costa S, Frezza C, Krawczyk CM, Friedman A, Manfredi M, Richer MJ, Hess C, Jones RG (2017) Serine is an essential metabolite for effector T cell expansion. Cell Metab 25(2):345–357. https://doi.org/10.1016/j.cmet.2016.12.011

32. Blagih J, Coulombe F, Vincent EE, Dupuy F, Galicia-Vazquez G, Yurchenko E, Raissi TC,

van der Windt GJ, Viollet B, Pearce EL, Pelletier J, Piccirillo CA, Krawczyk CM, Divangahi M, Jones RG (2015) The energy sensor AMPK regulates T cell metabolic adaptation and effector responses in vivo. Immunity 42(1):41–54. https://doi.org/10.1016/j.immuni.2014.12.030

33. Chang CH, Qiu J, O'Sullivan D, Buck MD, Noguchi T, Curtis JD, Chen Q, Gindin M, Gubin MM, van der Windt GJ, Tonc E, Schreiber RD, Pearce EJ, Pearce EL (2015) Metabolic competition in the tumor microenvironment is a driver of cancer progression. Cell 162(6):1229–1241. https://doi.org/10.1016/j.cell.2015.08.016

34. Romero-Garcia S, Moreno-Altamirano MM, Prado-Garcia H, Sanchez-Garcia FJ (2016) Lactate contribution to the tumor microenvironment: mechanisms, effects on immune cells and therapeutic relevance. Front Immunol 7:52. https://doi.org/10.3389/fimmu.2016.00052

35. Angelin A, Gil-de-Gomez L, Dahiya S, Jiao J, Guo L, Levine MH, Wang Z, Quinn WJ III, Kopinski PK, Wang L, Akimova T, Liu Y, Bhatti TR, Han R, Laskin BL, Baur JA, Blair IA, Wallace DC, Hancock WW, Beier UH (2017) Foxp3 reprograms T cell metabolism to function in low-glucose, high-lactate environments. Cell Metab 25:1282. https://doi.org/10.1016/j.cmet.2016.12.018

36. Klein-Wieringa IR, Andersen SN, Kwekkeboom JC, Giera M, de Lange-Brokaar BJ, van Osch GJ, Zuurmond AM, Stojanovic-Susulic V, Nelissen RG, Pijl H, Huizinga TW, Kloppenburg M, Toes RE, Ioan-Facsinay A (2013) Adipocytes modulate the phenotype of human macrophages through secreted lipids. J Immunol 191(3):1356–1363. https://doi.org/10.4049/jimmunol.1203074

37. Ioan-Facsinay A, Kwekkeboom JC, Westhoff S, Giera M, Rombouts Y, van Harmelen V, Huizinga TW, Deelder A, Kloppenburg M, Toes RE (2013) Adipocyte-derived lipids modulate CD4+ T-cell function. Eur J Immunol 43(6):1578–1587. https://doi.org/10.1002/eji.201243096

38. Gistera A, Hansson GK (2017) The immunology of atherosclerosis. Nat Rev Nephrol 13:368. https://doi.org/10.1038/nrneph.2017.51

39. Tam VC (2013) Lipidomic profiling of bioactive lipids by mass spectrometry during microbial infections. Semin Immunol 25(3):240–248. https://doi.org/10.1016/j.smim.2013.08.006

40. Husted AS, Trauelsen M, Rudenko O, Hjorth SA, Schwartz TW (2017) GPCR-mediated signaling of metabolites. Cell Metab 25(4):777–796. https://doi.org/10.1016/j.cmet.2017.03.008

41. Lin L, Zhang J (2017) Role of intestinal microbiota and metabolites on gut homeostasis and human diseases. BMC Immunol 18(1):2. https://doi.org/10.1186/s12865-016-0187-3

42. Furusawa Y, Obata Y, Fukuda S, Endo TA, Nakato G, Takahashi D, Nakanishi Y, Uetake C, Kato K, Kato T, Takahashi M, Fukuda NN, Murakami S, Miyauchi E, Hino S, Atarashi K, Onawa S, Fujimura Y, Lockett T, Clarke JM, Topping DL, Tomita M, Hori S, Ohara O, Morita T, Koseki H, Kikuchi J, Honda K, Hase K, Ohno H (2013) Commensal microbe-derived butyrate induces the differentiation of colonic regulatory T cells. Nature 504(7480):446–450. https://doi.org/10.1038/nature12721

43. Arpaia N, Campbell C, Fan X, Dikiy S, van der Veeken J, deRoos P, Liu H, Cross JR, Pfeffer K, Coffer PJ, Rudensky AY (2013) Metabolites produced by commensal bacteria promote peripheral regulatory T-cell generation. Nature 504(7480):451–455. https://doi.org/10.1038/nature12726

44. Song H, Yoo Y, Hwang J, Na YC, Kim HS (2016) Faecalibacterium prausnitzii subspecies-level dysbiosis in the human gut microbiome underlying atopic dermatitis. J Allergy Clin Immunol 137(3):852–860. https://doi.org/10.1016/j.jaci.2015.08.021

45. Amiot A, Dona AC, Wijeyesekera A, Tournigand C, Baumgaertner I, Lebaleur Y, Sobhani I, Holmes E (2015) (1)H NMR spectroscopy of fecal extracts enables detection of advanced colorectal neoplasia. J Proteome Res 14(9):3871–3881. https://doi.org/10.1021/acs.jproteome.5b00277

46. Machiels K, Joossens M, Sabino J, De Preter V, Arijs I, Eeckhaut V, Ballet V, Claes K, Van Immerseel F, Verbeke K, Ferrante M, Verhaegen J, Rutgeerts P, Vermeire S (2014) A decrease of the butyrate-producing species Roseburia hominis and Faecalibacterium prausnitzii defines dysbiosis in patients with ulcerative colitis. Gut 63(8):1275–1283. https://doi.org/10.1136/gutjnl-2013-304833

47. Guo Z, Zhang J, Wang Z, Ang KY, Huang S, Hou Q, Su X, Qiao J, Zheng Y, Wang L, Koh E, Danliang H, Xu J, Lee YK, Zhang H (2016) Intestinal microbiota distinguish gout patients from healthy humans. Sci Rep 6:20602. https://doi.org/10.1038/srep20602

48. Donia MS, Fischbach MA (2015) HUMAN MICROBIOTA. Small molecules from the human microbiota. Science 349

(6246):1254766. https://doi.org/10.1126/science.1254766

49. Saric J (2010) Interactions between immunity and metabolism - contributions from the metabolic profiling of parasite-rodent models. Parasitology 137(9):1451–1466. https://doi.org/10.1017/S0031182010000697

50. Munshi SU, Rewari BB, Bhavesh NS, Jameel S (2013) Nuclear magnetic resonance based profiling of biofluids reveals metabolic dysregulation in HIV-infected persons and those on anti-retroviral therapy. PLoS One 8(5): e64298. https://doi.org/10.1371/journal.pone.0064298

51. Alonso A, Julia A, Vinaixa M, Domenech E, Fernandez-Nebro A, Canete JD, Ferrandiz C, Tornero J, Gisbert JP, Nos P, Casbas AG, Puig L, Gonzalez-Alvaro I, Pinto-Tasende JA, Blanco R, Rodriguez MA, Beltran A, Correig X, Marsal S (2016) Urine metabolome profiling of immune-mediated inflammatory diseases. BMC Med 14(1):133. https://doi.org/10.1186/s12916-016-0681-8

52. Zhang A, Sun H, Wang P, Han Y, Wang X (2012) Modern analytical techniques in metabolomics analysis. Analyst 137(2):293–300. https://doi.org/10.1039/c1an15605e

53. Sergushichev AA, Loboda AA, Jha AK, Vincent EE, Driggers EM, Jones RG, Pearce EJ, Artyomov MN (2016) GAM: a web-service for integrated transcriptional and metabolic network analysis. Nucleic Acids Res 44(W1): W194–W200. https://doi.org/10.1093/nar/gkw266

Part II

LC-MS-Based Metabolomics

## Chapter 3

LC-MS-Based Metabolomics of Biofluids Using All-Ion Fragmentation (AIF) Acquisition

Romanas Chaleckis, Shama Naz, Isabel Meister, and Craig E. Wheelock

Abstract

The field of liquid chromatography-mass spectrometry (LC-MS)-based nontargeted metabolomics has advanced significantly and can provide information on thousands of compounds in biological samples. However, compound identification remains a major challenge, which is crucial in interpreting the biological function of metabolites. Herein, we present a LC-MS method using the all-ion fragmentation (AIF) approach in combination with a data processing method using an in-house spectral library. For the purposes of increasing accuracy in metabolite annotation, up to four criteria are used: (1) accurate mass, (2) retention time, (3) MS/MS fragments, and (4) product/precursor ion ratios. The relative standard deviation between ion ratios of a metabolite in a biofluid vs. its analytical standard is used as an additional metric for confirming metabolite identity. Furthermore, we include a scheme to distinguish co-eluting isobaric compounds. Our method enables database-dependent targeted as well as nontargeted metabolomics analysis from the same data acquisition, while simultaneously improving the accuracy in metabolite identification to increase the quality of the resulting biological information.

Key words Metabolomics, Liquid chromatography-mass spectrometry (LC-MS), All-ion fragmentation (AIF), Metabolite annotation

Abbreviations

ACN	Acetonitrile
AIF	All-ion fragmentation
AM	Accurate mass
CID	Collision-induced dissociation
EIC	Extracted ion chromatogram
HILIC	Hydrophilic interaction liquid chromatography
LC-MS	Liquid chromatography-mass spectrometry
MeOH	Methanol
RP	Reverse phase
RT	Retention time

Romanas Chaleckis and Shama Naz contributed equally to this work.

Martin Giera (ed.), *Clinical Metabolomics: Methods and Protocols*, Methods in Molecular Biology, vol. 1730, https://doi.org/10.1007/978-1-4939-7592-1_3, © Springer Science+Business Media, LLC 2018

1   Introduction

Mass spectrometry (MS)-based metabolomics has been used extensively to investigate biological processes and perform biomarker discovery work [1, 2]. In addition, it is considered an important part of precision medicine initiatives [3]. However, metabolite annotation is still a major challenge and significant bottleneck in translating metabolomics data into biochemical context [4, 5]. Following the identification criteria proposed by the Metabolomics Standard Initiative (MSI), the highest level of metabolite identification is based upon matching two or more orthogonal properties [e.g., accurate mass (AM), retention time (RT)/index, isotopic pattern, MS/MS spectrum] of an authentic reference standard analyzed under the same condition as the metabolite of interest [6].

Liquid chromatography-mass spectrometry (LC-MS) is one of the most commonly applied techniques in the field of metabolomics, offering high sensitivity and wide metabolite coverage [7]. In LC-MS-based nontargeted metabolomics approaches, metabolites are generally reported based on AM and RT matching; however, these criteria are often insufficient to conclusively identify a metabolite due to co-elution and/or the presence of metabolites of similar m/z and molecular formula. Accordingly, these data are frequently complemented with MS/MS spectra to increase the certainty in the identification of metabolites of interest. While useful, this approach relies heavily upon the use of spectral databases to confirm metabolite structure. The number and quality of spectral databases has steadily increased, improving the confidence in metabolite identification; however, additional resources and efforts are needed for comprehensive structural confirmation.

Herein, we describe a LC-MS metabolomics method using an all-ion fragmentation (AIF) approach. An AIF experiment uses the low-energy collision-induced dissociation (CID) to acquire precursor ion mass spectra, whereas the high-energy CID is used to obtain product ion information by tandem mass spectrometry [8]. In the subsequent data analysis, extracted ion chromatograms (EIC) from any precursor or associated product ions of interest are extracted from the low- or high-energy scan data. One EIC is chosen for relative quantification (the quantifier ion) of the metabolite, and further product ions from the same compound are used as qualifier ions. This approach has the advantage of simultaneously acquiring extensive fragmentation data on the sample, which can then be searched against spectral databases for MS/MS-based confirmation. For the purposes of the reported method, an in-house spectral database was constructed using commercially available analytical standards [9].

To further increase the metabolite identification specificity, the product/precursor ion ratio was calculated, and the % relative error was calculated between each analytical standard and the corresponding compound in the biological matrix. The use of specific quantifier and qualifier ions is also used to distinguish co-eluting isobaric compounds. The described data acquisition strategy enables a simultaneous combination of database-dependent targeted and nontargeted metabolomics in combination with improved accuracy in metabolite identification, thus increasing the quality of the biological information acquired in a metabolomics experiment.

2   Materials

2.1   Samples

1. Plasma and/or serum samples kept at −80 °C until the day of analysis (*see* **Note 1**).

2. Urine samples kept at −80 °C until the day of analysis.

2.2   Chemicals and Standards

The analytical standards for the construction and expansion of the compound spectral database, as well as the internal standards, can be purchased from Sigma-Aldrich (St. Louis, USA), Cayman Chemical Company (Michigan, USA), Toronto Research Chemicals (Ontario, Canada), Zhejiang Ontores Biotechnologies Co., Ltd. (Zhejiang, China), and Avanti Polar Lipids, Inc. (Alabama, USA) depending upon availability. The standards are prepared at 1 mM concentrations in the appropriate solvent for dissolution, stored at −20 °C, and diluted appropriately on the day of analysis.

2.3   Solutions and Solvents

All solutions are prepared at room temperature (25 °C) using LC-MS grade solvents and analytical grade reagents.

1. ESI-low concentration tuning mix (Agilent Technologies, Santa Clara, USA) for calibrating the TOF-MS: For 100 mL of solution, mix 10 mL of ESI-low concentration tuning mix, 85.5 mL acetonitrile (ACN), 4.5 mL water, and 3 μL HP-321 (Agilent Technologies, Santa Clara, USA).

2. An internal lock mass mixture (Agilent Technologies, Santa Clara, USA) for continuous mass calibration during data acquisition: For 500 mL of solution, mix 475 mL ACN, 25 mL water, 0.2 mL purine [5 mM in (9:1, *v/v*) ACN/water], and 0.5 mL of HP-0921 [2.5 mM in (9:1, *v/v*) ACN/water].

3. Solvents for hydrophilic interaction liquid chromatography (HILIC).

   Solvent A (water with 0.1% formic acid): prepare by adding 1 mL formic acid to 1 L of water.

   Solvent B (ACN with 0.1% formic acid): prepare by adding 1 mL formic acid to 1 L of ACN.

4. Solvents for the reversed phase (RP) chromatography.

Solvent A (water with 0.1% formic acid): prepare by adding 1 mL formic acid to 1 L of water.

Solvent B [(90:10, *v/v*) 2-propanol:ACN]: prepare by adding 1 mL formic acid to 1 L of (90:10, *v/v*) 2-propanol:ACN.

5. A methanol (MeOH) crash solution for plasma/serum metabolite extraction or urine nonpolar metabolite extraction containing internal standards (Table 1) (*see* **Note 2**).

6. An ACN crash solution for urine polar metabolite extraction containing internal standards (Table 1) (*see* **Note 2**).

2.4 LC-MS System

1. Ultrahigh performance liquid chromatography (UHPLC) 1290 Infinity II system (including in-line filter 0.3 μm, Agilent Technologies, Santa Clara, USA) with 1260 Infinity II isocratic pump (including 1:100 splitter) coupled to a 6550 iFunnel quadrupole-time of flight (Q-TOF) mass spectrometer with a dual AJS electrospray ionization source (Agilent Technologies, Santa Clara, USA).

2. For separation of polar metabolites, HILIC SeQuant® ZIC®-HILIC column (100 mm × 2.1 mm, 100 Å, 3.5 μm particle size, Merck, Darmstadt, Germany) coupled to a guard column (2.1 mm × 2 mm, 3.5 μm particle size, Merck, Darmstadt, Germany).

3. For separation of nonpolar metabolites, reversed phase (RP) Zorbax Eclipse Plus C18, RRHD column (100 mm × 2.1 mm, 1.8 μm particle size, Agilent Technologies, Santa Clara, USA) coupled to a guard column (5 mm × 2 mm, 1.8 μm Agilent Technologies, Santa Clara, USA).

2.5 Software

1. Agilent MassHunter Acquisition for Q-TOF (version B.06.01, Agilent Technologies, Santa Clara, USA) for acquiring the data.

2. Agilent PCDL (version B.07.00, Agilent Technologies, Santa Clara, USA) for managing the compound database (AM, RT, and MS/MS spectra).

3. Agilent MassHunter TOF-Quant software (version B.07.00, Agilent Technologies, Santa Clara, USA) for construction of the data processing method and processing the acquired data for targeted metabolite screening.

2.6 Other Equipments

1. Pipettes and tips (2–20, 20–200, 100–1000, 500–5000 μL).

2. Specific gravity refractometer (Atago, Tokyo, Japan).

3. Eppendorf 1.5 mL tubes.

4. Vortex mixer.

5. Pyrex measuring cylinders (100 mL, 1000 mL).

Table 1
Composition of the crash solution (total volume of 500 mL in MeOH or ACN)

Compound	Stock concentration (μg/mL or ppm)	Solvent for the stock solution	Volume for crash solution (μL) in 500 mL	Concentration in crash solution (μg/mL or ppm)	Formula	Monoisotopic mass
L-phenylalanine-$^{13}C_9$,^{15}N	200	Water	1250	0.5	$C_6H_{14}[^{15}N]4O_2$	175.1062
L-arginine-$^{15}N_4$	500	Water	2600	2.6	$C_6H_9[^{15}N]3O_2$	178.0998
Uracil-$^{15}N_2$	500	Warm water/ ammonia water	500	0.5	$C_5H_{11}[^{15}N]O_2$	114.0213
L-valine-$^{15}N_3$	1200	Water	8333	20	$C_3[^{13}C]H_3D_2O_3$	118.0760
L-tyrosine-$^{13}C_9$,^{15}N	133	Water	3759	1	$[^{13}C]2H_7NO_3S$	191.1011
Taurine-$^{13}C_2$	1000	Water	1250	2.5	$C_{11}D_5H_7N_2O_2$	127.0214
L-asparagine-$^{15}N_2$	1000	Water	1250	2.5	$C_9H_{14}D_3NO_4$	134.0476
Acetyl-d_3-carnitine	1240	MeOH	202	0.5	$C_6H_5D_7O_6$	206.1346
Allantoin ^{13}C,^{15}N	1300	Water	385	1	$C_3[^{13}C]H_6N_3[^{15}N]O_3$	160.0444
Arachidonic acid-d_8	100	Ethanol	5000	1	$C_{20}H_{24}D_8O_2$	312.2904
Linoleic acid-d_{11}	100	Ethanol	5000	1	$C_{18}H_{21}D_{11}O_2$	291.3093
Docosahexaenoic acid-d_5	100	Ethanol	5000	1	$C_{22}H_{27}D_5O_2$	333.2716
Eicosapentaenoic acid-d_5	100	Ethanol	5000	1	$C_{20}H_{25}D_5O_2$	307.2560
S1P(d18:1/17:0)	456	MeOH	1000	0.186	$C_{17}H_{36}NO_5P$	365.2331
SM(d18:1/17:0)	1000	MeOH	5000	10	$C_{40}H_{81}N_2O_6P$	716.5832
Cer(d18:1/17:0)	500	MeOH	1000	1	$C_{35}H_{69}NO_3$	551.5277
Palmitoyl-L-carnitine (N-methyl-d_3)	100	MeOH	250	0.05	$C_{23}D_3H_{42}NO_4$	402.3537
Chenodeoxycholic acid-2,2,4,-d_4	1000	MeOH	200	0.4	$C_{24}D_4H_{36}O_4$	396.3178

6. Centrifuge.

7. Evaporator-concentrator.

8. HPLC vials (fixed insert 300 μL amber, Agilent Technologies, Santa Clara, USA) and screw caps (Agilent Technologies, Santa Clara, USA).

3   Methods

3.1   Processing of Plasma/Serum Samples

1. Thaw the samples on ice (~1 h, depending on volume), followed by vortexing for 30 s.

2. Place 150 μL of plasma or serum sample in a 1.5 mL Eppendorf tube (*see* **Note 3**).

3. Add 450 μL of ice-cold MeOH crash solution (sample/solvent 1:3, *v/v*) (*see* **Note 4**).

4. Vortex 10 s.

5. Centrifuge the sample to pellet the precipitate (13,000 × *g*, 10 min, 4 °C).

6. Transfer 70 μL of the supernatant containing extracted metabolites to a new 1.5 mL Eppendorf tube (*see* **Note 5**).

7. Evaporate the samples using an evaporator-concentrator. Set temperature at 30 °C, and dry for 45 min (drying time will vary depending on unit/volume) (*see* **Note 6**).

8. Reconstitute the evaporated samples on the day of the analysis in 50 μL (8:2, *v/v*) ACN/water for HILIC and 50 μL MeOH for RP metabolomics.

9. Centrifuge the reconstituted samples (13,000 × *g*, 2 min, 4 °C).

10. Transfer 35 μL to HPLC vial for analysis (*see* **Note 7**).

3.2   Processing of Urine Samples

1. Thaw the samples on ice (~1 h, depending on volume).

2. Vortex 30 s.

3. Centrifuge the sample (13,000 × *g*, 10 min, 4 °C), and use the supernatant in the following steps.

4. Measure the specific gravity by pipetting 100 μL of urine on the specific gravity refractometer (*see* **Note 8**).

5. Normalize all the urine samples to the sample with the lowest measured specific gravity using LC-MS grade water [10] (*see* **Note 9**).

6. Place 20 μL of normalized urine sample in a 1.5 mL Eppendorf tube.

7. Add 180 μL of crash solution containing internal standards (sample/solvent 1:9, *v/v*).

8. Vortex 5 s.

9. Centrifuge the sample ($13,000 \times g$, 10 min, 4 °C).

10. Transfer 40 µL of supernatant to HPLC vial for analysis (*see* **Note 10**).

3.3 **LC Parameters**

To obtain consistently high-quality data, regularly inspect and maintain the system (*see* **Note 11**).

1. Maintain the samples in the autosampler module at 10 °C (*see* **Note 12**).

2. Injection volume 2 µL (injection loop 20 µL) (*see* **Note 13**).

3. The mobile phase gradient for both HILIC and RP chromatography is presented in Tables 2 and 3, respectively (*see* **Note 14**).

Table 2
Gradient settings for HILIC chromatography

Time (min)	Solvent A (%)	Solvent B (%)
0	5	95
1.5	5	95
12	60	40
14	60	40
14.2	75	25
17	75	25
18	5	95
25	5	95

Table 3
Gradient settings for RP chromatography

Time (min)	Solvent A (%)	Solvent B (%)
0	95	5
3	95	5
5	70	30
18.5	2	98
20	2	98
20.5	95	5
25	95	5

4. The mobile phase gradient flow rate is 0.3 mL/min for HILIC (*see* **Note 15**) and 0.4 mL/min for RP (*see* **Note 16**).

5. The column oven temperature is maintained at 25 °C for HILIC chromatography and 50 °C for RP chromatography.

3.4 MS Parameters

1. Tune the MS system before each project using the ESI-low concentration tuning mix solution.

2. The internal lock mass mixture is constantly infused at a flow rate of 1 mL/min using an isocratic pump (split ESI spray/ return to the stock bottle 1:100; *see* **Note 17**) together with the LC eluent for constant mass correction. Positive ionization mode: purine ($[M+H]^+$ m/z 121.0509), HP-0921 ($[M+H]^+$ m/z 922.0098). Negative ionization mode: purine ($[M-H]^-$ m/z 119.0363), HP-0921 ($[M+COOH]^-$ m/z 966.0007). Use a detection window of 100 ppm and minimum height of 1000 counts.

3. Set the sheath and drying gas (nitrogen purity > 99.999%) flows to 8 L/min and 15 L/min, respectively. Set the temperature of the drying and sheath gas at 250 °C, with the nebulizer pressure at 35 psig. Set the voltages for positive and negative ionization modes at +3000 V and −3000 V, respectively.

4. Set the fragmentor voltages to 380 V at 0 eV, 185 V at 10 eV, and 410 V at 30 eV.

5. Acquire the data in centroid mode with a mass range of 40–1200 m/z for HILIC and 40–1200 m/z for RP.

6. Perform MS acquisition in AIF mode, where full-scan high-resolution data are acquired at three alternating collision energies (0 eV – full scan, 10 eV, and 30 eV) (*see* **Note 18**).

7. Set the data acquisition rate at six scans/s.

3.5 Data Processing and Metabolite Identification

A data processing method for database-dependent metabolite screening was constructed in the Agilent TOF-Quant software (version B.07.00, Agilent Technologies, *see* **Note 19**) using precursor and product ion information. The manually constructed data processing method is useful in improving metabolite identification as well as deconvoluting isobaric compounds. A list of 413 compounds with AM, RT (HILIC and RP chromatography), fragmentation, and ions ratios is provided in Naz et al. [9]. The level of confidence in the identification is reflected by the ranking: Rank 1, AM and RT; Rank 2, AM, RT, and MS/MS; and Rank 3, AM, RT, MS/MS, and ion ratio. Of the 413 compounds, 229 metabolites (Rank 1, 40; Rank 2, 86; Rank 3, 99) were detected in plasma. The steps for adding a compound to the data processing method are explained using the example of *N*-acetylcarnosine (Fig. 1).

1. Inject and measure appropriately diluted analytical standards using HILIC and/or RP chromatography in positive and/or negative ionization modes.

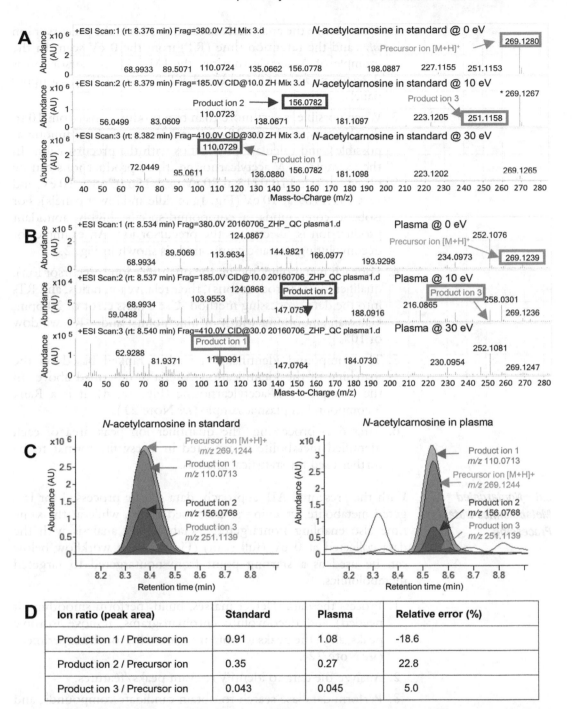

Fig. 1 Metabolite identification with MS/MS fragments and ion ratio confirmation using all-ion fragmentation (AIF) data. (**a**) *N*-acetylcarnosine spectra obtained from an analytical standard (HILIC column, positive ionization mode) at 0 eV (upper panel), 10 eV (middle panel), 30 eV (lower panel). (**b**) *N*-acetylcarnosine spectra obtained in plasma (HILIC column, positive ionization mode) at the elution time of 8.37 min at 0 eV (upper panel), 10 eV (middle panel), 30 eV (lower panel). (**c**) Extracted ion chromatograms of *N*-acetylcarnosine precursor and MS/MS product ions overlaid in standard (left panel) and plasma (right panel). (**d**) Ion ratios based on the peak areas in standard and plasma samples and the respective relative errors

2. Characterize the analytical standard to obtain the precursor ion m/z and the retention time (RT) from the 0 eV scan. In the example of N-acetylcarnosine, the $[M+H]^+$ precursor ion is detected at m/z 269.1244 and RT 8.37 min (Fig. 1a upper panel).

3. When possible, select more than two product ions (from 10 to 30 eV scans; *see* **Note 20**) for each compound (as unique as possible), and calculate their ratios with the precursor ion. In the example of N-acetylcarnosine, three product ions can be observed: m/z 110.0713 at 30 eV, m/z 156.0768 at 10 eV, and m/z 251.1139 at 10 eV (Fig. 1a middle and lower panels). For isobaric compounds, a compound-specific highly abundant product ion is selected as the precursor ion. An example for deconvoluting isobaric compounds is shown in Fig. 2.

4. For each metabolite, merge the quantifier ion (precursor ion), qualifier ion(s) (product ions), their relative ion ratios, and RTs into the data processing method. Use a mass error of 20 ppm, Gaussian smoothing width of nine points, and a RT window of 10%.

5. The compound identification is then ranked based on the matching of the identification criteria as described above. In the example of N-acetylcarnosine (Fig. 1c, d), it is a Rank 3 compound in plasma sample (*see* **Note 21**).

6. After data processing, the quantifier ion peak area of each identified metabolite is exported in a .csv file format to be further used for statistical analysis.

3.6 Nontargeted Metabolomics Data Processing

With the presented AIF approach, data can be processed for targeted metabolite screening as described above, while at the same time also enabling nontargeted metabolomics analysis with the data collected at 0 eV (full scan) (Fig. 3). The workflow below can be used as a starting point from nontargeted to targeted metabolomics.

1. Process the data: Detect masses, build (perform smoothing if needed) and deconvolute chromatograms, remove isotopic peaks, align the peaks to obtain a feature list for the experiment (*see* **Note 22**).

2. Analyze the data to identify relevant peaks/features.

3. Perform database search to obtain candidate compounds, and if MS/MS spectra are available, search for fragments in the 10 eV and 30 eV scans (*see* **Note 23**).

4. Acquire an analytical standard, and add it to the database-dependent data processing method as described in Subheading 3.5.

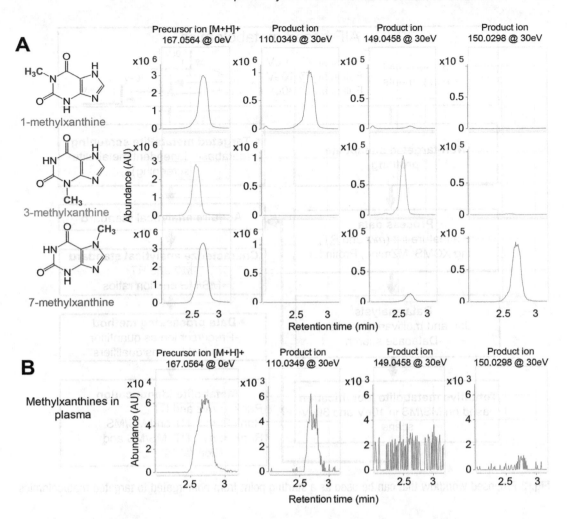

Fig. 2 Improving selectivity for co-eluting isobaric compounds using all-ion fragmentation (AIF) data. (**a**) Standards of 1-, 3-, and 7-methylxanthines (HILIC method, positive ionization mode) have very close retention times for the precursor [M+H]$^+$ ion at 0 eV (first column). However, each of the methylxanthines has a distinct product ion at 30 eV, m/z 110.0349, m/z 149.0458, and m/z 150.0298 for 1-, 3-, and 7-methylxanthines, respectively (second to fourth columns). (**b**) In plasma (HILIC column, positive ionization mode), the precursor ion for methylxanthines [M+H]$^+$ m/z 167.0564 is detected at 2.7 min, 0 eV (first column). In the 30 eV scans, product ion for 1-methylxanthine (m/z 110.0349, second column) and 7-methylxanthine (a low level of m/z 150.0298, fourth column), but no product ion for 3-methylxanthine, is observed, consistent with the literature [11]. Integration of the specific fragment ions enables the relative quantification of each co-eluting isobaric metabolite

4   Notes

1. While beyond the scope of the current protocol, sample collection and processing procedures are vital for data quality.

2. Compounds can be excluded/included as well as concentrations adjusted as required by the metabolomics experiment.

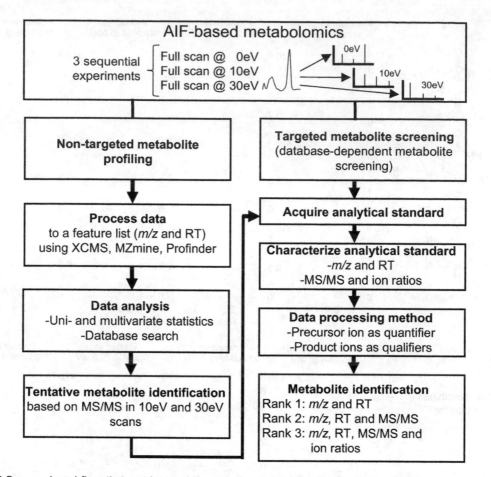

Fig. 3 Proposed workflow that can be used as a starting point from nontargeted to targeted metabolomics

3. This volume can be adjusted depending upon how many aliquots are needed (e.g., 100 μL are sufficient for four aliquots).

4. Crash solution containing standards should be stored at −80 °C until needed.

5. A maximum of six 70 μL aliquots can be obtained from 150 μL plasma or serum sample.

6. If required, the sample can be stored after this step at −80 °C.

7. Sometimes after reconstitution, a precipitate may form; avoid transferring it to the HPLC vial.

8. Wipe off the prism with lint-free tissue between measurements.

9. Create a dilution scheme by using the urine sample with highest specific gravity and subsequent 2-, 5-, 10-, and 50-fold dilutions.

10. Sometimes after adding crash solution, a precipitate (very little) may form; avoid transferring it to HPLC vial. A maximum of four 40 μL aliquots can be obtained from 20 μL urine sample.

11. Before running a new project, we recommend to clean the autosampler needle, change the in-line filter, and prepare fresh solvents.

12. For the strong needle wash, the current method uses ACN/water ratios of 9:1(v/v) and 1:9 (v/v) for HILIC and RP chromatographic methods, respectively.

13. The draw speed for the syringe is 100 μL/min, the eject speed is 400 μL/min, and the wait time after the draw is 1–2 s.

14. For the seal wash of the pumps, use (9:1, v/v) water/MeOH.

15. Expected back pressure at initial conditions 65–70 bars.

16. Expected back pressure at initial conditions 365 bars.

17. Expected back pressure ~23 bars.

18. To increase sensitivity, use two instead of three alternating scans (e.g., 0 eV and 10 eV).

19. Alternatively, other metabolomics packages such as MS-DIAL [12] can be adopted to the workflow.

20. Alternative energies can be used if desired.

21. We consider compounds confirmed by ion ratio if the % relative error of ion ratio (peak area) for at least one product/product pair is <25%.

22. The parameters for data processing (mass detection, deconvolution, smoothing, etc.) are project and matrix dependent.

23. A list with a short description of online metabolomics databases is available under http://metabolomicssociety.org/resources/metabolomics-databases.

Acknowledgments

We acknowledge the support of the Swedish Heart Lung Foundation (HLF 20140469), the Swedish Research Council (2016-02798), the Swedish Foundation for Strategic Research, the Karolinska Institutet and AstraZeneca Joint Research Program in Translational Science (ChAMP; Centre for Allergy Research Highlights Asthma Markers of Phenotype), the Novo Nordisk Foundation (TrIC NNF15CC0018486 and MSAM NNF15CC0018346), and Gunma University Initiative for Advanced Research (GIAR). This work was supported in part by The Environment Research and Technology Development Fund (ERTDF) (Grant No 5-1752). CEW was supported by the Swedish Heart Lung Foundation (HLF 20150640).

References

1. Dunn WB, Broadhurst DI, Atherton HJ et al (2011) Systems level studies of mammalian metabolomes: the roles of mass spectrometry and nuclear magnetic resonance spectroscopy. Chem Soc Rev 40:387–426. https://doi.org/10.1039/b906712b

2. Kell DB, Oliver SG (2016) The metabolome 18 years on: a concept comes of age. Metabolomics 12:148. https://doi.org/10.1007/s11306-016-1108-4

3. Beger RD, Dunn W, Schmidt MA et al (2016) Metabolomics enables precision medicine: "a white paper, community perspective". Metabolomics 12:149. https://doi.org/10.1007/s11306-016-1094-6

4. Viant MR, Kurland IJ, Jones MR, Dunn WB (2017) How close are we to complete annotation of metabolomes? Curr Opin Chem Biol 36:64–69. https://doi.org/10.1016/j.cbpa.2017.01.001

5. Dias DA, Jones OAH, Beale DJ et al (2016) Current and future perspectives on the structural identification of small molecules in biological systems. Meta 6:46. https://doi.org/10.3390/metabo6040046

6. Sumner LW, Amberg A, Barrett D et al (2007) Proposed minimum reporting standards for chemical analysis chemical analysis working group (CAWG) metabolomics standards initiative (MSI). Metabolomics 3:211–221. https://doi.org/10.1007/s11306-007-0082-2

7. Dunn WB, Broadhurst D, Begley P et al (2011) Procedures for large-scale metabolic profiling of serum and plasma using gas chromatography and liquid chromatography coupled to mass spectrometry. Nat Protoc 6:1060–1083. https://doi.org/10.1038/nprot.2011.335

8. Plumb RS, Johnson KA, Rainville P et al (2006) UPLC/MS(E); a new approach for generating molecular fragment information for biomarker structure elucidation. Rapid Commun Mass Spectrom 20:1989–1994. https://doi.org/10.1002/rcm.2550

9. Naz S, Gallart-Ayala H, Reinke SN et al (2017) Development of an LC-HRMS metabolomics method with high specificity for metabolite identification using all ion fragmentation (AIF) acquisition. Anal Chem 89(15):7933–7942. https://doi.org/10.1021/acs.analchem.7b00925.

10. Edmands WMB, Ferrari P, Scalbert A (2014) Normalization to specific gravity prior to analysis improves information recovery from high resolution mass spectrometry metabolomic profiles of human urine. Anal Chem 86:10925–10931. https://doi.org/10.1021/ac503190m

11. Martínez-López S, Sarriá B, Baeza G et al (2014) Pharmacokinetics of caffeine and its metabolites in plasma and urine after consuming a soluble green/roasted coffee blend by healthy subjects. Food Res Int 64:125–133. https://doi.org/10.1016/j.foodres.2014.05.043

12. Tsugawa H, Cajka T, Kind T et al (2015) MS-DIAL: data-independent MS/MS deconvolution for comprehensive metabolome analysis. Nat Methods 12:523–526. https://doi.org/10.1038/nmeth.3393

Chapter 4

Lipid Mediator Metabolomics Via LC-MS/MS Profiling and Analysis

Jesmond Dalli, Romain A. Colas, Mary E. Walker, and Charles N. Serhan

Abstract

Solid-phase extraction coupled with liquid chromatography tandem mass spectrometry provides a robust and sensitive approach for the identification and quantitation of specialized pro-resolving mediators (lipoxins, resolvins, protectins, and maresins), their pathway markers and the classic eicosanoids. Here, we provide a detailed description of the methodologies employed for the extraction of these mediators from biological systems, setup of the instrumentation, sample processing, and then the procedures followed for their identification and quantitation.

Key words Lipid mediator metabololipidomics, Flux analysis, Eicosanoids, Resolvin, Protectin, Maresin, Profiling, Liquid chromatography-tandem mass spectrometry

1 Introduction

The role of lipid mediators in regulating distinct aspects of the body's functions in humans and experimental systems is well appreciated [1–4]. In this chapter, we shall detail the methodologies pioneered in the Serhan laboratory to obtain a snapshot of the dynamic pathways for the four major bioactive metabolomes that include the arachidonic acid, eicosapentaenoic acid, and docosahexaenoic acid metabolomes [5–12]. These methodologies provide insights into mechanisms activated during inflammation as well as new leads into underlying causes of disease by measuring the flux down each of the major bioactive metabolomes [5–11]. The methodologies that will be discussed herein employ C18-based extraction (Fig. 1) to enrich for lipid mediators in biological systems. These are then coupled with reversed-phase liquid chromatography electrospray tandem mass spectrometry that allows for the separation, identification, and quantitation of these molecules [6, 10] (Figs. 2 and 3). Given that the structures of these mediators are conserved throughout evolution, these methodologies are applicable to biological material from

Martin Giera (ed.), *Clinical Metabolomics: Methods and Protocols*, Methods in Molecular Biology, vol. 1730,
https://doi.org/10.1007/978-1-4939-7592-1_4, © Springer Science+Business Media, LLC 2018

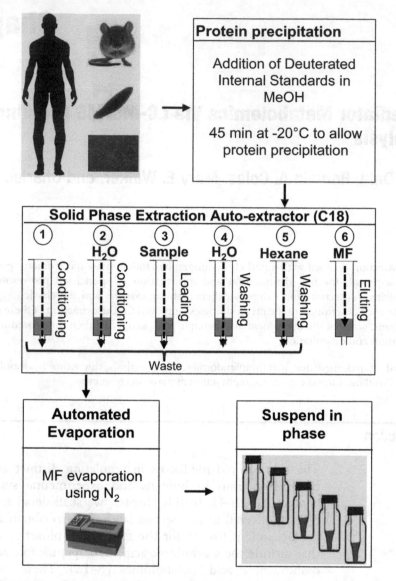

Fig. 1 Sample preparation. Tissues from various origins (human or animal tissues or fluids, mice, planaria, cell culture, etc.) should be homogenized in ice-cold MeOH containing deuterated internal standards. Samples should always be kept on ice to prevent mediator isomerization. For biological fluids 4 volumes of methanol should be added to the samples. Homogenized tissues and biological fluids should then be kept at $-20\ ^\circ C$ for 45 min to allow for protein precipitation. Samples should then be centrifuged and supernatant collected and acidified to pH 3.5 with HCl prior to solid-phase extraction. The eluting fraction containing the bioactive lipid mediators is dried under a gentle flux of nitrogen (N_2) and resuspended in phase prior injection. *MeOH* methanol, H_2O water, *MF* methyl formate, N_2 nitrogen

experimental systems such as tunicates [13], mice [5, 7], and baboons [9] as well as humans [6, 8, 11, 14, 15]. Thus, they facilitate the direct translation of findings made in experimental systems to humans and vice versa.

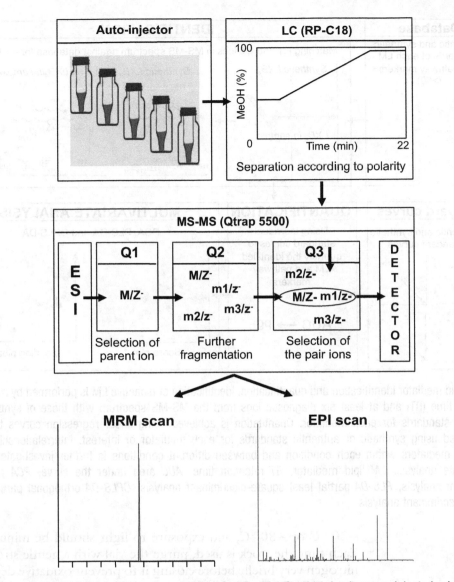

Fig. 2 Schematic of instrument setup for mediator identification. Samples in phase are injected using a HPLC-MS-MS system; no more than 40 μL should be injected. *LC* liquid chromatography, *RP* reverse phase, *MeOH* methanol, *ESI* electron spay ionization, *m/z* mass to charge ratio, *MRM* multiple reaction monitoring, *EPI* enhanced product ion

2 Materials

All solvents should be LC-MS or analytical grade. All stocks should be handled using zero dead volume Hamilton syringes. Ensure that the Hamilton syringes are appropriately cleaned using isopropanol and methanol; this should be done by aliquoting at least 20 syringe volumes with each solvent. Please note that these compounds are sensitive to light, oxygen, and heat. They should be stored at

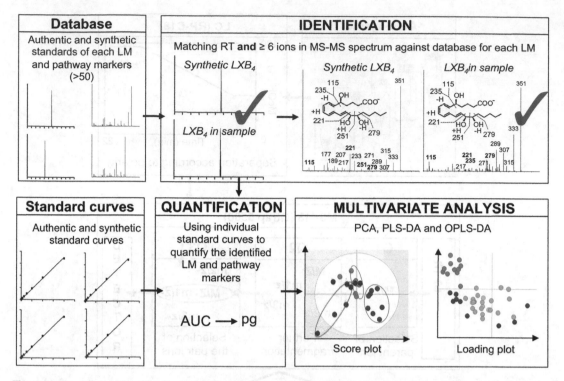

Fig. 3 Lipid mediator identification and quantification. Identification of bioactive LM is performed by matching retention time (RT) and at least six diagnostic ions from the MS-MS spectrum with those of synthetic or authentic standards for each mediator. Quantitation is achieved using linear regression curves that are constructed using synthetic or authentic standards for each mediator of interest. Interrelationship(s) for identified mediators within each condition and between different conditions is further investigated using multivariate analysis. *LM* lipid mediator, *RT* retention time, *AUC* area under the curve, *PCA* principal component analysis, *PLS-DA* partial least square-discriminant analysis, *OPLS-DA* orthogonal partial least square-discriminant analysis

−20 °C or −80 °C, and exposure to light should be minimized. Each time the stock is used, purge the vial with a gentle stream of nitrogen very briefly before closing it to prevent oxidative degradation. DO NOT use these compounds with DMSO because the SPM is sensitive to isomerization and oxidation in this solvent. Please *see* **Notes 1–9** in the notes section for important precautions to be taken in handling lipid mediators and materials.

2.1 Sample Preparation and Solid-Phase Extraction

1. Methanol (MeOH) is used for protein precipitation and sample extraction as well as solid-phase extraction (SPE) cartridge equilibration.

2. For washing of SPE cartridges, use *n-hexane*.

3. Methyl formate is used as SPE elution solvent.

4. Isolute 500 mg/3 mL C18 SPE columns are used (Biotage).

5. Deuterium-labelled internal standards used for quantification can be found in Table 1.

Table 1
List of deuterated internal standards, their complete stereochemistries, and source

Abbreviation	Trivial name	Full stereochemistry	Source
5S-HETE-d_8	Deuterium-labelled 5S-HETE	5S-hydroxy-eicosa-6E,8Z,11Z, 14Z-tetraenoic-5,6,8,9,11,12, 14,15-d_8 acid	Cayman Chemical
LTB$_4$-d_4	Deuterium-labelled leukotriene B$_4$	5S,12R-dihydroxy-eicosa-6Z,8E, 10E,14Z-tetraenoic-6,7,14,15-d_4 acid	Cayman Chemical
LXA$_4$-d_5	Deuterium-labelled lipoxin A$_4$	5S,6R,15S-trihydroxy-eicosa-7E,9E,11Z, 13E-tetraenoic-19,19,20,20,20-d5 acid	Cayman Chemical
RvD2-d_5	Deuterium-labelled resolvin D2	7S,16R,17S-trihydroxy-docosa-4Z,8E,10Z,12E,14E,19Z-hexaenoic-21,21,22,22,22-d_5 acid	Cayman Chemical
PGE$_2$-d_4	Deuterium-labelled prostaglandin E$_2$	9-oxo-11α,15S-dihydroxy-prosta-5Z,13E-dien-1-oic-3,3,4,4-d_4 acid	Cayman Chemical

2.2 Liquid Chromatography Tandem Mass Spectrometry

1. Water containing 0.01% acetic acid is used as solvent A.
2. MeOH containing 0.01% acetic acid is used as solvent B.
3. LC-MS standards (*see* Tables 2, 3, 4, and 5).
4. A Poroshell 120 EC-18 4.6 mm × 100 mm × 2.7 μm reversed-phase column is employed (Agilent).

3 Methods

3.1 Preparing Standard Mixes for Internal Standards and Standard Curves

3.1.1 Determining the Concentration of Synthetic Material

1. Blank the UV spectrophotometer using the solvent of interest.
2. Aliquot a known volume of the stock using a Hamilton syringe.
3. Measure the UV absorbance ensuring that a characteristic chromophore is observed for the molecule of interest (i.e., diene, triene, or tetraene) [6].
4. Calculate absorption using a three-point dropline.
5. Calculate concentration using the Beer-Lambert law. Concentration = absorbance/extinction coefficient*dilution (*see* Table 6 for λ_{max}^{MEOH} *and extinction coefficients for each conjugated double-bond system*).

3.2 Tuning of the LC-MS-MS for Lipid Mediator Profiling

1. Using synthetic standards, first tune the probe position to maximize signal.
2. Next, adjust electrode position.
3. Afterward, in the instrument parameter window, adjust curtain gas, collisionally activated dissociation (CAD) gas, electrode

Table 2
List of arachidonic acid metabolome, their complete stereochemistries, and source

Abbreviation	Trivial name	Full stereochemistry	Source
AA	Arachidonic acid	eicosa-5Z,8Z,11Z,14Z-tetraenoic acid	Cayman Chemical
PGD$_2$	Prostaglandin D$_2$	9α,15S-dihydroxy-11-oxo-prosta-5Z,13E-dien-1-oic acid	Cayman Chemical
PGE$_2$	Prostaglandin E$_2$	9-oxo-11α,15S-dihydroxy-prosta-5Z,13E-dien-1-oic acid	Cayman Chemical
PGF$_{2a}$	Prostaglandin F$_{2a}$	9α,11α,15S-trihydroxy-prosta-5Z,13E-dien-1-oic acid	Cayman Chemical
TxB$_2$	Thromboxane B$_2$	9α,11,15S-trihydroxy-thromba-5Z,13E-dien-1-oic acid	Cayman Chemical
LTB$_4$	Leukotriene B$_4$	5S,12R-dihydroxy-eicosa-6Z,8E,10E,14Z-tetraenoic acid	Cayman Chemical
LXA$_4$	Lipoxin A$_4$	5S,6R,15S-trihydroxy-eicosa-7E,9E,11Z,13E-tetraenoic acid	Cayman Chemical
LXB$_4$	Lipoxin B$_4$	5S,14R,15S-trihydroxy-eicosa-6E,8Z,10E,12E-tetraenoic acid	Cayman Chemical
15-HETE		15S-hydroxy-eicosa-5Z,8Z,11Z,13E-tetraenoic acid	Cayman Chemical
12-HETE		12-hydroxy-eicosa-5Z,8Z,10E,14Z-tetraenoic acid	Cayman Chemical
5-HETE		5S-hydroxy-eicosa-6E,8Z,11Z,14Z-tetraenoic acid	Cayman Chemical
Δ6-trans-LTB$_4$	Δ6-trans-leukotriene B$_4$	5S,12R-dihydroxy-eicosa-6E,8E,10E,14Z-tetraenoic acid	Cayman Chemical
5S,12S-diHETE		5S,12S-dihydroxy-eicosa-6E,8E,10E,14Z-tetraenoic acid	Biogenic Synthesis

voltage, source temperature, and the source gases one at a time to identify optimal source settings.

4. Subsequently tune the compound parameters (declustering potential, entrance potential, collision energy, and collision cell exit potential) for each compound and for each transition.

5. It is suggested that for each compound (both pathway markers and mediators) at least two transitions are used.

3.3 Automated Lipid Mediator Extraction

3.3.1 Sample Preparation (Fig. 1)

1. Ensure that samples have not been freeze-thawed; fresh samples are preferred.

2. For frozen samples allow to defrost on ice.

3. For tissues dissociate gently in ice-cold MeOH ensuring that the samples remain cold throughout the process.

Table 3
List of eicosapentaenoic acid metabolome, their complete stereochemistries, and source

Abbreviation	Trivial name	Full stereochemistry	
EPA		eicosa-5Z,8Z,11Z,14Z, 17Z-pentaenoic acid	Cayman Chemical
RvE1	Resolvin E1	5S,12R,18R-trihydroxy-eicosa- 6Z,8E,10E,14Z,16E-pentaenoic acid	Cayman Chemical
RvE2	Resolvin E2	5S,18R-dihydroxy-eicosa- 6Z,8E,11E,14E,16Z-pentaenoic acid	Biogenic Synthesis
RvE3	Resolvin E3	17R,18R/S-dihydroxy-eicosa- 5Z,8Z,11Z,13E,15E-pentaenoic acid	Custom Synthesis (Dr Makoto Arita, Riken Institute Japan)
18-HEPE		18-hydroxy-eicosa- 5Z,8Z,11Z,14Z,16E-pentaenoic acid	Cayman Chemical
15-HEPE		15-hydroxy-eicosa- 5Z,8Z,11Z,13E,17Z-pentaenoic acid	Cayman Chemical
12-HEPE		12-hydroxy-eicosa- 5Z,8Z,10E,14Z,17Z-pentaenoic acid	Cayman Chemical
5-HEPE		5-hydroxy-eicosa-6E,8Z,11Z,14Z, 17Z-pentaenoic acid	Cayman Chemical

Table 4
List of docosahexaenoic acid metabolome, their complete stereochemistries, and source

Abbreviation	Trivial name	Full stereochemistry	
DHA		docosa-4Z,7Z,10Z,13Z,16Z, 19Z-hexaenoic acid	Cayman Chemical
RvD1	Resolvin D1	7S,8R,17S-trihydroxy-docosa- 4Z,9E,11E,13Z,15E, 19Z-hexaenoic acid	Cayman Chemical
RvD2	Resolvin D2	7S,16R,17S-trihydroxy-docosa- 4Z,8E,10Z,12E,14E, 19Z-hexaenoic acid	Cayman Chemical
RvD3	Resolvin D3	4S,7R,17S-trihydroxy-docosa- 5Z,7E,9E,13Z,15E, 19Z-hexaenoic acid	Cayman Chemical
RvD4	Resolvin D4	4S,5R,17S-trihydroxy-docosa- 6E,8E,10Z,13Z,15E, 19Z-hexaenoic acid	Custom Synthesis (Dr Charles Serhan
RvD5	Resolvin D5	7S,17S-dihydroxy-docosa- 4Z,8E,10Z,13Z, 15E,19Z-hexaenoic acid	Cayman Chemical

(continued)

Table 4
(continued)

Abbreviation	Trivial name	Full stereochemistry	
RvD6	Resolvin D6	4S,17S-dihydroxy-docosa-5E,7Z,10Z,13Z,15E,19Z-hexaenoic acid	Biogenic Synthesis
MaR1	Maresin 1	7R,14S-dihydroxy-docosa-4Z,8E,10E,12Z,16Z,19Z-hexaenoic acid	Cayman Chemical
4S,14S-diHDHA		4S,14S-dihydroxy-docosa-5Z,7E,10E,12Z,16E,19E-hexaenoic acid	Biogenic Synthesis
7S,14S-diHDHA		7S,14S-dihydroxy-docosa-4Z,8E,10E,12Z,16E,19E-hexaenoic acid	Biogenic Synthesis
PD1	Protectin D1	10R,17S-dihydroxy-docosa-4Z,7Z,11E,13E,15Z,19Z-hexaenoic acid	Custom Synthesis (Dr Charles Serhan)
10S,17S-diHDHA	Protectin Dx	10S,17S-dihydroxy-docosa-4Z,7Z,11E,13Z,15E,19Z-hexaenoic acid	Cayman Chemical
22-OH-PD1	22-OH-protectin D1	10R,17S,20-trihydroxy-docosa-4Z,7Z,11E,13E,15Z,19Z-hexaenoic acid	Custom Synthesis (Dr Trond V. Hansen, University of Oslo)
17-HDHA		17-hydroxy-docosa-4Z,7Z,10Z,13Z,15E,19Z-hexaenoic acid	Cayman Chemical
14-HDHA		14S-hydroxy-docosa-4Z,7Z,10Z,12E,16Z,19Z-hexaenoic acid	Cayman Chemical
13-HDHA		13-hydroxy-docosa-4Z,7Z,10Z,14E,16Z,19Z-hexaenoic acid	Cayman Chemical
7-HDHA		7-hydroxy-docosa-4Z,8E,10Z,13Z,16Z,19Z-hexaenoic acid	Cayman Chemical
4-HDHA		4-hydroxy-docosa-5E,7Z,10Z,13Z,16Z,19Z-hexaenoic acid	Cayman Chemical

Table 5
List of aspirin-triggered mediators, their complete stereochemistries, and source

Abbreviation	Trivial name	Full stereochemistry	
AT-LXA$_4$	Aspirin-triggered lipoxin A$_4$	5S,6R,15R-trihydroxy-eicosa-7E,9E,11Z,13E-tetraenoic acid	Cayman Chemical
AT-LXB$_4$	Aspirin-triggered lipoxin B$_4$	5S,14R,15R-trihydroxy-eicosa-6E,8Z,10E,12E-tetraenoic acid	Cayman Chemical
AT-RvD1	Aspirin-triggered resolvin D1	7S,8R,17R-trihydroxy-docosa-4Z,9E,11E,13Z,15E,19Z-hexaenoic acid	Cayman Chemical
AT-RvD3	Aspirin-triggered resolvin D3	4S,7R,17R-trihydroxy-docosa-5Z,7E,9E,13Z,15E,19Z-hexaenoic acid	Custom Synthesis (Dr Charles Serhan)
AT-PD1	Aspirin-triggered protectin D1	10R,17R-dihydroxy-docosa-4Z,7Z,11E,13E,15Z,19Z-hexaenoic acid	Custom Synthesis (Dr Charles Serhan)

Table 6
Extinction coefficient and λ_{max}^{MEOH} for distinct conjugated double-bond systems

Double-bond system	Extinction coefficient	λ_{max}^{MEOH}
Conjugated diene (e.g., monohydroxy acids)	25,000	235
Two conjugated dienes (e.g., RvD5)	25,000	240
Diene-triene conjugated system (e.g., RvE1)	40,000	271
Conjugated triene (e.g., PD1)	40,000	269
Conjugated tetraene (e.g., RvD1)	50,000	301

4. For liquid samples (e.g., cell culture preparations and plasma) add four equal volumes of ice-cold methanol containing deuterium-labelled internal standards to each sample.

5. Place at −20 °C for 45 min to allow for protein precipitation.

6. Centrifuge samples at 2000 × g for 10 min at 4 °C.

7. Collect supernatant.

8. Transfer tubes to TurboVap Evaporator.

9. Ensure that the water bath is set to 37 °C.

10. Turn nitrogen feed on and set the flow rate to no more than 15 psi.

11. Ensure that the lid is closed.

12. Methanol volume should be evaporated to less than 1 mL using a steady nitrogen stream.

13. Centrifuge samples at $2000 \times g$ for 10 min at 4 °C.

14. Samples are now ready for extraction. At this stage samples should not be stored.

3.3.2 Lipid Mediator Extraction

1. Add new C18 columns on the rack.

2. Add the collection plates containing elution tubes.

3. Make sure run-through plate is at position D.

4. Click "run single method."

5. Select the extraction method.

6. Click "prepare run."

7. Go to solvent feeder (5), and prime each individually (make sure, the containers are filled, especially the MF).

8. Go to extraction media (3) and select how many columns you will use.

9. Go to solvent tips (1) and sample tips (2), and check tips are filled as displayed on screen. If necessary, add or delete missing columns (or add tips).

10. Click "run method."

11. Monitor that the extraction procedure is initiated correctly.

12. At the end of the extraction, transfer the eluted samples (i.e., the MF fraction) to glass tubes, rinse the collection tube or the collecting well once with MeOH, and add to the sample.

13. Samples are ready for drying on TurboVap.

3.3.3 Solvent Evaporation

1. Switch the TurboVap Evaporator on.

2. Ensure that the water bath is set to 37 °C.

3. Turn nitrogen feed on and set the flow rate to no more than 15 psi.

4. Place 10 mL conical borosilicate tubes containing the methyl formate fraction or methanol fraction obtained during solid-phase extraction in the TurboVap.

5. Ensure that the lid is closed.

6. When the solvent is more than 95% evaporated, rinse the walls of the tube using methyl formate.

7. When the solvent is more than 95% evaporated, rinse the walls of the tube using methanol.

8. When the solvent is completely evaporated, add 40 μL of methanol/water (1:1).

9. Centrifuge the tube at $2000 \times g$ for 2 min.

10. Carefully transfer the supernatant to an injection vial insert.

11. Place the insert in a clearly labelled 1.5 mL Eppendorf tube.

12. Centrifuge the insert at $10,000 \times g$ for no more than 2 min.

13. Transfer the supernatant to an injection vial insert, and place the insert in a clearly labelled sample vial.

14. Samples are now ready for LC-MS-MS profiling.

3.4 Chromatography

1. Using a Poroshell 120 EC-18 (4.6 mm $\times$ 100 mm $\times$ 2.7 μm) and water containing 0.01% acetic acid as solvent A, and methanol containing 0.01% acetic acid as solvent B, the following gradient should be used to chromatographically separate the mediators from their isomers and pathway markers.

2. Column temperature should be set at 50 °C.

3. Equilibrate the column with mobile phase at 80:20 (A:B).

4. This should be ramped to 50:50 (A:B) over 12 s.

5. The gradient should be maintained for 2 min.

6. Then to 80:20 over the subsequent 9 min.

7. This should be maintained for the next 3.5 min.

8. Then ramped to 98:2 (A:B).

9. Finally maintain this for 5.4 min to wash the column.

10. The follow rate should be maintained at 0.5 mL/min throughout the experiment.

3.5 Data Analysis Using Analyst Quantitate Tool

3.5.1 Setting Up a Quantitation Method

1. In analyst click on build quantitation method.

2. Select a file that contains a standard mix.

3. Assign which transitions correspond to the internal standards.

4. Assign the internal standards that will be used for identification and quantitation of each molecule.

5. In the integrate tab, select a retention time window of not more than 5 s.

6. Save the method.

3.5.2 Analyzing the Data

1. Click the quantitation wizard, and follow the instructions to load your samples and the appropriate analysis method obtained as detailed above.

2. Start by integrating the deuterium-labelled internal standards making a note of any drifts from the expected retention times.

3. If retention time drifts are observed, correct the expected retention times for each of the analyte according to the retention time drift observed in the respective internal standards being used for identification (e.g., if d_5-LXA$_4$ is observed to elute 2 s after the anticipated retention time, then LXA$_4$ expected retention time will also shift by 2 s).

4. After determining the expected retention times for each of the samples, integrate each of the mediators. Here all peaks that are clearly visible even if not more than three times higher than baseline should be integrated.

5. After integrating all the peaks in explore mode, search for MS-MS spectra for molecules where a peak was obtained in at least one transition in the same region of the chromatogram where the peak was recorded.

6. Compare the tandem mass spectrum obtained in the sample with that obtained from the synthetic/authentic compound matching at least six ions in the MS-MS spectrum, with one of the ions being derived from a backbone break. These can be compared and matched to the Spectra Book available for all of these mediators and pathways on the Serhan lab website.

7. For those molecules where the MS-MS spectrum is a positive match, proceed to quantitation using a standard curve constructed using either synthetic or authentic standard mixes with concentrations determined as detailed above.

8. For molecules where standards are not available, molecules that carry similar physical properties may be used.

9. If an MS-MS spectrum is not obtained for a molecule or group of molecules and additional sample is available, this could be rerun looking specifically for the molecule(s) where a spectrum was not obtained.

10. If no additional sample is available then the identification criteria are not fulfilled.

4 Notes

1. Concentrated HCl and acetic acid should be handled in a fumehood. Gloves and eye protection need to be worn at all times when handling these acids. Also always add the acid to water to avoid injury.

2. All standard stocks should be stored under nitrogen, shielded from light, and at $-20\,^\circ C$ for short-term storage and $-80\,^\circ C$ for long-term storage.

3. Each time the stock is used, purge the vial with a gentle stream of nitrogen very briefly before closing it to prevent oxidative degradation.

4. All solvents are flammable; they should be handled with care in a fumehood and kept away from open flames.

5. All compound stocks should be handled using zero dead volume Hamilton syringes.

6. Ensure that the Hamilton syringes are appropriately cleaned using isopropanol and methanol; this should be done by aliquoting at least 20 syringe volumes with each solvent.

7. Only use nitrogen or an inert gas to evaporate solvents; do not use air since this will lead to mediator oxidation.

8. Upon solvent evaporation and suspension of samples in water/methanol, samples need to be profiled within 24 h to avoid isomerization of the mediators. At all times samples should be kept at 4 °C and not frozen to maximize mediator integrity.

9. DO NOT DMSO at any stage since SPM is sensitive to isomerization and oxidation in this solvent.

Acknowledgments

This work was supported by funding from the European Research Council (ERC) under the European Union's Horizon 2020 research and innovation program (Grant number: 677542), a Sir Henry Dale Fellowship jointly funded by the Wellcome Trust and the Royal Society (Grant number: 107613/Z/15/Z), and the Barts Charity (Grant number: MGU0343). CN Serhan is supported by the National Institutes of Health, USA Grant Number P01GM095467.

References

1. Basil MC, Levy BD (2016) Specialized pro-resolving mediators: endogenous regulators of infection and inflammation. Nat Rev Immunol 16(1):51–67. https://doi.org/10.1038/nri.2015.4

2. Haworth O, Buckley CD (2015) Pathways involved in the resolution of inflammatory joint disease. Semin Immunol 27(3):194–199. https://doi.org/10.1016/j.smim.2015.04.002

3. Miyata J, Arita M (2015) Role of omega-3 fatty acids and their metabolites in asthma and allergic diseases. Allergol Int 64(1):27–34. https://doi.org/10.1016/j.alit.2014.08.003

4. Serhan CN, Chiang N, Dalli J (2015) The resolution code of acute inflammation: novel pro-resolving lipid mediators in resolution. Semin Immunol 27(3):200–215. https://doi.org/10.1016/j.smim.2015.03.004

5. Chiang N, Fredman G, Backhed F, Oh SF, Vickery T, Schmidt BA, Serhan CN (2012) Infection regulates pro-resolving mediators that lower antibiotic requirements. Nature 484(7395):524–528. https://doi.org/10.1038/nature11042

6. Colas RA, Shinohara M, Dalli J, Chiang N, Serhan CN (2014) Identification and signature profiles for pro-resolving and inflammatory lipid mediators in human tissue. Am J Physiol Cell Physiol 307(1):C39–C54. https://doi.org/10.1152/ajpcell.00024.2014

7. Dalli J, Colas RA, Arnardottir H, Serhan CN (2017) Vagal regulation of group 3 innate lymphoid cells and the Immunoresolvent PCTR1 controls infection resolution. Immunity 46(1):92–105. https://doi.org/10.1016/j.immuni.2016.12.009

8. Dalli J, Colas RA, Quintana C, Barragan-Bradford D, Hurwitz S, Levy BD, Choi AM, Serhan CN, Baron RM (2017) Human sepsis eicosanoid and proresolving lipid mediator temporal profiles: correlations with survival and clinical outcomes. Crit Care Med 45(1):58–68. https://doi.org/10.1097/CCM.0000000000002014

9. Dalli J, Kraft BD, Colas RA, Shinohara M, Fredenburgh LE, Hess DR, Chiang N, Welty-Wolf K, Choi AM, Piantadosi CA, Serhan CN (2015) The regulation of proresolving lipid mediator profiles in baboon pneumonia by

inhaled carbon monoxide. Am J Respir Cell Mol Biol 53(3):314–325. https://doi.org/10.1165/rcmb.2014-0299OC

10. Dalli J, Serhan CN (2012) Specific lipid mediator signatures of human phagocytes: microparticles stimulate macrophage efferocytosis and pro-resolving mediators. Blood 120(15): e60–e72. https://doi.org/10.1182/blood-2012-04-423525

11. Weiss GA, Troxler H, Klinke G, Rogler D, Braegger C, Hersberger M (2013) High levels of anti-inflammatory and pro-resolving lipid mediators lipoxins and resolvins and declining docosahexaenoic acid levels in human milk during the first month of lactation. Lipids Health Dis 12:89. https://doi.org/10.1186/1476-511X-12-89

12. Serhan CN, Petasis NA (2011) Resolvins and protectins in inflammation resolution. Chem Rev 111(10):5922–5943. https://doi.org/10.1021/cr100396c

13. Knight J, Taylor GW, Wright P, Clare AS, Rowley AF (1999) Eicosanoid biosynthesis in an advanced deuterostomate invertebrate, the sea squirt (Ciona intestinalis). Biochim Biophys Acta 1436(3):467–478

14. Rathod KS, Kapil V, Velmurugan S, Khambata RS, Siddique U, Khan S, Van Eijl S, Gee LC, Bansal J, Pitrola K, Shaw C, D'Acquisto F, Colas RA, Marelli-Berg F, Dalli J, Ahluwalia A (2017) Accelerated resolution of inflammation underlies sex differences in inflammatory responses in humans. J Clin Invest 127 (1):169–182. https://doi.org/10.1172/JCI89429

15. Titos E, Rius B, Lopez-Vicario C, Alcaraz-Quiles J, Garcia-Alonso V, Lopategi A, Dalli J, Lozano JJ, Arroyo V, Delgado S, Serhan CN, Claria J (2016) Signaling and immunoresolving actions of resolvin D1 in inflamed human visceral adipose tissue. J Immunol 197 (8):3360–3370. https://doi.org/10.4049/jimmunol.1502522

Chapter 5

UHPSFC/ESI-MS Analysis of Lipids

Miroslav Lísa and Michal Holčapek

Abstract

This new analytical approach for high-throughput and comprehensive lipidomic analysis of biological samples using ultrahigh-performance supercritical fluid chromatography (UHPSFC) with electrospray ionization–mass spectrometry (ESI-MS) is based on lipid class separation using 1.7 μm particle bridged ethylene hybrid silica columns and a gradient of methanol–water–ammonium acetate mixture as a modifier. The method enables a fast separation of 30 nonpolar and polar lipid classes within 6-min analysis time covering six main lipid categories including fatty acyls, glycerolipids, glycerophospholipids, sphingolipids, sterols, and prenols. Individual lipid species within lipid classes are identified based on positive- and negative-ion full scan and tandem mass spectra measured with high mass accuracy and high resolving power. The method is used for the quantitative analysis of lipid species in biological tissues using internal standards for each lipid class. This high-throughput, comprehensive, and accurate UHPSFC/ESI-MS method is suitable for the lipidomic analysis of large sample sets in clinical research.

Key words Supercritical fluid chromatography, Ultrahigh-performance supercritical fluid chromatography, Mass spectrometry, Electrospray ionization, Lipids, Lipidomics, Glycerolipids, Glycerophospholipids, Sphingolipids

1 Introduction

Supercritical fluid chromatography (SFC) is a chromatographic technique, which uses a supercritical fluid heated above critical temperature and pressurized above critical pressure as the mobile phase [1]. The supercritical fluid combines some unique properties of gases and liquids, such as low viscosity similarly to gases, high density similarly to liquids, and diffusivity between gases and liquids. These properties enable fast analyses with high resolving power in SFC. Any fluid above the critical point can be used as the mobile phase in SFC, but carbon dioxide is almost exclusively employed due to its favorable physicochemical properties, such as relatively low values of critical characteristics (7.4 MPa and 31 °C), which can be easily achieved using conventional chromatographic instrumentation. Organic modifiers are added to carbon dioxide to enhance its polarity, solvating power, and elution strength, such as

Martin Giera (ed.), *Clinical Metabolomics: Methods and Protocols*, Methods in Molecular Biology, vol. 1730,
https://doi.org/10.1007/978-1-4939-7592-1_5, © Springer Science+Business Media, LLC 2018

methanol, ethanol, acetonitrile, 2-propanol, or their mixtures. Conventional LC columns with 3–5 μm particle size have been mainly used in SFC, but in recent years, sub-2 μm particles providing ultrahigh-performance SFC (UHPSFC) separations demonstrate their benefits, such as faster analyses and higher efficiency.

The UHPSFC/ESI-MS coupling is a powerful tool for the sensitive identification/quantitation in complex mixtures. The organic solvent is usually added after the column to the mobile phase as a makeup liquid to ensure ESI ionization.

SFC of lipids can be performed in two separation modes, i.e., the lipid class separation according to the class polarity or the separation of individual lipid species based on the fatty acyl chain composition [2]. The lipid class separation is used for the analysis of total lipid extracts, where lipids are separated into lipid classes according to their polarities, while species within the class differing only in fatty acyl chain composition are not separated, and they elute together in one chromatographic peak of particular lipid class [3, 4]. Lipid classes are identified based on retention times and class characteristic fragment ions in ESI mass spectra. Averaged mass spectra of lipid class chromatographic peaks are used for the determination of lipid species level, i.e., the number of carbon atoms and double bonds of attached fatty acyl/alkyls. For the quantitative analysis, nonendogenous lipid species for each lipid class are added before the extraction step as internal standards (IS) [5, 6] to cover the extraction efficiency and possible fluctuation of MS signal. The lipid class separation provides the coelution of lipid species with the lipid class IS in one peak under identical mobile phase and matrix composition, which is especially important for accurate quantitative measurements.

This protocol presents a new analytical strategy for the high-throughput and comprehensive lipidomic analysis of biological samples applicable for large-scale clinical studies. High-throughput UHPSFC/ESI-MS method using sub-2 μm particle bridged ethylene hybrid silica column is used for the lipid class separation of a wide range of nonpolar and polar lipids in one analysis including the identification and quantitation of individual lipid species using ESI-MS.

2   Materials

Solutions and samples are prepared using acetonitrile, 2-propanol, methanol, *n*-hexane, and chloroform stabilized with 0.5–1% ethanol (all HPLC/MS or HPLC grade). Ammonium acetate, deionized water, methanol (all HPLC/MS grade), and carbon dioxide 4.5 grade (99.995%) are used for UHPSFC/ESI-MS analysis.

2.1 Lipid Standards

Qualitative standards of lipid class representatives used for UHPSFC/ESI-MS method development, i.e., CE, TG, FA, 1,3-DG, 1,2-DG, and 1-MG containing oleoyl acyls and cholesterol, can be purchased from Sigma-Aldrich, and Cer, PG, PE, LPG, PI, LPE, CL, LPI, PA, PC, pPC, ePC, PS, LPA, SM, LPC, and LPS containing oleoyl acyls, sphingosine, sphinganine, GlcCer d18:1/16:0, LacCer d18:1/16:0, S1P d17:1, desmosterol, and DHEA can be purchased from Avanti Polar Lipids (Alabaster, AL, USA). Internal standards (IS) used for quantitation, i.e., CE 19:0, TG 19:0/19:0/19:0, FA 14:0, DG 19:0/0:0/19:0, and MG 19:0/0:0/0:0, are purchased from Nu-Chek Prep (Elysian, MN, USA); D7-cholesterol, Cer d18:1/17:0, GlcCer d18:1/12:0, PG 14:0/14:0, LacCer d18:1/12:0, PE 14:0/14:0, LPG 14:0/0:0, LPE 14:0/0:0, PC 22:1/22:1, PC 14:0/14:0, SM d18:1/17:0, and LPC 17:0/0:0 are purchased from Avanti Polar Lipids.

2.2 Solutions

1. Stock solutions of individual qualitative and quantitative standards: prepare individual stock solutions at a concentration of 2 mg/mL using a (1:4, v/v) chloroform–2-propanol mixture.

2. The solution of all qualitative standards is prepared by mixing of 20 μL of each standard diluted to the appropriate concentration with a (7:3, v/v) hexane–chloroform mixture for UHPSFC/ESI-MS analysis.

2.3 Instruments

UHPSFC experiments were performed on an Acquity UPC² instrument (Waters, Milford, MA, USA). HPLC 515 pump (Waters) was used as a makeup pump. The hybrid quadrupole–travelling wave ion mobility–time-of-flight mass spectrometer Synapt G2Si (Waters) was used for MS experiments.

3 Methods

3.1 Sample Preparation

Total lipid extracts of tissues (typically 25 mg) or plasma (typically 25 μL) are prepared according to the modified Folch procedure [7] using a chloroform–methanol–water system, subsequently evaporated by a gentle stream of nitrogen and redissolved in 1 mL of a (1:1, v/v) chloroform–2-propanol mixture. 10 μL of the extract stock solution is diluted into 1 mL of a (7:3, v/v) hexane–chloroform mixture for nontargeted UHPSFC/ESI-MS analysis. The same procedure is used for the preparation of the total lipid extract for UHPSFC/ESI-MS quantitation with the addition of IS into the sample before the extraction. 2 μL of LPE and DG; 4 μL of TG, Cer, PG, LPC, and MG; and 40 μL of FA, GlcCer, CE, D7-cholesterol, sulfatide, PE, PI, PC, and SM stock solutions of IS were added.

3.2 UHPSFC Conditions

1. Prepare the modifier by mixing methanol and water in the ratio of 99:1 (v/v) with the addition of 30 mM of ammonium acetate.

2. Condition the Acquity BEH UPC^2 column (100 × 3 mm, 1.7 μm, Waters) using a flow rate of 1.9 mL/min consisting of 99% carbon dioxide and 1% of the modifier, column temperature 60 °C, and the active back pressure regulator (ABPR) 1800 psi for 20 min before the first injection (*see* **Note 1**).

3. Inject 1 μL of sample with the gradient of the modifier: 0 min, 1%; 5 min, 51%; 6, min 51%; 7 min, 1%; and 10 min, 1% (*see* **Note 2**) (Fig. 1).

4. Wash the injector needle with a (2:2:1, v/v/v) hexane–2-propanol–water mixture after each injection.

3.3 UHPSFC/ESI-MS Analysis

1. Prepare a mixture of (99:1, v/v) methanol–water as a makeup liquid, flush a makeup pump, and set up the flow rate of 0.25 mL/min.

2. Calibrate the MS instrument using a sodium formate solution prepared by mixing 10 μL 1 M NaOH, 20 μL formic acid, and 20 mL (1:1, v/v) water–acetonitrile.

3. Connect the UHPSFC instrument with the mass spectrometer via the commercial interface kit (Waters) composed of two T-pieces enabling the backpressure control and mixing of column effluent with a makeup liquid.

4. Collect MS data in the centroid high-resolution mode for 6 min using 0.15 s scan speed in the mass range *m/z* 50–1600 (*see* **Note 3**) with the following setting of tuning parameters: capillary voltages 3.0 kV and 2.5 kV for positive-ion and negative-ion modes (*see* **Note 4**), respectively, the sampling cone 20 V, the source offset 90 V, the source temperature 150 °C, the drying temperature 500 °C, the cone gas flow 0.8 L/min, the drying gas flow 17 L/min, and the nebulizer gas flow 4 bar. Use 10 μL/min of leucine enkephalin as the lock mass for all experiments with the acquisition of lock mass spectra each 30 s for 0.1 s.

5. Identify lipid classes and individual lipid species (*see* **Note 5**) based on their retention times (Fig. 2) (*see* **Note 6**), observed ions (*see* **Note 7**), and characteristic fragments (*see* **Note 8**) in ESI mass spectra.

3.4 UHPSFC/ESI-MS Quantitation of Lipids

1. Perform UHPLC/ESI-MS analysis of prepared lipidomic extracts.

2. Process data using MarkerLynx XS software (*see* **Note 9**), i.e., combine together individual ESI scans corresponding to lipid

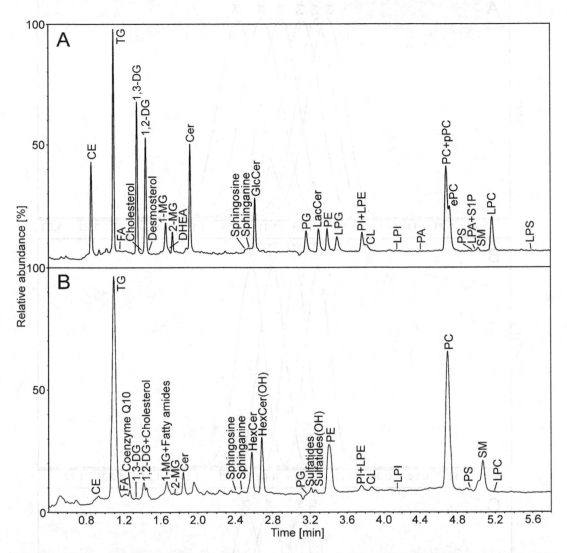

Fig. 1 Positive-ion UHPSFC/ESI-MS chromatograms of the mixture of lipid class standards (**a**) and a total lipid extract of porcine brain (**b**). UHPSFC conditions: Acquity BEH UPC^2 column (100 × 3 mm, 1.7 μm, Waters), the flow rate 1.9 mL/min, the column temperature 60 °C, the ABPR pressure 1800 psi, and the gradient of methanol–water mixture (99:1, v/v) containing 30 mM of ammonium acetate as the modifier: 0 min, 1%, 5 min, 51%, 6 min, 51%. Peak annotation: CE, cholesteryl esters; TG, triacylglycerols; FA, fatty acids; DG, diacylglycerols; MG, monoacylglycerols; DHEA, dehydroepiandrosterone; Cer, ceramides; GlcCer, glucosyl-ceramides; HexCer, hexosylceramides; PG, phosphatidylglycerols; LacCer, lactosylceramides; pPE, 1-alke-nyl-2-acyl phosphatidylethanolamines (plasmalogens); ePE, 1-alkyl-2-acyl phosphatidylethanolamines (ethers); PE, phosphatidylethanolamines; LPG, lysophosphatidylglycerols; PI, phosphatidylinositols; LPE, lysophosphatidylethanolamines; CL, cardiolipins; LPI, lysophosphatidylinositols; PA, phosphatidic acids; PC, phosphatidylcholines; pPC, 1-alkenyl-2-acyl phosphatidylcholines; ePC, 1-alkyl-2-acyl phosphatidyl-cholines; PS, phosphatidylserines; LPA, lysophosphatidic acids; S1P, sphingosine-1-phosphate; SM, sphin-gomyelins; LPC, lysophosphatidylcholines; LPS, lysophosphatidylserines. Reprinted with permission from (Lísa M, Holčapek M (2015). Analytical Chemistry 87 (14):7187–7195). Copyright (2015) American Chemical Society

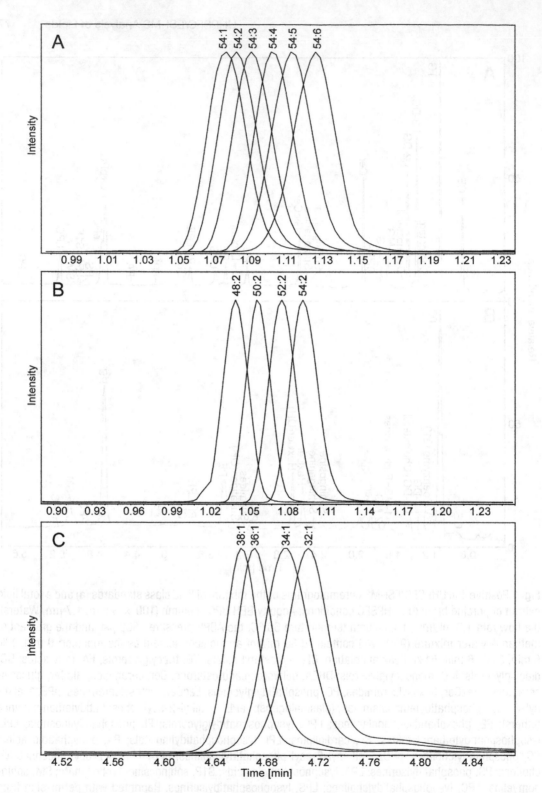

Fig. 2 Effects of DB number and fatty acyl chain length on the retention behavior of lipids using UHPSFC/ESI-MS. Reconstructed ion chromatograms from the analysis of porcine brain extract: (**a**) TG with 54 carbon atoms and the different number of DB, (**b**) TG with 2 DB and the different number of carbon atoms, and (**c**) PC with 1 DB and the different number of carbon atoms. Reprinted with permission from (Lísa M, Holčapek M (2015). Analytical Chemistry 87 (14):7187–7195). Copyright (2015) American Chemical Society

class peaks with the peak separation set to 50 mDa and an intensity threshold of 3000 for markers extraction.

3. Detect lipid species within extracted markers according to accurate m/z values, and perform isotopic correction of ion intensities (*see* **Note 10**).

4. Calculate concentrations of lipids from corrected ion intensities related to the intensity of lipid class IS.

4   Notes

1. A properly conditioned column is necessary before the first injection to achieve appropriate and stable retention times. Stable column temperature and system pressure have to be achieved as well. The UHPSFC/ESI-MS method provides an excellent intraday stability of retention times in the range ± 0.01 min. A small continuous reduction of retention times of mainly polar lipids can be observed during multiple consecutive days. This phenomenon of small retention shift on SFC columns is probably given by the formation of silyl ethers on the particle surface with alcohols from the mobile phase leading to the decrease of hydrophilicity [7], but it does not affect the identification and quantitation of lipids. The process is reversible and silanols can be hydroxylated again using water. For this reason, the column is regenerated daily by five consecutive injections of 10 µL of water and acetonitrile into 100% of CO_2 at 0.5 mL/min flow rate or occasionally is flushed with a (1:1, v/v) water–acetonitrile mixture, which minimizes this effect.

2. In this UHPSFC method, lipids are mainly separated according to their polarity into lipid classes. Retention times increase with the increased lipid polarity. In total, up to 30 lipid classes from six different lipid categories covering a wide range of nonpolar and polar lipids are separated within 6 min analysis. Under optimized UHPSFC conditions, most of the lipid classes are baseline or at least partially separated. Similar to LC separations, the strong peak tailing is observed for UHPSFC of acidic lipids, such as PA, PS, LPA, S1P, and LPS, which complicates their precise quantitation.

3. The employed m/z range should be defined by the molecular weights of lipid molecules analyzed.

4. Polarity of acquisition depends on the monitored lipid classes. Most of the lipid classes provide appropriate signal in the positive-ion mode. The combination of positive- and negative-ion modes is used for the identification of lipids with a higher confidence.

5. Retention times from the UHPSFC analysis and ESI mass spectra measured with high mass accuracy and high resolving power in both positive- and negative-ion modes are used for the unambiguous identification of individual lipid species. The total lipid extract is separated into lipid classes using UHPSFC enabling the direct identification of lipid class based on the comparison of retention times with standards. And then, the averaged mass spectrum of a lipid class chromatographic peak is used for the determination at lipid species level, i.e., the number of carbon atoms and DB of attached fatty acyl/alkyls. The partial separation of individual species within lipid classes according to fatty acyl lengths and the number of DB provide another supporting information for their identification. The fatty acyl composition of individual species can be determined using $[RCOO]^-$ ions in negative-ion MS/MS spectra [8].

6. In addition to the class separation, lipid species within individual lipid classes are partially separated according to the fatty acyl composition (Fig. 2). Retention times of all lipids increase with increased DB number, as demonstrated on reconstructed ion chromatograms of TG with 54 carbon atoms and different number of DB (Fig. 2a). A different situation is observed for species differing in fatty acyl lengths, where the retention behavior of lipids differs according to the individual lipid classes. Retention times of TG increase with the fatty acyl length (Fig. 2b), which is observed also for other nonpolar lipid classes, such as CE, FA, DG, MG, fatty amides, and Cer. On the other hand, retention times of polar lipids decrease for longer fatty acyls, which can be demonstrated on the separation of PC species (Fig. 2c).

7. The most abundant ions in positive-ion UHPSFC/ESI-MS full-scan mass spectra are protonated molecules $[M+H]^+$ (base peaks for fatty amides, sphinganine, LacCer, PE, LPE, CL, PC, SM, and LPC), adducts with ammonium ion $[M+NH_4]^+$ (TG and coenzyme Q10) and neutral losses of water $[M+H-H_2O]^+$ (DG, cholesterol, MG, Cer, sphingosine, and LacCer), attached fatty acyls $[M+H-acyl]^+$ (CE) or phosphoglycerol $[M+H- H_2PO_4CH_2CHOHCH_2OH]^+$ (PG and LPG), and sulfo $[M+H-SO_3]^+$ (sulfatides) groups. UHPSFC/ESI-MS mass spectra provide relatively low abundance of sodium adduct ions, which reduces the risk of incorrect identification between $[M+H]^+$ and $[M+Na]^+$ ions, because the difference $\Delta m/z$ 22 also corresponds to an additional two methylene units minus 3 DB. The data are also correlated with negative-ion mass spectra and retention times of lipid species within lipid classes to unambiguously confirm the identification of all reported lipids.

8. Positive-ion MS/MS spectra of identified classes provide well-known characteristic fragment ions and neutral losses observed in HPLC/MS, such as the phosphocholine fragment ion m/z 184 ($[H_2PO_4CH_2CH_2N(CH_3)_3]^+$) for moieties containing choline (PC, LPC, and SM), $m/z369$ ($[M+H-H_2O]^+$ or $[M+H-acyl]^+$) for cholesterol-containing lipids (cholesterol and CE), or fragment ions corresponding to ceramide bases (Cer, GlcCer, and LacCer). The neutral loss of phosphoethanolamine $\Delta m/z = 141$ ($H_2PO_4CH_2CH_2NH_2$) is observed for PE and neutral losses of fatty acyls for TG, DG, and MG. In the negative-ion mode, base peaks of spectra are mostly deprotonated molecules $[M-H]^-$, except for DG, PC, SM, and LPC, providing mainly adduct ions with acetate $[M+CH_3COO]^-$. Relatively high abundances of $[M+CH_3COO]^-$ ions are also observed for Cer, GlcCer, and LacCer (70–95%). CL species exhibit $[M-2H]^{2-}$ ions, which can be used for their identification even in the lower mass range, such as $m/z = 50$–1000 usually used in lipidomic analyses. Negative-ion MS/MS spectra show mainly $[RCOO]^-$ ions corresponding to the fatty acyl/alkyl composition.

9. Ion intensities on ESI full-scan mass spectra of lipid class peaks are used for the data evaluation of UHPSFC/ESI-MS experiments.

10. A contribution of M+2 isotope intensity of species with one more double bond ($\Delta m/z = 2$) has to be subtracted to achieve more accurate quantitative results. We use our Excel macro script LipidQuant for detection of lipid species based on accurate m/z values, isotopic correction, and calculation of concentrations.

Acknowledgments

This work was supported by the grant project LL1302 sponsored by the Ministry of Education, Youth and Sports of the Czech Republic.

References

1. Nováková L, Perrenoud AG-G, Francois I, West C, Lesellier E, Guillarme D (2014) Modern analytical supercritical fluid chromatography using columns packed with sub-2 micrometer particles: a tutorial. Anal Chim Acta 824:18–35. https://doi.org/10.1016/j.aca.2014.03.034

2. Čajka T, Fiehn O (2014) Comprehensive analysis of lipids in biological systems by liquid chromatography-mass spectrometry. Trends Anal Chem 61:192–206. https://doi.org/10.1016/j.trac.2014.04.017

3. Lesellier E, Gaudin K, Chaminade P, Tchapla A, Baillet A (2003) Isolation of ceramide fractions from skin sample by subcritical chromatography with packed silica and evaporative light scattering detection. J Chromatogr A 1016(1):111–121. https://doi.org/10.1016/s0021-9673(03)01323-2

4. Lísa M, Holčapek M (2015) High-throughput and comprehensive lipidomic analysis using ultrahigh-performance supercritical fluid chromatography-mass spectrometry. Anal Chem 87(14):7187–7195. https://doi.org/10.1021/acs.analchem.5b01054

5. Schuhmann K, Almeida R, Baumert M, Herzog R, Bornstein SR, Shevchenko A (2012) Shotgun lipidomics on a LTQ Orbitrap mass spectrometer by successive switching between acquisition polarity modes. J Mass Spectrom 47 (1):96–104. https://doi.org/10.1002/jms.2031

6. Wiesner P, Leidl K, Boettcher A, Schmitz G, Liebisch G (2009) Lipid profiling of FPLC-separated lipoprotein fractions by electrospray ionization tandem mass spectrometry. J Lipid Res 50(3):574–585. https://doi.org/10.1194/jlr.D800028-JLR200

7. Fairchild JN, Brousmiche DW, Hill JF, Morris MF, Boissel CA, Wyndham KD (2015) Chromatographic evidence of Silyl ether formation (SEF) in supercritical fluid chromatography. Anal Chem. https://doi.org/10.1021/ac5035709

8. Lísa M, Cífková E, Holčapek M (2011) Lipidomic profiling of biological tissues using off-line two-dimensional high-performance liquid chromatography mass spectrometry. J Chromatogr A 1218(31):5146–5156. https://doi.org/10.1016/j.chroma.2011.05.081

Chapter 6

LC-MS/MS Analysis of Lipid Oxidation Products in Blood and Tissue Samples

Yiu Yiu Lee and Jetty Chung-Yung Lee

Abstract

Oxygenated lipid products of non-cyclooxygenase derivatives, namely, prostanoids such as, isoprostanes and isofurans, are formed in vivo through lipid autoxidation. Insofar it has been marked as novel biomarkers of oxidative stress in the biological systems. Elevations of these oxidized products are associated with several diseases. This chapter describes the preparation and measurement of the products, including newly identified F_2-dihomo-isoprostanes and dihomo-isofurans, from plasma and tissue samples using the liquid chromatography-tandem mass spectrometry approach.

   Key words Lipid peroxidation products, Isoprostanes, Isofurans, Liquid chromatography-tandem mass spectrometry

1   Introduction

Oxidative stress is highly related to various metabolic diseases associated to cardiovascular, neurodegenerative, pulmonary disorders, and cancer. The measurement of eicosanoid-based oxidized lipid products is claimed to be the most reliable indicator of oxidative damage in diseases [1–4]. These oxidized lipid products are generated nonenzymatically and typically initiated by the action of free radicals and reactive oxygen species on polyunsaturated fatty acids (PUFA) and in certain diseases catalyzed by elements such as iron. Among the oxidized PUFA products, F_2-isoprostanes (F_2-IsoPs) from arachidonic acid are the most commonly measured due to their specificity and sensitivity [5]. These compounds are non-cyclooxygenase derivatives of prostanoids that are formed in situ, esterified to phospholipid, and released as free form into circulation via phospholipase A_2-catalyzed hydrolysis [6, 7].

   Depending on the parent PUFA, different types of IsoPs have been discovered and quantified in numerous pathological conditions [8]. These include F_2-IsoPs and isofurans (IsoFs) from arachidonic acid, F_2-dihomo-IsoPs and dihomo-IsoFs from adrenic

Martin Giera (ed.), *Clinical Metabolomics: Methods and Protocols*, Methods in Molecular Biology, vol. 1730,
https://doi.org/10.1007/978-1-4939-7592-1_6, © Springer Science+Business Media, LLC 2018

acid, and F_4-neuroprostanes (NeuroPs) and neurofurans (NeuroFs) from docosahexaenoic acid [8]. Among them, 15-F_{2t}-IsoP is the most represented and measured IsoP [9] and has been corroborated by the European Food Safety Authority (EFSA) as biomarkers for oxidative damage in cardiovascular health [10]. Besides being oxidative stress biomarkers, some autoxidized products have been reported to be bioactive, but the research progression in their bioactivities has been slow, partly due to the lack of chemical standards available in the commercial market.

The quantitation of oxidized PUFA products originates from the measurement of 15-F_{2t}-IsoP using gas chromatography-mass spectrometry (GC-MS) in negative chemical ionization (NCI). The merit of GC-NCI-MS over many other analytical methods is owed to its superior sensitivity and is recognized as the "gold standard" in targeted lipid profiling (*see* preceding chapter by Barden and Mori). However, isomeric compounds or different compounds with the exact mass and elution time cannot be effectively differentiated in GC-NCI-MS. With recent advancements in MS, the use of LC-MS/MS in multiple reaction monitoring (MRM) mode has received great attention for the ease of sample preparation, relatively lesser run time, and high selectivity which enables the separation of isomeric compounds.

In this chapter, the preparation procedure and analysis of oxidized PUFA products in plasma and tissue samples for LC-MS/MS measurement are described. The method is applicable for the measurement of all IsoPs and IsoFs; however, we will focus on the measurement of F_2-dihomo-IsoPs and dihomo-IsoFs. These are newly identified oxidized lipid products and potentially a robust biomarker for neurological disorders such as Rett syndrome and epilepsy [10–12].

2   Materials

Prepare all reagents using ultrapure water. All solvents should be at least high-performance liquid chromatography (HPLC) grade. All solvents and buffers prepared should be stored at room temperature, unless otherwise stated. It is not encouraged to use sodium azide as buffer preservative for the potential of ion suppression in the LC-MS/MS measurement.

2.1   Lipid Extraction

1. Ethylenediaminetetraacetic acid (EDTA)-primed syringe: expose the syringe inner wall with 5 M EDTA (pH 8.0) and subsequently empty the syringe before blood withdrawal.

2. Phosphate-buffered saline (PBS): dissolve 8 g sodium chloride (NaCl), 0.2 g KCl, 1.44 g Na_2PO_4, and 0.24 g KH_2PO_4 in 800 mL of ultrapure water. Then adjust to pH 7.4 with HCl and top up to 1 L.

3. 20 ml EPA glass vials with PTFE septa screw cap (VWR).

4. 1 M methanolic potassium hydroxide (KOH): dissolve 5.6 g of KOH pellets per 100 mL of absolute methanol.

5. 5% (w/v) butylated hydroxytoluene (BHT): dissolve 5 g of BHT per 100 mL of absolute ethanol.

6. 1% (w/v) indomethacin: dissolve 1 g of indomethacin per 100 mL of ethanol.

7. Folch solution: mix chloroform and methanol in a ratio of 2:1 and spike with 0.005% (w/v) BHT (i.e., 1 mL 5% BHT in 1 L Folch solution). Store at −20 °C.

8. A cocktail of internal standards (ISTD): accurately dilute concentrated stock ISTD (e.g., 15-F_{2t}-IsoP-d_4, 4-F_{4t}-NeuroP-d_4, 10-F_{4t}-NeuroP-d_4) to 1 ng/μL in absolute methanol. Store at −80 °C.

9. 0.9% NaCl: dissolve 0.9 g of NaCl per 100 mL ultrapure water.

10. 0.45 μm PTFE syringe filters (13 mm diameter) (Membrane Solutions).

2.2 Solid-Phase Extraction (SPE)

1. 5 M HCl: add 41.059 mL of 37% stock HCl to ultrapure water per 100 mL final volume.

2. 40 mM formic acid: add 0.159 mL of stock formic acid to ultrapure water per 100 mL final volume.

3. 20 mM formic acid: 1:1 dilute from 40 mM formic acid with ultrapure water.

4. Mixed-mode reversed-phase/strong anion-exchange (MAX) cartridges (Oasis).

5. Visiprep™ SPE Vacuum Manifold (Supelco).

6. 2% ammonium hydroxide (NH_4OH): add 3.5 mL of 56.6% stock NH_4OH per 100 mL final volume of ultrapure water.

7. Elution solvent: for every 100 mL, add 70 mL of n-hexane, 29.4 mL of ethanol, and 0.4 mL of acetic acid. Store at room temperature in amber bottle.

2.3 LC-ESI-MS/MS Analysis

1. LC solvent A: add 1 mL of formic acid to ultrapure water per 1 L final volume.

2. LC solvent B: add 1 mL of formic acid to absolute acetonitrile per 1 L final volume.

3. Kinetex reverse phase C18 column (2.6 μ, 100 Å, 150 × 2.1 mm) (Phenomenex).

4. Ultrahigh-performance liquid chromatography (UHPLC) system (Agilent) (1290 Infinity).

5. Hybrid triple quadrupole/linear ion trap (QqQLIT) mass spectrometer (MS) (AB Sciex (QTRAP 3200)).

3 Methods

3.1 Extraction of Lipid Autoxidized Products from Blood and Tissue Samples

3.1.1 Blood Samples

1. Use 22 G needle with a 10 mL EDTA-primed syringe tube for blood collection in rats via cardiac puncture, followed by immediate euthanization (*see* **Note 1**).

2. Transfer the collected blood into an EDTA-primed polypropylene Falcon™ tube and gently shake the tube with hand (*see* **Note 1**). Separate the plasma (top layer) from whole blood by centrifugation at 3000 × *g* for 15 min, 4 °C.

3. Following centrifugation, dispense 1 mL aliquots of blood plasma into prechilled microcentrifuge tubes and spike with 20 µL of 5% (w/v) BHT and 1% (w/v) indomethacin (*see* **Note 2**). Briefly vortex the samples and proceed to hydrolysis step immediately or store at −80 °C until analysis (*see* **Note 3**).

4. On the day of analysis, thaw 1 mL (one vial) of frozen plasma on ice and transfer to a clean 20 mL glass vial containing 1 mL of 1 M methanolic KOH and add 50 µL of IS cocktail. Mix the plasma mixture by aspiration, and incubate in 50 °C oven for 1 h.

3.1.2 Tissue Samples

1. Briefly rinse the freshly harvested organs with ice-cold PBS containing BHT (0.005%, w/v) and (0.005%, w/v) indomethacin (*see* **Note 2**). Cleaned tissue samples not analyzed immediately should be snap frozen with liquid nitrogen and stored at −80 °C (*see* **Note 3**).

2. On the day of tissue lipid extraction, excise a portion of tissues from the frozen organ, and grind into powder using mortar and pestle in the presence of liquid nitrogen. Weigh about 0.2–0.5 g of tissue powder into a clean 15 mL Falcon™ tube before it becomes moist, and add 10 mL of ice-cold Folch solution and 50 µL of ISTD cocktail (*see* **Note 4**).

3. Homogenize the tissue using a Polytron benchtop homogenizer at 24,000 rpm for four bursts of 10 s, while placing the Falcon tube on ice. The homogenizer probe should pass through multiple washes of ultrapure water and methanol to prevent cross contamination between samples.

4. Firmly screw cap the Falcon™ tubes and place them on ice in a rocking motion for at least 30 min, and vortex every 10 min to enhance extraction efficiency.

5. Introduce phase separation by adding 2 mL of 0.9% NaCl to the homogenate, shake vigorously, and centrifuge at 3000 × *g* for 10 min, 4 °C.

6. Transfer the lower phase using a glass pipette and filter through a 0.45 µm PTFE filter membrane into a new 20 mL glass vial (*see* **Note 5**).

7. Further add 6 mL of chloroform to the remaining homogenates, vortex, and repeat the centrifugation and filtration steps (**steps** 5 and 6).

8. Pool the lower phases together and dry under a gentle stream of nitrogen, on a heat block set to 37 °C (*see* **Note 6**).

9. To the dried lipid extracts, add 1 mL of 1 M methanolic KOH, and sonicate until all extracts dissociate from the bottom of the vial. Then, add 1 mL of PBS, mix thoroughly by aspiration, flush the vial with nitrogen, and incubate in darkness at room temperature overnight (*see* **Note 7**).

3.2 Solid-Phase Extraction (SPE)

1. Ensure the sample is cooled to room temperature, and add 0.5 mL methanol, 0.2 mL 5 M HCl, 2.7 mL 40 mM formic acid, and 4 mL 20 mM formic acid, in succession, vortex, and transfer to a new 15 mL Falcon™ tube. Centrifuge at $3000 \times g$ for 10 min to remove any protein precipitates (*see* **Note 8**).

2. Mount Oasis MAX 3 cc extraction cartridges onto a SPE Vacuum Manifold and precondition with 2 mL of methanol and 2 mL of 20 mM formic acid.

3. Load the acidified samples and allow passing the solution at a flow of about one drop per second to maximize the interaction of analytes with the cartridge sorbents.

4. Wash the unwanted and weakly bound compounds by adding 2 mL of 2% NH_4OH, followed by 2 mL of 20 mM formic acid. Dry the cartridges by increasing air flow for at least 5 min to remove all water content (*see* **Note 9**).

5. Elute all lipid analytes by passing through 2 mL of n-hexane and twice with 2 mL of elution solvent into a clean 20 mL glass vial. Dry with nitrogen gas to approximately 1 mL of volume before transferring to the autosampler vial.

6. Continue drying the SPE extracts in the autosampler vials until complete dryness and reconstitute with 200 µL of acetonitrile (*see* **Note 10**).

7. Allow at least 5 min of incubation on bench in the dark before transferring to autosampler spring insert (*see* **Note 11**). Firmly screw the cap and store at 4 °C until ready for analysis within the day.

3.3 LC-ESI-MS/MS

1. Prepare solvent A (0.1% formic acid in ultrapure water) and B (0.1% formic acid in acetonitrile), filtered and degassed and store at room temperature.

2. Clean the LC C18 RP column by flushing 98% solvent B at 0.2 mL/min until the bin pump pressure is stabilized. Then, equilibrate the column for at least 30 min with 20% solvent A immediately before the start of the analysis.

3. The elution gradient program begins with 20% solvent B for 2 min and ramp to 98% solvent B over an 8 min period and hold for 5 min. Then reduce solvent B to 20% for another 5 min to equilibrate for the next run. Set the flow rate to 0.3 mL/min and the injection volume to 10 µL. Fit the C18 column in the column oven and maintain at 30 °C.

4. QTRAP 3200 hybrid triple quadrupole-linear ion trap mass spectrometry is equipped with a Turbo V ESI source operated in negative (−) mode with the following settings: curtain gas (CUR) = 10 psi, ion spray voltage (IS) = −4500 V, CAD = HIGH, temperature (TEM) = 500 °C, ion source gas 1 (GS1, nebulizer) = 30 psi (nitrogen), GS2 (heater) = 30 psi (nitrogen), interface heater (ihe) = on, CEP = −10 V, and CXP = −10 V.

5. To achieve maximum selectivity, QTRAP 3200 MS is operated in *scheduled* multiple reaction monitoring mode (MRM) with a retention time window of 1 min. The transitions of each analyte are chosen by infusing their pure standards into MS and generate MS/MS spectra from which the values are selected and optimized. Optimal declustering potential (DP) and collision energy (CE) of each transition are listed in Table 1. For illustration, only F_2-dihomo-IsoPs and dihomo-IsoFs that were first characterized by our laboratory are shown (Fig. 1).

6. The amount of endogenous autoxidation products is then calculated from normalized response using relative response factors (RFFs) with respect to internal standards. The RFFs are usually predetermined using peak areas and masses of deuterated standards versus analyte standards. This is often determined by the construction of a calibration line with at least five varying analyte concentrations (typically 0.01–1 ng/µL), each having a fixed concentration of internal standard (*see* **Note 12**). The calibration line slope plotting $Area_{analyte}/Area_{IS}$ against $Mass_{analyte}/Mass_{IS}$ represents the RRF.

7. The final concentration of an autoxidized product from the sample is then calculated as $(Area_{analyte} \times Mass_{IS})/(Area_{IS} \times RRF)$ (*see* **Note 13**).

---

4   Notes

1. Blood drawn into the syringe must be accompanied with appropriate needle size to minimize rupturing of the red blood cells that can cause hemolysis. Plasma contaminated with hemolyzed red blood cells will lead to IsoPs artifacts [13]. Similarly, the needle must be removed before the transfer

Table 1
The availability of some isomeric autoxidized products currently in the market

Autoxidized products	Transition (m/z)	DP (eV)	CE (eV)	ISTD	Supplier
Arachidonic acid					
5-F_{2t}-IsoP	$353 \rightarrow 115$	−50	−25	5-F_{2t}-IsoP-d_4	Cayman
8-F_{2t}-IsoP	$353 \rightarrow 127$	−50	−25	15-F_{2t}-IsoP-d_4	Cayman
12-F_{2t}-IsoP	$353 \rightarrow 151$	−50	−25	15-F_{2t}-IsoP-d_4	NA
15-F_{2t}-IsoP	$353 \rightarrow 193$	−50	−25	15-F_{2t}-IsoP-d_4	Cayman
Isofurans	$369 \rightarrow 115$	−45	−20	15-F_{2t}-IsoP-d_4	Na
Adrenic acid					
7-F_{2t}-Dihomo-IsoP	$381 \rightarrow 143$	−35	−34	15-F_{2t}-IsoP-d_4	Durand
17-F_{2t}-Dihomo-IsoP	$381 \rightarrow 237$	−65	−22	15-F_{2t}-IsoP-d_4	Durand
7(RS)-ST-Δ^8-11-Dihomo-IsoF	$397 \rightarrow 201$	−60	−32	15-F_{2t}-IsoP-d_4	Durand
17(RS)-10-*epi*-SC-Δ^{15}-11-Dihomo-IsoF	$397 \rightarrow 155$	−70	−30	15-F_{2t}-IsoP-d_4	Durand
Eicosapentaenoic acid					
5-F_{3t}-IsoP	$351 \rightarrow 115$	−50	−25	15-F_{2t}-IsoP-d_4	Durand
8-F_{3t}-IsoP	$351 \rightarrow 127$	−55	−35	15-F_{2t}-IsoP-d_4	Durand
11-F_{3t}-IsoP	$351 \rightarrow 167$	−50	−25	15-F_{2t}-IsoP-d_4	NA
12-F_{3t}-IsoP	$351 \rightarrow 151$	−50	−25	15-F_{2t}-IsoP-d_4	NA
15-F_{3t}-IsoP	$351 \rightarrow 193$	−50	−25	15-F_{2t}-IsoP-d_4	Cayman
18-F_{3t}-IsoP	$351 \rightarrow 233$	−50	−25	15-F_{2t}-IsoP-d_4	NA
Docosahexaenoic acid					
4-F_{4t}-NeuroP	$377 \rightarrow 101$	−60	−30	4-F_{4t}-NeuroP-d_4	Durand
7-F_{4t}-NeuroP	$377 \rightarrow 141$	−60	−30	4-F_{4t}-NeuroP-d_4	NA
10-F_{4t}-NeuroP	$377 \rightarrow 153$	−55	−40	10-F_{4t}-NeuroP-d_4	Durand
11-F_{4t}-NeuroP	$377 \rightarrow 178$	−60	−30	4-F_{4t}-NeuroP-d_4	NA
13-F_{4t}-NeuroP	$377 \rightarrow 221$	−60	−30	4-F_{4t}-NeuroP-d_4	NA
14-F_{4t}-NeuroP	$377 \rightarrow 138$	−60	−30	4-F_{4t}-NeuroP-d_4	Durand
17-F_{4t}-NeuroP	$377 \rightarrow 98$	−60	−30	4-F_{4t}-NeuroP-d_4	NA
20-F_{4t}-NeuroP	$377 \rightarrow 58$	−60	−30	4-F_{4t}-NeuroP-d_4	NA
4(RS)-ST-Δ^5-8-NeuroF	$393 \rightarrow 325$	−55	−32	4-F_{4t}-NeuroP-d_4	Durand
Neurofurans	$393 \rightarrow 193$	−35	−20	4-F_{4t}-NeuroP-d_4	NA

(continued)

Table 1
(continued)

Autoxidized products	Transition (*m/z*)	DP (eV)	CE (eV)	ISTD	Supplier
ISTD					
5-F$_{2t}$-IsoP-d$_4$	357 → 115	−50	−25	−	Cayman
15-F$_{2t}$-IsoP-d$_4$	357 → 197	−50	−25	−	Cayman
4-F$_{4t}$-IsoP-d$_4$	381 → 101	−60	−30	−	Durand
10-F$_{4t}$-IsoP-d$_4$	381 → 157	−55	−40	−	Durand

DP declustering potential, *CE* collision energy, *ISTD* internal standard, *NA* not available, *Cayman Chemicals*, USA, *IBMM* Dr. Thierry Durand, Institut des Biomolécules Max Mousseron, Montpellier, France

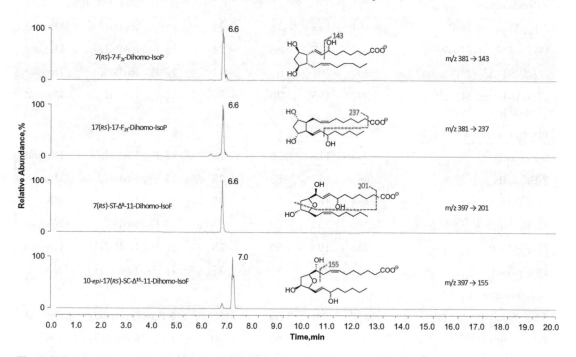

Fig. 1 Chromatograms of F$_2$-dihomo-IsoPs and F$_2$-dihomo-IsoFs in a typical LC-MS/MS analytical run. *Dotted red line* indicates the proposed site of fragmentation

of the blood into the 15 mL Falcon™ tube as failure to remove the needle may induce severe hemolysis.

2. It is known that improper handling of blood and tissue samples can lead to artificial increase of IsoPs and IsoFs; therefore, it is important to treat the samples appropriately using anti-cyclooxygenase (indomethacin) and antioxidants (BHT) before storing away.

3. For long-term storage, samples are recommended to be kept at $-80\ °C$ as higher temperature, like $-20\ °C$, has shown to induce significant lipid autoxidation.

4. Wear personal protective equipment when handling liquid nitrogen. To avoid the tissue powder from building up moisture, carefully add in more liquid nitrogen and grind until dry.

5. Sample matrices filtered through PTFE membranes eliminate the inclusion of sample tissues and allow easier sample cleanups at later stages of sample preparation.

6. Adding heat to the glass vial while drying enhances the rate of solvent volatility. A higher temperature, e.g., $40\ °C$, may be applied to the glass vial when large volume of solvent is present. It is essential to make sure that the vial is completely dry before further processing.

7. Hydrolysis of lipid compounds is achievable by placing the vials at $50–60\ °C$ for a period of 30 min to 2 h. It is found that the best hydrolyzing method for tissues is to incubate the sample overnight at room temperature to allow longer time for the release of the products embedded in the tissue matrix.

8. Acidification of sample is essential to unionize the analytes to maximize the reversed-phase retention of the MAX SPE cartridge. Also, note that it is necessary to perform centrifugation prior to loading onto the SPE cartridge to prevent clots.

9. It is advisable to remove all remaining small amount of water after washes as failure to do so may lead to the elution of contaminants in the subsequent steps.

10. Complete dryness of the lipid extracts is critical, or the final analyte concentrations may be affected.

11. In some instances, precipitates may form during mobile phase resuspension. Passing through a small diameter PTFE membrane (low holdup volume) can help to resolve this problem and prevent clotting of HPLC needle and column.

12. It is advisable to use the deuterated form of the exact analyte as the respective internal standard, i.e., $4\text{-}F_4\text{-NeuroP-}d_4$ as the IS for $4\text{-}F_4\text{-NeuroP}$ detection and quantitation. However, the availability of deuterated IS is lacking in the commercial market for many compounds; therefore, it is common to use the most available one, $15\text{-}F_{2t}\text{-IsoP-}d_4$, as the IS.

13. The calculated concentration represents only single injection and is subjected to technical repeats ($n = 3$). Also, the calculated concentration requires normalization to the total volume of lipid extract and the weight of tissue samples used when expressing the amount of analyte per gram of tissue.

References

1. Mallat Z, Nakamura T, Ohan J et al (1999) The relationship of hydroxyeicosatetraenoic acids and F2-isoprostanes to plaque instability in human carotid atherosclerosis. J Clin Invest 103:421–427

2. Practico D, Lee VMY, Trojanowski JQ et al (1998) Increased F_2-isoprostanes in Alzheimer's disease: evidence for enhanced lipid peroxidation *in vivo*. FASEB J 12:1777–1783

3. Lee CYJ, Seet RCS, Huang SH et al (2009) Different patterns of oxidized lipid products in plasma and urine of dengue fever, stroke, and Parkinson's disease patients: cautions in the use of biomarkers of oxidative stress. Antioxid Redox Signal 11:407–420

4. Basu S, Harris H, Wolk A et al (2016) Inflammatory F_2-isoprostane, prostaglandin-$F_{2\alpha}$, pentraxin 3 levels and breast cancer risk: The Swedish Mamography Cohort. Prostaglandins Leukot Essent Fatty Acids 113:28–32

5. Milne GL, Gao B, Terry ES et al (2013) Measurement of F_2-isoprostanes and isofurans using gas chromatography-mass spectrometry. Free Radic Biol Med 59:36–44

6. Morrow JD, Awad JA, Boss HJ et al (1992) Non-cyclooxygenase-derived prostanoids (F_2-isoprostanes) are formed *in situ* on phospholipids. Proc Natl Acad Sci U S A 89:10721–10725

7. Awad JA, Morrow JD, Takahashi K et al (1993) Identification of non-cyclooxygenase-derived prostanoid (F_2-Isoprostane) metabolite in human urine and plasma. J Biol Chem 268:4161–4169

8. Galano JM, Lee JCY, Gladine C et al (2015) Non-enzymatic cyclic oxygenated metabolites of adrenic, docosahexaenoic, eicosapentaenoic and a-linolenic acids; bioactivities and potential use as biomarkers. Biochim Biophys Acta 1851:446–455

9. Kadiiska MB, Gladen BC, Baird DD et al (2005) Biomarkers of oxidative stress study II. Are oxidation products of lipids, proteins, and DNA markers of CCl_4 poisoning? Free Radic Biol Med 38:698–710

10. De Felice C, Signorini C, Durand T et al (2011) F_2-dihomo-isoprostanes as potential early biomarkers of lipid oxidative damage in Rett syndrome. J Lipid Res 52:2287–2297

11. Medina S, De Miguel-Elízaga I, Oger C et al (2015) Dihomo-isoprostanes—nonenzymatic metabolites of AdA – are higher in epileptic patients compared to healthy individuals by a new ultrahigh pressure liquid chromatography–triple quadrupole–tandem mass spectrometry method. Free Radic Biol Med 79:154–163

12. De La Torre A, Lee YY, Oger C et al (2014) Synthesis, discovery, and quantitation of dihomo-isofurans: biomarkers for in vivo adrenic acid peroxidation. Angew Chem Int Ed 53:6249–6252

13. Gruber J, Tang SY, Jenner AM et al (2009) Allantoin in human plasma, serum, and nasal-lining fluids as a biomarker of oxidative stress: avoiding artifacts and extasblishing real *in vivo* concentrations. Antioxid Redox Signal 11:1767–1776

## Chapter 7

Serum Testosterone by Liquid Chromatography Tandem Mass Spectrometry for Routine Clinical Diagnostics

Lennart J. van Winden, Olaf van Tellingen, and Huub H. van Rossum

Abstract

In clinical diagnostics, samples containing low testosterone cannot be analyzed by random access immunoassays normally available at clinical laboratories. For these samples, sensitive and specific LC-MS/MS-based testosterone methods are required. An LC-MS/MS-based testosterone assay is described that was developed and validated for routine clinical application.

Key words Testosterone, LC-MS/MS, Steroids, Hormone

1 Introduction

Testosterone is the most important male sex hormone (androgen) in terms of potency and amounts secreted [1]. In clinical diagnostics, circulating testosterone concentrations are used for the diagnosis and follow-up of various diseases such as hypogonadism, polycystic ovary syndrome, and precocious or delayed puberty [2]. Furthermore, androgen deprivation therapy aimed at suppressing testosterone is the most successful first-line treatment for advanced prostate cancer [3]. Monitoring proper androgen suppression and identification of resistance to androgen deprivation therapy requires measurement of circulating testosterone concentrations.

In clinical laboratories, analysis of testosterone is performed mainly by automated, random access immunoassays. The advantages of these routine assays are their high throughput and low turnaround time properties. Furthermore, they do not require specialized machinery (HPLC and mass spectrometer systems) and expertise [3]. However, for low concentrations of testosterone as observed in pediatric, female and castrated male samples, these immunoassays lack sensitivity and specificity [3]. For such samples, application of liquid chromatography-tandem mass spectrometry

Martin Giera (ed.), *Clinical Metabolomics: Methods and Protocols*, Methods in Molecular Biology, vol. 1730,
https://doi.org/10.1007/978-1-4939-7592-1_7, © Springer Science+Business Media, LLC 2018

(LC-MS/MS)-based testosterone assays is required for accurate testosterone analysis. In addition, recognizing the limitations of the testosterone immunoassays, leading journals now require testosterone analysis by LC-MS when testosterone is used as a major end point in studies [4].

Our hospital serves a relatively large population of prostate cancer patients. For routine application of testosterone analysis by LC-MS/MS in clinical diagnostics, we developed and validated a liquid chromatography-tandem mass spectrometry-based testosterone assay. Besides the analytical characteristics, the practical aspects of minimization of the hands-on time and minimization of potential errors were taken into account when the assay was designed.

2 Materials

2.1 Stock Solutions and Calibration Curve

1. Preparation of testosterone stock solution: dissolve 10 mg testosterone (Sigma-Aldrich, St. Louis, MO, USA) in 3.473 mL DMSO to a concentration of approximately 10 mmol/L.

2. Preparation of internal standard (IS) working solution: dissolve 10 mg testosterone-d5 (CDN isotopes, Pointe-Claire, QC, Canada) in 3.473 mL DMSO to a concentration of approximately 10 mmol/L.

3. For a concentration of 10 μmol/L, 10 μL stock solution is mixed with 9990 μL methanol. Subsequently, final testosterone concentrations of 1, 3.3, 10, 33, 100, 333, and 1000 nmol/L are established by serial dilution in methanol (*see* **Notes 1** and **2**).

4. The calibration curve is standardized against the NIST standard reference material (SRM) 971 (*see* **Note 3**).

2.2 Other Reagents and Materials

All solutions prepared in this method are of analytical grade or better, and ultrapure water (purified deionized water, with a sensitivity of 15 MΩ/cm at 25 °C) is used.

1. In total, three serum-based control levels are used. The quality control (QC) samples are made from leftover patient materials, and pools are aimed at levels representing (1) (chemically) castrated men, (2) female normal range, and (3) adult male concentrations.

2. Storage: patient serum samples are stored for up to 1 week at 4 °C before analysis. QC samples are stored long term at −20 °C (*see* **Note 4**).

3. Preparation of system suitability test (SST) working solution: for a concentration of 10 μmol/L, 10 μL stock solution is mixed with 9990 μL methanol. Subsequently, a final testosterone concentration of 5 nmol/L is established by dilution.

4. Extraction solvent: methyl *tert*-butyl ether (MTBE) is used to extract steroids from human serum.

5. Injection solution: methanol and water are mixed to a 7:3 ratio (*see* **Note 5**).

2.3 Liquid Chromatography-Tandem Mass Spectrometry System

1. HPLC system: liquid chromatography is performed on a Dionex Ultimate 3000 HPLC system consisting of a vacuum degasser, pump, and an autosampler or equivalent. The autosampler needle can pierce the cap of plastic microvials (*see* **Note 6**) and has an injection volume range of 1–100 μL.

2. Mobile phase: the mobile phase is composed of water (containing 0.1% formic acid) (hereafter referred to as mobile phase A) and methanol (hereafter referred to as mobile phase B).

3. Separation column: reversed-phase C-18 column (5 μm particle size, 2.1 × 100 m internal diameter, Agilent Zorbax Extend C-18) (*see* **Note 7**).

4. Ion source: Turbo V electrospray ionization (ESI).

5. Mass spectrometer: API4000 (AB Sciex, Concord, ON, Canada) triple quadrupole mass spectrometer or equivalent or higher-end MS should be used.

3 Methods

3.1 Preparation of Samples and Controls

1. Add 250 μL human serum to a 2 mL Safe-Lock microvial (*see* **Notes 8** and **9**).

2. Add 25 μL calibration standard (*see* **Note 10**) in 1.5 mL tubes.

3. Add 100 μL SST working solution in 1.5 mL tubes.

4. Add 10 μL IS working solution to calibrators, blanks, and serum samples. The SST, calibrator, blank, and double blank samples are excluded from **steps 4–8** (*see* **Note 11**).

5. Add 1 mL of extraction solvent, MTBE, to the serum (*see* **Notes 12** and **13**).

6. Mix for 15 min on an orbital shaker (*see* **Note 14** and **15**).

7. Centrifuge at 3000 × g for 5 min at room temperature (RT).

8. Snap freeze the human serum using dry ice and ethanol (*see* **Notes 13** and **16**).

9. Decant the extraction solvent into a 1.5 mL soft cap microtube (*see* **Note 6**).

10. Dry all samples in a SpeedVac concentrator in combination with a vapor trap (*see* **Note 17**).

11. Reconstitute the sample extract in 100 μL injection solution.

12. Mix (400 rpm) for 10 min at RT (*see* **Note 18**).

13. Centrifuge the samples at 18,213 × *g* for 5 min at RT (*see* **Note 19**).

3.2 Liquid Chromatography

1. Purge the system with 50% mobile phase A and 50% mobile phase B at 1 mL/min (*see* **Note 20**).

2. Equilibrate the column at least 10 min prior to run with 20% mobile phase A and 80% mobile phase B (*see* **Note 21**).

3. Apply a run time of 5 min with an isocratic mobile phase flow.

4. Set the injection volume to 50 μL.

5. Program an autosampler wash cycle between each sample to minimize cross-contamination.

3.3 Tandem Mass Spectrometry Settings

1. Set ESI settings according to the values listed in Table 1.

2. Measure in multiple reaction monitoring (MRM) mode. Select the quantifier and qualifier transitions for testosterone and testosterone-d5 (Table 2).

3. Tune for optimal MRM settings. The obtained optimal MRM settings for the collision energy (CE), declustering potential (DP), entrance potential (EP), and cell exit potential (CXP) for both the mass transitions of our system are listed in Table 2 (*see* **Note 22**).

4. Set the dwell time to 100 ms.

Table 1
General settings of the electrospray ionization (ESI) source

General settings	
Nebulizer gas (psi)	50
Curtain gas (psi)	20
Ion spray (V)	5500
ESI temp. (°C)	500

Table 2
Multiple reaction monitoring (MRM) settings

		MRM transition (*m/z*)		CE (V)	DP (V)	EP (V)	CXP (V)	Dwell time (ms)
		Q1	Q3					
Quantifier	Testosterone	289.5	109.1	35	90	10	7	100
	Testosterone-d5	294.5	113.1	35	90	10	7	100
Qualifier	Testosterone	289.5	97.1	35	110	10	7	100
	Testosterone-d5	294.5	100.1	35	110	10	7	100

3.4 LC-MS Run and Batch Design

1. The analytical run starts with three system suitability tests (SST) containing 2 nmol/L testosterone standard and IS (*see* **Note 23**).

2. Next, eight calibrators including a blank are measured in duplicate.

3. Double blank samples in duplicate are introduced after the calibrators (*see* **Note 24**).

4. Two sets of QC samples (three levels, singular samples) are used: one set placed before and one set placed after the patient samples (*see* **Notes 25** and **26**).

5. Patient samples are scheduled in between the sets of QC samples and run in duplicate (*see* **Note 27**).

3.5 Data Analysis

1. Starting data analysis, chromatography of the quantifier and qualifier transitions are checked for their similarity. Aberrant peak shapes between quantifier and qualifier transitions, such as shouldering or twin peaks, are an indicator of interference (*see* **Note 28**).

2. Next, in quantification mode, integration of peaks is automatically performed and manually reviewed and corrected when inaccurate automated integration was observed. Integration of the analyte should match integration of the corresponding IS (*see* **Note 29**).

3. For calibration, the testosterone/IS ratio is calculated using the quantifier transition.

4. A calibration curve is generated using a linear regression, and a $1/x^2$ weighting is applied (*see* **Notes 3** and **30**).

5. Patient and QC results are individually calculated using the calibration curve.

6. QCs are individually reviewed and a 2SD control rule is applied for every control.

7. For patient samples, the duplicate results obtained are averaged, and the average concentrations are reported as testosterone concentration.

8. The difference between the duplicate sample results relative to the mean concentration was calculated and used to control for sample handling errors (*see* **Note 31**).

4 Notes

1. The calibration curve range is 0.1–100 nmol/L. By using a concentrated calibrator stock, the standard volumes added to the calibrator samples can be decreased accordingly. This method prolongs the usage of one set of standard stock.

2. Generally, a calibration curve is made in the same matrix as the QC and patient samples. However, a human serum pool with undetectable levels of testosterone is difficult to obtain in the required volumes. For this reason, we verified whether the methanol calibration matrix was suitable as an alternative to the serum matrix and chose to use methanol as calibration matrix.

3. The assay was standardized against the serum-based NIST (National Institute of Standards and Technology) reference material SRM 971. The two reference material samples were analyzed in four independent runs, and both SRM 971 standards were diluted in various concentrations to confirm trueness throughout the measuring range.

4. Serum testosterone stability was tested as part of the method validation. Testosterone stability in collected serum was at least 1 week when stored at room temperature, at least 2 weeks when stored at 4 °C, and at least 2 months when stored at −20 °C. Furthermore, when serum was not separated from the separator gel containing collection tube after centrifugation, testosterone was stable for at least 1 week when stored at 4 °C [5, 6]. Stability was defined as a $\leq \pm 6\%$ bias from the fresh testosterone sample.

5. Injection solution can be stored at RT. However, to avoid contamination we prepared the injection solution on a weekly basis.

6. Specifically, we used Brand soft cap micro-tubes (Wertheim, Germany) in combination with the cap piercing of the autosampler. Therefore, no certified glass sample vials are needed.

7. Our assay was validated with an Agilent Zorbax Extend C-18 column, since this column is also used for other applications run on the LC-MS/MS system.

8. A 250 μL sample size was chosen as this offers a sufficiently low lower limit of quantification (LLOQ) of 0.17 nmol/L with acceptable sample consumption. At this level, a total CV of 5.88% and a mean signal-to-noise ratio (S/N) of 14 are obtained. The application of higher-end mass spectrometers [2, 7] or incorporation of derivatization methods [8, 9] allows higher assay sensitivity, although derivatization processes are often more time-consuming and therefore not preferred for routine clinical diagnostics.

9. For this step, we used 2 mL Safe-Lock Eppendorf tubes (Hamburg, Germany) that provide additional protection against spilling during the mixing step. We find that normal tubes often lack cap security, resulting in sample loss.

10. Note that by reconstitution of sample extract of 250 μL serum in 100 μL injection solution, the samples are concentrated 2.5 times. This is accounted for by adding a volume of 25 μL standard working solution into the calibrator samples.

11. SST, calibrator, blank, and double blank samples do not contain serum. Therefore, no extraction is needed for these samples.

12. MTBE is highly volatile: this results in inaccurate pipetting when using air displacement pipettes. We find that using a repeater pipette significantly increases the accuracy of distributing equal volumes.

13. Precautionary measures are accounted for by working in a fume hood.

14. As a reference, we performed sample mixing with an IKA KS 130 basic orbital shaker (IKA, Staufen im Breisgau, Germany).

15. We find that by placing the sample vials sideways in the shaker, the mixing is performed with improved efficiency. By increasing the surface area, the immiscible liquids distribute more equally inside the tube.

16. For snap freezing a Styrofoam container is used. First, a permeable micro-tube rack is placed in the container. Subsequently, dry ice is placed around the rack. Ethanol is then poured into the container. Importantly, the ethanol should cover the aqueous phase in the micro-tubes. Freezing the aqueous phase usually takes 20–40 s, depending on the volume of ethanol and the amount of dry ice. A good indicator is the formation of a projection on the surface of the frozen serum.

17. As a reference, we used a Savant SC210A SpeedVac concentrator (Thermo Fisher Scientific, Waltham, MA, USA) in combination with refrigerated vapor trap RVT4104 (Thermo Fisher Scientific, Waltham, MA, USA) to dry the organic phase after extraction.

18. The samples should not be mixed sideways as this step focuses on reconstitution of the dried sample, which is mainly positioned at the bottom of the tube.

19. By briefly centrifuging the samples, droplets of dissolved sample located in the upper part of the tube are repositioned at the bottom of the tube. Additionally, in the case of increased turbidity in the samples, which is often present in human serum, this step acts as purification of the sample. After intensive centrifugation, the turbidity extract is pelleted at the bottom of the tube. As a result, the turbidity extract remains in the tube during injection of the sample.

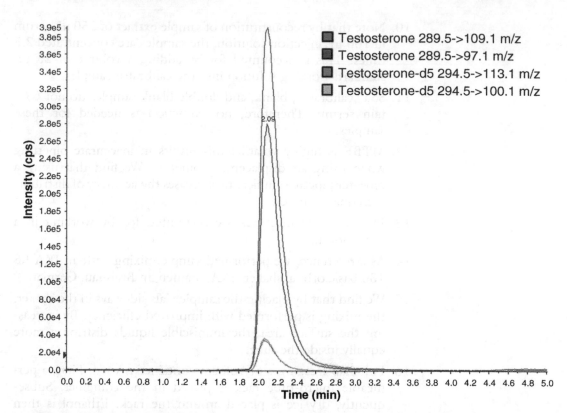

Fig. 1 Testosterone chromatogram. Chromatogram of a calibrator sample containing 25.4 nmol/L testosterone. The sample is spiked with 2 nmol/L internal standard (IS). Blue represents the quantifier testosterone transition, red represents the qualifier testosterone transition, green represents the quantifier testosterone IS transition, and gray represents the qualifier testosterone IS transition

20. Generally, the system is purged for at least 1 min. This results in clearance of any air bubbles present in the tubing leading to the pumps.

21. Testosterone and the corresponding IS have a tested retention time around 2.1 min using a 80:20 mobile phase B:A ratio in our system setup (Fig. 1).

22. Details on the general settings and MRM settings of testosterone and testosterone-d5 used in our method are given in Tables 1 and 2, respectively. Note that the parameters of the IS corresponding to the analyte are identical. Firstly, variations in parameter values between analyte and IS could result in signal differences troubling quantification. Secondly, replacement of hydrogens by deuterium yields a molecule that has virtually identical chemical-physical properties. Therefore, tuning both molecules should result in similar values for all parameters.

23. An SST is performed in triplicate, prior to the analytical run, to confirm proper system functioning before starting an analytical run.

24. Normally, double blank samples are used to check the matrix and LC-MS/MS system for contaminants and interferences. In this case, however, the calibration curve is generated in methanol, and double blank samples are introduced to identify a possible contamination in the IS working solution or the injection solution.

25. The QC samples serve as a measure of the quality for each run. When a significant aberration of QC concentration is detected (exceeding a 2SD rule), the run and obtained results require additional evaluation before the obtained results can be released.

26. Assay precision was tested by measuring three QC levels in quadruplicate for ten individual runs. Total coefficient of variation (CV) was 4.7, 4.1, and 3.3% for testosterone concentrations of 0.4, 3.1, and 7.6 nmol/L, respectively.

27. We find that increasing the number of samples reduces the practicality of the sample pretreatment. Therefore, we recommend to divide large patient serum sample sets (i.e., > 40 patient serum samples) over multiple runs.

28. No interference was detected from DHT, androstenedione, DHEA, 17-OHP, cortisol, hemolysate, bilirubin, and Intralipid. Only epitestosterone tested positive for interference. However, to date, no epitestosterone interference (i.e., a peak at a retention time of 2.4 min) has been observed in any patient sample analyzed.

29. Generally, the Analyst® software package (Version 1.6.2) integrates peaks automatically based on predetermined settings. In general, these settings should provide correct integration in more than 90% of samples, but at low concentrations (smaller peaks), fluctuation may result in poor automatic recognition of peak start and ending. Correct where needed, and ascertain that the peaks of the analyte and the corresponding IS are integrated similarly.

30. The applied $1/x^2$ weighting was selected based on overall trueness and linearity results obtained during assay validation. We determined that a $1/x^2$ weighting outperformed a $1/x$ weighting assay performance, especially for lower testosterone concentrations that we considered most relevant. A similar observation has recently been reported [2].

31. Patient serum samples are measured in duplicate to detect laboratory handling errors. In the case of inconsistent results (difference of duplicate results > 15%), the sample should be considered for reanalysis.

References

1. Marshall SK, Bangert WJ (2008) Clinical chemistry, 6th edn. Elsevier Science, Edinburgh

2. Wang HW, Gay Y, Botelho GD, Caudill JC, Vesper SP (2014) Total testosterone quantitative measurement in serum by LC-MS/MS. Clin Chim Acta 8(5):583–592

3. van Rossum HH, Bergman AM, Lentjes E (2015) Analytical challenges and potential applications of sex steroid hormone analysis in breast and prostate cancer patients. Chromatographia 78(5–6):359–365

4. Handelsman DJ, Wartofsky L (2013) Requirement for mass spectrometry sex steroid assays in the journal of clinical endocrinology and metabolism. J Clin Endocrinol Metab 98(10):3971–3973

5. Hepburn S et al (2016) Sex steroid hormone stability in serum tubes with and without separator gels. Clin Chem Lab Med 54(9):1451–1459

6. Shi RZ, van Rossum HH, Bowen RAR (2012) Serum testosterone quantitation by liquid chromatography-tandem mass spectrometry: interference from blood collection tubes. Clin Biochem 45(18):1706–1709

7. Methlie P et al (2013) Multisteroid LC-MS/MS assay for glucocorticoids and androgens, and its application in Addison's disease. Endocr Connect 2(12):125–136

8. Bui HN et al (2010) Serum testosterone levels measured by isotope dilution-liquid chromatography-tandem mass spectrometry in postmenopausal women versus those in women who underwent bilateral oophorectomy. Ann Clin Biochem 47:248–252

9. Kushnir MM et al (2010) Liquid chromatography-tandem mass spectrometry assay for androstenedione, dehydroepiandrosterone, and testosterone with pediatric and adult reference intervals. Clin Chem 56(7):1138–1147

## Chapter 8

LC-MS/MS Analysis of Bile Acids

Sabrina Krautbauer and Gerhard Liebisch

Abstract

Besides their role as lipid solubilizers, bile acids (BAs) are increasingly appreciated as bioactive molecules. They bind to G-protein-coupled receptors and nuclear hormone receptors. So they control their own metabolism and act on lipid and energy metabolism. Here we describe a simple, accurate, and fast liquid chromatography-tandem mass spectrometry (LC-MS/MS) method for the quantification of BAs in human plasma/serum.

Key words Tandem mass spectrometry, Electrospray ionization, Bile acids, Quantification, Internal standards, Method, Liquid chromatography, Isomer separation

1 Introduction

Bile acids (BAs) are primarily synthesized from cholesterol in the liver. Cholesterol is catabolized by oxidation, shortening of the side chain to form BAs. The free BAs are finally conjugated by glycine and taurine, respectively [1, 2]. Amphiphatic BAs are essential to solubilize dietary lipids and vitamins to promote their absorption. In humans most abundant BAs comprise the primary BAs cholic acid (CA) and chenodeoxycholic acid (CDCA) and the secondary BAs deoxycholic acid (DCA), lithocholic acid (LCA), and ursodeoxycholic acid (UDCA) generated by deconjugation and dehydroxylation by intestinal bacteria. BAs are effectively reabsorbed and transported back to the liver to again enter enterohepatic circulation [3].

Besides their role as lipid solubilizers, BAs are increasingly studied due to their signaling functions. They can bind to G-protein-coupled receptors and nuclear hormone receptors to control their own metabolism and to influence lipid and energy metabolism [4]. To study BA function in detail, it is necessary to use methods for their quantification covering the structural diversity of this group. For quantification of BAs including also their

Martin Giera (ed.), *Clinical Metabolomics: Methods and Protocols*, Methods in Molecular Biology, vol. 1730, https://doi.org/10.1007/978-1-4939-7592-1_8, © Springer Science+Business Media, LLC 2018

conjugates, liquid chromatography-tandem mass spectrometry (LC-MS/MS) is considered as the method of choice.

In this chapter we describe a simple, accurate, and fast LC-MS/MS method for the analysis of the main BAs in human plasma/serum. This approach uses protein precipitation as simple sample preparation. For accurate quantification stable isotope-labeled standards are included for the major species. The method provides baseline separation of isomeric BAs within a run time of 6 min due to application of core-shell material at basic pH of the mobile phase [5, 6].

2   Materials

Only highest purity solvents and standards should be used. Milli-Q water purified to a specific resistance of 18.2 MΩ is used.

2.1   Standards

Dissolve the following BA standards (Table 1) at a concentration of 20 mg/mL (deuterated standards at 1 mg/mL) in methanol in 10 mL screw cap glass vessels. Store standards at −80 °C.

2.2   Reagents

1. 1 M hydrochloric acid: Fill 100 mL water into a screw cap bottle and add 6.35 mL of concentrated hydrochloric acid (37%) and store in 100 mL reagent bottle (see **Note 1**).

2. Mobile phase A: Dissolve 771 mg ammonium acetate in 900 mL water and 100 mL methanol (measure volume in appropriate graduated cylinder), add 1000 µL ammonium hydroxide (25%), and store in 1 L reagent bottle.

3. Mobile phase B: Dissolve 771 mg ammonium acetate in 1000 mL methanol (measure volume in appropriate graduated cylinder), add 1000 µL ammonium hydroxide (25%), and store in 1 L reagent bottle.

4. Combined solution of internal standards: Pipette the following volume of 1 mg/mL standard solutions in a 10 mL volumetric flask, and add methanol to a volume of 10 mL. Store standards below −18 °C when used within 6 months (for long-term storage −80 °C). 15 µL each of D4-LCA, D4-GUDCA, and D5-TLCA, 20 µL each of D4-GDCA and D5-TDCA, 50 µL each of D4-CA and D4-DCA, 75 µL each of D4-CDCA and D4-UDCA, and 200 µL each of D4-GCA, D4-GCDCA, D4-GLCA, D5-TCA, D5-TCDCA, and D5-TUDCA.

5. Matrix calibrator: Human serum double charcoal-stripped, delipidized Golden West Biologicals (Temecula, CA, USA) is used as level **0**, and pooled serum of healthy donors is used as level **1**. Levels **2–5** are spiked up to the following concentrations (level 5 concentration; see **Notes 2** and **3**): 30 µM TCA,

Table 1
Analytical standards

Analyte	Abbreviation	Internal standard	Possible supplier
Cholic acid	CA		Sigma-Aldrich (Taufkirchen, Germany)
Chenodeoxycholic acid	CDCA		
Deoxycholic acid	DCA		
Lithocholic acid	LCA		
Ursodeoxycholic acid	UDCA		
Hyodeoxycholic acid	HDCA		
Glycocholic acid	GCA		
Glycochenodeoxycholic acid	GCDCA		
Glycodeoxycholic acid	GDCA		
Taurocholic acid	TCA		
Taurochenodeoxycholic acid	TCDCA		
Taurodeoxycholic acid	TDCA		
Taurolithocholic acid	TLCA		
Tauroursodeoxycholic acid	TUDCA		
Taurohyodeoxycholic acid	THDCA		
D4-CDCA		x	
D4-LCA		x	
D4-CA		x	
D4-DCA		x	
Glycolithocholic acid	GLCA		Steraloids (Newport, USA)
Glycoursodeoxycholic acid	GUDCA		
Glycohyodeoxycholic acid	GHDCA		
D4-UDCA		x	
D4-GCDCA		x	
D4-GCA		x	
D4-GLCA		x	CDN Isotopes (Pointe-Claire, QC, Canada)
D4-GUDCA		x	
D4-GDCA		x	
D5-TCA		x	Toronto Research Chemicals (Toronto, Canada)
D5-TUDCA		x	
D5-TCDCA		x	
D5-TDCA		x	
D5-TLCA		x	

30 μM TUDCA, 20 μM THDCA, 40 μM TCDCA, 16 μM TDCA, 4 μM TLCA, 70 μM GCA, 100 μM GUDCA, 70 μM GHDCA, 40 μM GCDCA, 16 μM GDCA, 4 μM GLCA, 16 μM CA, 22 μM UDCA, 8 μM HDCA, 8 μM CDCA, 6 μM DCA, 2 μM LCA. Store in 5 mL aliquots at −80 °C.

6. Working aliquots of 0.5 mL are kept at −18 °C for up to 6 months.

7. Serum quality controls (QCs): As low control, pooled serum of healthy donors; as high control, pooled serum of healthy

donors spiked with 1/10 of the highest calibrator level. Store in 5 mL aliquots at −80 °C. Working aliquots of 0.5 mL are kept at −18 °C for up to 6 months.

2.3 Liquid Chromatography-Tandem Mass Spectrometry (LC-MS/MS)

1. MS/MS: Hybrid triple quadrupole/linear ion trap mass spectrometer, 4000 QTRAP® (Concord, Ontario, Canada) equipped with Turbo V ion spray source (*see* **Note 4**).

2. LC: Autosampler, HTC PAL autosampler (CTC Analytics, Zwingen, Switzerland) with additional 6-port valve, Agilent 1200 binary pump, 1200 isocratic pump, degasser (Waldbronn, Germany).

3. LC column: Core-shell column NUCLEOSHELL RP18, 50 × 2 mm, 2.7 μm (Macherey-Nagel, Düren, Germany) equipped with a 0.5 μm prefilter (Upchurch Scientific, Oak Harbor, WA, USA).

4. Glass autosampler vials (1.5 mL) with 200 μL glass insert.

3 Methods

3.1 Sample Preparation

1. Place 100 μL sample (patient material, QCs, calibrators, internal standard blank = water) in 1.5 mL Eppendorf tube.

2. Add 20 μL of internal standard mix.

3. Mix briefly.

4. Add 30 μL 1 M HCl.

5. Add 1 mL acetonitrile.

6. Mix thoroughly about 1 min.

7. Centrifuge 15 min at 14,000 × g.

8. Transfer 1 mL supernatant into 1.5 mL Eppendorf tube.

9. Remove solvent in a vacuum concentrator.

10. Reconstitute in 100 μL (3/7; v/v) methanol/water.

11. Mix thoroughly about 1 min.

12. Place for 10 min into ultrasonic bath.

13. Centrifuge 15 min at 14,000 × g.

14. Transfer clear supernatant into autosampler vial (1.5 mL) with 200 μL glass insert (*see* **Note 5**).

3.2 Liquid Chromatography-Tandem Mass Spectrometry (LC-MS/MS)

Samples are analyzed by LC-MS/MS together with calibrators, QCs including double blank ((3/7; v/v) methanol/water), and internal standard blank (*see* Subheading 3.1) in each batch.

1. LC settings: Purge gradient pump with mobile phases A and B. Wash solvent 1: 50/50 (v/v) methanol/2-propanol. Wash solvent 2: 50/50 (v/v) methanol/water.

2. Install core-shell column NUCLEOSHELL RP18, 50 × 2 mm, 2.7 μm (*see* **Note 6**).

3. Set column oven temperature: 50 °C.

4. The following gradient is used: 80% A for 0.05 min, a stepwise linear decrease to 53% A at 0.1 min, to 49% A at 2.0 min and to 28% A at 4.5 min. The flow rate was set to 0.5 mL/min. A column wash at 100% B for 0.5 min and re-equilibration at 100% A for 0.6 min were performed at a flow rate of 0.8 mL/min.

5. Isocratic pump delivers methanol at a flow rate of 250 μL/min.

6. Divert valve directs column flow only from 65 to 290 s into the mass spectrometer.

7. Inject volume: 5 μL.

8. Mass spectrometer settings (for 4000 QTRAP): Ensure the following parameters are set: ion source, Turbo V ion spray source; ionization mode, negative; ion spray voltage, −4500 V; ion source heater temperature, 450 °C; source gas 1, 40 psi; source gas 2, 35 psi; curtain gas, 20 psi; collision gas, 6 psi; entrance potential, −10 V; dwell time, 30 msec; operate both Q1 and Q3 at unit resolution; analysis mode, MRM (for mass transitions, *see* Table 2).

Table 2
Mass transitions and retentions times of the individual BAs including their internal standards

BA species	RT [min]	Q1 [*m/z*]	Q3 [*m/z*]	DP [V]	CE [V]	CXP [V]	IS
CA	2.63	407.3	407.3	−135	−30	−11	D4-CA
UDCA	1.96	391.3	391.3	−135	−30	−11	D4-UDCA
HDCA	2.19	391.3	391.3	−135	−30	−11	–
CDCA	3.64	391.3	391.3	−135	−30	−11	D4-CDCA
DCA	3.83	391.3	391.3	−135	−30	−11	D4-DCA
LCA	4.55	375.3	375.3	−135	−30	−11	D4-LCA
GCA	2.57	464.3	74	−105	−72	−1	D4-GCA
GUDCA	1.87	448.3	74	−115	−70	−1	D4-GUDCA
GHDCA	2.11	448.3	74	−115	−70	−1	–
GCDCA	3.57	448.3	74	−115	−70	−1	D4-GCDCA
GDCA	3.78	448.3	74	−115	−70	−1	D4-GDCA
GLCA	4.45	432.3	74	−105	−64	−1	D4-GLCA
TCA	2.54	514.3	80	−185	−116	−1	D5-TCA

(continued)

Table 2
(continued)

BA species	RT [min]	Q1 [m/z]	Q3 [m/z]	DP [V]	CE [V]	CXP [V]	IS
TUDCA	1.86	498.3	80	−185	−116	−1	D5-TUDCA
THDCA	2.10	498.3	80	−185	−116	−1	–
TCDCA	3.52	498.3	80	−185	−116	−1	D5-TCDCA
TDCA	3.73	498.3	80	−185	−116	−1	D5-TDCA
TLCA	4.41	482.3	80	−165	−108	−1	D5-TLCA
D4-CA	2.61	411.3	411.3	−135	−30	−11	
D4-CDCA	3.62	395.3	395.3	−135	−30	−11	
D4-DCA	3.81	395.3	395.3	−135	−30	−11	
D4-LCA	4.53	379.3	379.3	−135	−30	−11	
D4-UDCA	1.94	395.3	395.3	−135	−30	−11	
D4-GCA	2.55	468.3	74	−105	−72	−1	
D4-GCDCA	3.55	452.3	74	−115	−70	−1	
D4-GDCA	3.76	452.3	74	−115	−70	−1	
D4-GUDCA	1.85	452.3	74	−115	−70	−1	
D4-GLCA	4.43	436.3	74	−105	−64	−1	
D5-TCA	2.52	519.3	80	−185	−116	−1	
D5-TUDCA	1.84	503.3	80	−185	−116	−1	
D5-TCDCA	3.50	503.3	80	−185	−116	−1	
D5-TDCA	3.71	503.3	80	−185	−116	−1	
D5-TLCA	4.39	487.3	80	−165	−108	−1	

9. Data analysis: Integrate peaks by instrument software (here Analyst Software 1.4.2.; Applied Biosystems, Darmstadt, Germany) (*see* Fig. 1). Quantification is based on ratio of area counts of analyte to internal standard (*see* Table 2). Calibration lines were generated by linear regression with $1/x$ weighting (*see* **Notes** 7 and **8**).

10. Quality check: Check for baseline separation of isomeric BA species, e.g., CDCA and DCA including their conjugates (*see* Fig. 1). Check purity of the deuterated BAs: When interference on the analyte transitions is present, quantification needs to consider that. For calibration lines the coefficient of determination R^2 should be above 0.99, and individual calibrator

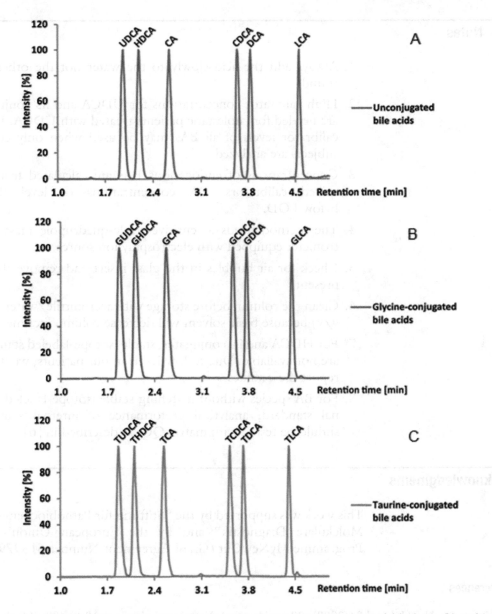

Fig. 1 Chromatogram of bile acid species. The chromatograms show unconjugated (**a**) glyco- (**b**) and tauro-conjugates (**c**). Displayed are chromatograms from a matrix calibrator after normalization to the highest peak

levels should not deviate more than 15% (more than 20% at LOQ) from the target value. Concentrations of QCs BA levels have to be in a defined range (recommended maximal deviation <15% from the target value; <20% at the LOQ).

4   Notes

1. Always add the acid slowly to the water not the other way round.

2. High calibrator concentrations for UDCA and its conjugates are needed for cholestatic patients treated with UDCA. Lower calibrator levels of all BAs may be used when only control subjects are analyzed.

3. Concentrations of endogenous BAs are calculated from the spiked calibrators. BA concentrations of level 0 are below LOD.

4. The method needs a sensitive triple-quadrupole mass spectrometer equipped with electrospray ion source.

5. Check for air bubbles in the glass insert and remove these if present.

6. Clean the column before storage with acetonitrile/water (3/1, v/v) because basic solvent will decrease column lifetime.

7. For HDCA and its conjugates, stable isotope-labeled standards are not available. Due to low levels in our patients, we do not report these values.

8. For BA species without matching stable isotope-labeled internal standard, analytical performance of internal standards should be tested with matrix QCs as described in [6].

Acknowledgments

This work was supported by the "Stiftung für Pathobiochemie und Molekulare Diagnostik" and by the European Union's FP7 Programme MyNewGut (Grant Agreement Number 613979).

References

1. Hofmann AF, Hagey LR (2008) Bile acids: chemistry, pathochemistry, biology, pathobiology, and therapeutics. Cell Mol Life Sci 65 (16):2461–2483

2. Thomas C, Pellicciari R, Pruzanski M, Auwerx J, Schoonjans K (2008) Targeting bile-acid signalling for metabolic diseases. Nat Rev Drug Discov 7(8):678–693. https://doi.org/10.1038/nrd2619

3. Dawson PA, Karpen SJ (2015) Intestinal transport and metabolism of bile acids. J Lipid Res 56 (6):1085–1099. https://doi.org/10.1194/jlr.R054114

4. Kuipers F, Bloks VW, Groen AK (2014) Beyond intestinal soap—bile acids in metabolic control.

Nat Rev Endocrinol 10(8):488–498. https://doi.org/10.1038/nrendo.2014.60

5. Scherer M, Gnewuch C, Schmitz G, Liebisch G (2009) Rapid quantification of bile acids and their conjugates in serum by liquid chromatography-tandem mass spectrometry. J Chromatogr B Analyt Technol Biomed Life Sci 877(30):3920–3925. https://doi.org/10.1016/j.jchromb.2009.09.038

6. Krautbauer S, Buechler C, Liebisch G (2016) Relevance in the use of appropriate internal standards for accurate quantification using LC-MS/MS: tauro-conjugated bile acids as an example. Anal Chem 88(22):10957–10961. https://doi.org/10.1021/acs.analchem.6b02596

Chapter 9

LC-MS/MS Analysis of Triglycerides in Blood-Derived Samples

Madlen Reinicke, Susen Becker, and Uta Ceglarek

Abstract

The increasing interest in the analysis of triglyceride (TG) species and the individual fatty acid (FA) composition requires expeditious and reliable quantification strategies. The utilization of flow injection analysis (FIA) coupled to quadrupole tandem mass spectrometry (MS/MS) for the simultaneous quantitation of TG and identification of FA composition facilitates the multiplexed verification of various biomarkers from small sample quantities. Enzymatic methods based on saponification and glycerol analysis are not suited for the determination of the FA distribution in TGs. This protocol proposes a procedure for the establishment of a relative quantitation method for middle- to high-abundance plasma TGs and the corresponding FA composition. Essential topics as FIA-MS/MS method development as well as sample preparation and validation strategies are described in detail.

Key words Triglyceride species, Fatty acid distribution, Neutral loss experiments, Relative quantification, Biomarker validation

1 Introduction

Reliable quantitation of triglyceride (TG) species and the corresponding fatty acid (FA) distribution is a basic requirement for biomarker verification studies in human body fluids. Routine diagnostic quantification methods of TGs are based on hydrolysis and spectrophotometric determination of total glycerol concentration. Thus, a differentiation of the TG molecular species and information about the FA distribution are not available. Due to the association of FAs with the development of cardiovascular diseases and diabetes, detailed information about the kind of FAs bound within the TG is required [1, 2]. The identification of TG molecular species in human body fluids is challenging because of the great variability of the FAs relating to the degree of saturation, the chain length, and the various combination possibilities of FAs at the glycerol backbone. The close resemblance of the TG species requires laborious chromatographic separation [3, 4]. We describe

Martin Giera (ed.), *Clinical Metabolomics: Methods and Protocols*, Methods in Molecular Biology, vol. 1730, https://doi.org/10.1007/978-1-4939-7592-1_9, © Springer Science+Business Media, LLC 2018

a flow injection analysis (FIA) coupled to quadrupole tandem mass spectrometry (MS/MS) method without chromatography benefitting from the advantages of fast measurement and mass separation analysis performed in neutral loss (NL) scan mode using a triple quadrupole mass spectrometer. In the first dimension, a defined range of mass-to-charge ratios (m/z) is scanned. The scan range of the second dimension is defined by a m/z offset corresponding to the FA of interest. Signals for TG species are detected for a limited number of NL-specific ions occurring in both scan dimensions. Combining and monitoring of such FA-specific mass transitions provide the basis for the high specificity in complex sample matrices [5–9]. Thereby, quantification is dependent on the implementation of internal standards to compensate for sample losses during sample preparation and matrix effects in the ionization process. For this purpose, mostly deuterated analogs of the TG species are used. Important for the deuterated analogs is a structurally stable position of the deuterium atoms to ensure stable labeled fragments during electrospray ionization.

The presented protocol for TG quantification by FIA-MS/MS can be easily implemented for middle- to high-abundance plasma TGs requiring only small sample quantities [10].

2   Materials

2.1   Reagents and Consumables

1. Solvents: only use high purity solvents, ideally of MS grade. For example, ULC-MS grade methanol, ULC-MS grade 2-propanol, and Rotisolv® HPLC grade toluene can be used.

2. Protein precipitation: use 200 µL glass micro-inserts with spring, 1.5 mL polypropylene microtube for centrifugation, and 10 mM ammonium acetate in (50/50, v/v) toluene/methanol.

3. LC-MS/MS analysis: use 200 µL glass micro-inserts in combination with 2.0 mL short thread autosampler vials with slotted cap.

2.2   HPLC and Mobile Phases

1. The HPLC instrument consists of an autosampler and a binary micro-pump system.

2. Isocratic conditions with a flow gradient of (10–500 µL/min) are used.

3. The mobile phase consists of (90/10, v/v) 2-propanol/methanol.

2.3   Mass Spectrometer

1. The employed mass spectrometer is triple quadrupole (e.g., API 4000, SCIEX).

2. Ionization source: electrospray ionization in the positive mode.

3. An analysis software Analyst (SCIEX) can be used.

Table 1
Standards and internal standards with abbreviations for total carbon/number of double bonds and the FA composition

Standard	Total carbon/number of double bonds	FA composition
Standards		
Trimyristin glyceride	TG 42:0	TG (14:0/14:0/14:0)
Tripalmitin glyceride	TG 48:0	TG (16:0/16:0/16:0)
1,3-palmitin-2-stearin glyceride	TG 50:0	TG (16:0/18:0/16:0)
Tristearin glyceride	TG 54:0	TG (18:0/18:0/18:0)
Triolein glyceride	TG 54:3	TG (18:1/18:1/18:1)
Triarachidonin glyceride	TG 60:12	TG (20:4/20:4/20:4)
(1,1,2,3,3)-deuterium-labeled (d_5) internal standards		
1,3-myristin-2-palmitolein glyceride	d_5-TG 44:1	d_5-TG (14:0/16:1/14:0)
1,3-pentadecanoin-2-olein glyceride	d_5-TG 48:1	d_5-TG (15:0/18:1/15:0)
1,3-palmitin-2-stearin glyceride	d_5-TG 50:0	d_5-TG (16:0/18:0/16:0)
1,3-nonadecanoin-2-laurin glyceride	d_5-TG 50:0	d_5-TG (19:0/12:0/19:0)
1,3-heptadecanoin-2-heptadecenoin glyceride	d_5-TG 51:1	d_5-TG (17:0/17:1/17:0)
1,3-eicosadienoin-2-linolenin glyceride	d_5-TG 58:7	d_5-TG (20:2/18:3/20:2)
1,3-linolein-2-linolein glyceride	d_5-TG 58:10	d_5-TG (20:4/18:2/20:4)
1,3-arachidin-2-eicosenoin glyceride	d_5-TG 60:1	d_5-TG (20:0/20:1/20:0)
1,3-eicosapentaenoin-2-docosahexaenoin glyceride	d_5-TG 62:16	d_5-TG (20:5/22:6/20:5)

2.4 Standards, Internal Standards, and Human Specimen

Standards and internal standards are given in Table 1. For method development, human serum or EDTA plasma is required.

3 Methods

The following protocol summarizes basic steps in the method development of a quantitative assay. Starting from the identification of characteristic mass transitions, all mandatory requirements for a relative quantification of middle- to high-abundance plasma TGs and the FA composition by MS/MS are described (*see* **Notes 1** and **2**).

Table 2
Possible NL transitions with the corresponding FA

Neutral loss [Da]	Corresponding FA	
245.24	Myristic acid	C14:0
259.25	Pentadecylic acid	C15:0
273.27	Palmitic acid	C16:0
271.25	Palmitoleic acid	C16:1
301.30	Stearic acid	C18:0
299.28	Oleic acid	C18:1
297.27	Linoleic acid	C18:2
295.25	Linolenic acid	C18:3
321.27	Arachidonic acid	C20:4

3.1 Selection of Mass Transitions and HPLC Optimization

In the presence of ammonium ions (10 mM ammonium acetate), TGs form the molecular ions $[M+NH_4]^+$ in positive mode ESI. The ammonium adducts were detected by neutral loss scans, characterized by the loss of the FA and ammonia (NH_3) as neutral molecules during the fragmentation process, leaving a positively charged molecule with the mass difference of the FA of interest. The respective NL transitions can be calculated by its monoisotopic mass in addition to the loss of NH_3, e.g., myristic acid C14:0, which has a monoisotopic mass of 228.21 Da and 17.03 Da for the loss of NH_3, yielding in a respective NL of 245.24 Da (*see* **Notes 2** and **3**). An example of possible NL transitions with the corresponding FA is given in Table 2.

After MS tuning and selection of the neutral loss transitions, the HPLC part of the method has to be optimized. Below a detailed procedure applying a triple quadrupole instrument is described:

1. Calculate potential NL fragments of the fatty acid residues in the TG species of interest.

2. Dissolve TG standards in toluene, to yield stock solutions of 1–5 mg/mL, and apply ultrasound if standards are less soluble (*see* **Note 4**).

3. Prepare working standards in (50/50, v/v) toluene/methanol.

4. Prepare tuning solutions of 250 ng/mL or lower in (50/50, v/v) toluene/methanol containing 10 mM ammonium acetate.

5. Activate the tune mode and directly infuse the tuning solution at 10 μL/min using the following parameters: curtain gas, 10 psi; collision gas, high; nebulizer gas, 20 psi; heater gas, 0 psi; ion spray voltage, 5500 V; source temperature, 20 °C; and polarity, positive.

6. Perform a Q1 scan to identify the precursor ion, e.g., m/z 350–1000.

7. Tune the declustering potential in a Q1 multiple ion scan, e.g., 0–200 V.

8. Identify fragment ions in a product ion scan performed with ramped collision energy in multichannel analysis (MCA) mode. Compare detected ions with the expected fragment ions identified before by the calculated neutral loss.

9. Tune the collision energy and the collision cell entrance and exit potential in a multiple reaction monitoring (MRM) scan by applying the ramped mode for each potential.

10. Optimize curtain gas, nebulizer gas, heater gas, ion spray voltage, source temperature, and collision gas by infusion of the TG standard solution in a constant flow of eluent operated under FIA conditions via a tee. Choose a flow rate which will be applied in the final method. If necessary, repeat this step during FIA optimization.

11. Create an analyzing method containing the optimized neutral loss and MS parameters. An optimized flow gradient and standard gas and temperature settings are given below.
The injection volume was 40 µL, and the mobile phase was 2-propanol/methanol (90/10, v/v) with a flow gradient of 10 µL/min (0–3.5 min) followed by 500 µL/min (3.5–5 min). The mass range was set from m/z 700 to m/z 1000 with a scan rate of 1.5 s. Experimental conditions for electrospray ionization of TGs in positive-ion mode with collisional activation by nitrogen gas were as follows: curtain gas, 20 psi; collision gas, high; declustering potential, 70 V; collision cell entrance potential, 10 V; collision energy, 35 V; collision cell exit potential, 7 V; nebulizer gas, 30 psi; heater gas, 50 psi; ion spray voltage, 5500 V; source temperature. 300 °C.

12. Optimize HPLC parameters and settings including analytical eluents, modifier, gradient, flow rate, column temperature, and sample volume. Refer to the basic literature on HPLC method development for guidance [11].

3.2 Internal Standardization and Quality Controls

One of the simplest approaches for the relative quantification of TG species by means of MS/MS is the FIA combined with internal standardization. Search the literature for reference values or published concentration levels of the target analyte in human plasma or serum, e.g., [10, 12, 13].

1. Internal standard mix
Prepare an internal standard mix in (50/50, v/v) toluene/methanol containing 10 mM ammonium acetate with concentrations similar to expected concentrations in human

specimens. Note that signal intensity and peak area of the internal standard should be comparable to the ones of the monitored analyte.

2. Quality controls

    For quality assessment and guarantee of batch comparability, quality controls should be carried along in the analytical process. Preferably, two to three concentration levels should be investigated (low, middle, high), and blank matrix should be preferred. TG standards can be used for the preparation of quality controls. Human specimens, which should be stored as aliquots to avoid freeze-thaw cycles, can be used as quality control. Ring trial materials or the like with known target values can also be used. Avoid multiple freeze-thaw cycles and store internal standard mix and quality controls aliquoted, ready for use.

3.3 Sample Pretreatment of Human Specimens and Control Material

1. Protein precipitation carried out in a glass micro-insert with spring which is placed in a 1.5 mL polypropylene microtube for later centrifugation.

2. Place 245 µL of precipitation solvent containing 150 ng/mL internal standard d_5-TG mixture and 10 mM ammonium acetate in the glass micro-insert with spring.

3. Add 5 µL of plasma/serum or standard or control material and close the tube tightly.

4. After mixing for 30 s, spin down the precipitate at $13,000 \times g$ for 10 min.

5. Transfer the clear supernatant in a new glass micro-insert placed in an autosampler vial.

3.4 Sample Analysis

1. Run the developed MS/MS method (Subheading 3.1).

2. For quality assessment of the analysis, several aliquots of a quality control should be carried along in the sample preparation procedure and be measured accordingly (*see* **Note 5**).

3.5 Data Analysis and Relative Quantification

TG species were relatively quantified using the intensity ratio of the summed NL intensities of each ion peak with that of the internal standard, e.g., TG d_5–50:0 (16:0/18:0/16:0), after correction for two ^{13}C isotope effects and the different fragmentation behavior of the cleaved neutral acyl residues in the positive-ion mode as the following [5, 10]:

1. Select a representative time window in the total ion current spectrum, e.g., 2 min.

2. Select "show spectrum" to get a NL spectrum of every neutral loss experiment included in the method.

3. To get the overall intensity of the TG species, the data of every single NL scan needs to be summed up. Use, e.g., the Analyst

(SCIEX) options "overlay" and "sum overlays" to sum all NL experiments. Use, e.g., "List data," and select the peak list to save the m/z and the corresponding peak intensity, e.g., as txt. file (*see* **Note 6**).

4. The m/z for each ion peak corresponds to a TG species. The peak list can be reduced to the m/z for the TG species of interest including the internal standards and the peak intensity of the corresponding [M-2]-peak for isotopic correction. An example peak list with TG species, ion peak, and corresponding [M-2]-peak is given in Table 3.

Table 3
Possible peak list with the *m/z* of the ion peak and corresponding [*M*-2]-peak for 19 TG species and 4 internal standards

Triglyceride	*m/z* [*M*+NH$_4$]$^+$	*m/z* [*M*-2]$^+$
TG 46:1	794.7	792.7
TG 46:0	796.7	794.7
TG 48:2	820.7	818.7
TG 48:1	822.7	820.7
TG 48:0	824.7	822.7
TG 50:3	846.8	844.8
TG 50:2	848.8	846.8
TG 50:1	850.8	848.8
TG 50:0	852.8	850.8
TG 52:5	870.8	868.8
TG 52:4	872.8	870.8
TG 52:3	874.8	872.8
TG 52:2	876.8	874.8
TG 52:1	878.8	876.8
TG 54:6	896.8	894.8
TG 54:5	898.8	896.8
TG 54:4	900.8	898.8
TG 54:3	902.8	900.8
TG 54:2	904.8	902.8
d$_5$-TG 44:1	771.7	769.7
d$_5$-TG 48:1	827.7	825.7
d$_5$-TG 50:0	857.8	855.8
d$_5$-TG 58:10	949.8	947.8

5. For further calculations, the summed NL intensities of each ion peak are corrected through multiplication by the following correction factors [5] (*see* **Note 7**): Z_1 is the first ^{13}C isotope correction factor, where the difference between the carbon number of a given TG species and the internal standard is considered with the total acyl carbon number n in the TG species of interest and the total acyl carbon number s of the internal standard, which is, e.g., 50 for d$_5$-TG 16:0/18:0/16:0:

$$Z_1 = \frac{1 + 0.011n + \frac{0.011^2 n(n-1)}{2}}{1 + 0.011s + \frac{0.011^2 s(s-1)}{2}}$$

Z_2 is the second ^{13}C isotope correction factor and results from the effect caused by the overlapping of the *[M+2]* isotope peak with the molecular ion peak of a species, which has a 2-Da higher mass. The general correction factor for this type of ^{13}C isotope effect is as follows:

$$Z_2 = 1 - \frac{I_{M-2}}{I_M} \times 0.011^2 \times \frac{m(m-1)}{2}$$

where m is the total acyl carbon number in the TG species with lower molecular mass and I_{M-2} and I_M are peak intensities of ions at molecular weight (M-2) and M, respectively. The signal intensity of the cleaved neutral acyl residue is influenced by the chain length and the degree of unsaturation. It has been shown that the longer the acyl chain lengths and the lower the unsaturation index, the lower the intensity of TG species in positive-ion ESI-MS. This effect is considered in the third correction factor (Z_3):

$$Z_3 = 4.4979 + 0.3441p - 0.1269q - 4.845 \times 10^{-3} p \times q$$
$$+ 9.9 \times 10^{-4} q^2$$

where p is the total double bond number in the TG species and q is the total acyl carbon number in the three acyl chains of the TG.

6. The m/z for each ion peak corresponds to a TG species. The relative amount of the TG species can now be calculated using the intensity ratio of the corrected summed NL intensities of each ion peak and the corrected intensity of the internal standard, e.g., d$_5$-TG 50:0 (16:0/18:0/16:0) with known concentration (*see* **Note 8**).

7. The relative FA composition for a specific TG species characterized by a corresponding m/z can be determined via the ratio of the relative ion peak intensity of each NL for a certain m/z and the corresponding summed NL intensity.

3.6 Validation of the Analytical Process

Assess the validity of the complete method including sample preparation with the following experiments:

1. Determine the limit of detection (LOD) and lower limit of quantification (LLOQ) in a serial dilution or more appropriately by spiking blank matrix according to concentration levels. By the use of a serial dilution, the method's linearity and the linear dynamic concentration range can be assessed simultaneously. A signal-to-noise ratio of 3 is commonly accepted for LOD estimation. LLOQ is defined as the analyte concentration which can be measured with a coefficient of variation <20% (*see* **Note 9**).

2. Assess within-day and between-day precision at two to three concentration levels to test the reliability of the entire method. Within-day precision: Prepare ten replicates per sample and measure them in one run. Between-day precision: Prepare and measure one replicate per sample in a single run on 10 consecutive working days.

3. Determine the recovery rate by spiking specified amounts of TG standard in blank matrix or human material. Analyze these samples as well as the starting material. Compare the difference in calculated concentration of spiked sample and concentration of starting material with the amount of analyte added.

4 Notes

1. The described method is a screening method with relative quantification of the TG species containing the FAs implemented in the method, independent of the frequency and position of the FA residue at the glyceride backbone. An absolute quantification for a limited selection of TGs with a distinct FA composition is possible if using an external calibration and setting up a different MS/MS experiment, e.g., multiple reaction monitoring.

2. An extension of the number of neutral loss experiments is possible according to the monoisotopic mass of the FA of interest, limited by the total scan time and the peak width.

3. All MS experiments can be performed in a modified way on other MS platforms, e.g., linear ion trap mass spectrometers or quadrupole time-of-flight mass spectrometers. However, specific experiments like enhanced product ion scans using the ion trap function for ion accumulation can only be performed by the use of hybrid triple quadrupole/linear ion trap mass spectrometers.

4. Toluene and toluene/methanol had been used for standard dilution and sample preparation due to the better solubility of

TGs and higher ionization efficiency during the ESI process than chloroform. The high vapor pressure of chloroform supports the evaporation process and negatively affects, e.g., the sample preparation, solvent handling, and eluent storage. Using tetrahydrofuran and n-hexane, we observed good TG solubility but worse ionization on account of its lacking protic properties.

5. If the quantification is performed via an external standard calibration set, run the calibration standards including a blank at the very beginning of each batch. The measurement of the external standard calibration set should be repeated at the end of each batch for quality assessment of the analysis.

6. For later calculations of the relative amount of each FA in a specific TG species, save every single NL experiment with the m/z and the peak intensity as txt.file, use, e.g., "List data," and select the peak list.

7. The selection of m/z for each ion peak including the [M-2]-peak and calculation of the correction factors can be done either manually using a common calculation program, e.g., Microsoft Office Excel, or program a software, e.g., RStudio.

8. Several internal standards for different mass ranges could be used.

9. Alternatively, a signal-to-noise ratio of 10 is commonly accepted for LLOQ estimation.

Acknowledgment

This publication is supported by LIFE—Leipzig Research Center for Civilization Diseases—Leipzig University. LIFE is funded by means of the European Union, by the European Regional Development Fund (ERDF), and by the Free State of Saxony within the framework of the excellence initiative.

References

1. Erkkilä A, de Mello VDF, Risérus U, Laaksonen DE (2008) Dietary fatty acids and cardiovascular disease: an epidemiological approach. Prog Lipid Res 47:172–187. https://doi.org/10.1016/j.plipres.2008.01.004

2. Rhee EP et al (2011) Lipid profiling identifies a triacylglycerol signature of insulin resistance and improves diabetes prediction in humans. J Clin Invest 121(4):1402–1411. https://doi.org/10.1172/JCI44442

3. Nagy K, Sandoz L, Destaillats F, Schafer O (2013) Mapping the regioisomeric distribution of fatty acids in triacylglycerols by hybrid mass spectrometry. J Lipid Res 54(1):290–305. https://doi.org/10.1194/jlr.D031484

4. Baiocchi C, Medana C, Dal Bello F (2015) Analysis of regioisomers of polyunsaturated triacylglycerols in marine matrices by HPLC/HRMS. Food Chem 166:551–560. https://doi.org/10.1016/j.foodchem.2014.06.067

5. Han X, Gross RW (2001) Quantitative analysis and molecular species fingerprinting of triacylglyceride molecular species directly from lipid extracts of biological samples by electrospray

ionization tandem mass spectrometry. Anal Biochem 295(1):88–100. https://doi.org/10.1006/abio.2001.5178

6. Krank J, Murphy RC, Barkley RM, Duchoslav E, McAnoy A (2007) Qualitative analysis and quantitative assessment of changes in neutral glycerol lipid molecular species within cells. In: Brown HA (ed) Methods in enzymology, vol 432. Academic Press, USA, pp 1–20. https://doi.org/10.1016/S0076-6879 (07)32001-6.

7. Fernandez C, Sandin M, Sampaio JL, Almgren P, Narkiewicz K, Hoffmann M, Hedner T, Wahlstrand B, Simons K, Shevchenko A, James P, Melander O (2013) Plasma lipid composition and risk of developing cardiovascular disease. PLoS One 8(8):e71846

8. Schwudke D, Oegema J, Burton L, Entchev E, Hannich JT, Ejsing CS, Kurzchalia T, Shevchenko A (2005) Lipid profiling by multiple precursor and neutral loss scanning driven by the data-dependent acquisition. Anal Chem 78 (2):585–595. https://doi.org/10.1021/ac051605m.

9. Hsu F-F, Turk J (2010) Electrospray ionization multiple-stage linear ion-trap mass spectrometry for structural elucidation of triacylglycerols: assignment of fatty acyl groups on the glycerol backbone and location of double bonds. J Am Soc Mass Spectrom 21 (4):657–669. https://doi.org/10.1016/j.jasms.2010.01.007

10. Sander M, Becker S, Thiery J, Ceglarek U (2015) Simultaneous identification and quantification of triacylglycerol species in human plasma by flow-injection electrospray ionization tandem mass spectrometry. Chromatographia 78(5):435–443. https://doi.org/10.1007/s10337-014-2782-x

11. Meyer VR (2010) Practical high-performance liquid chromatography. John Wiley & Sons, Ltd, Chichester, UK

12. Leiker TJ, Barkley RM, Murphy RC (2011) Analysis of diacylglycerol molecular species in cellular lipid extracts by normal-phase LC-electrospray mass spectrometry. Int J Mass Spectrom 305:103–108. https://doi.org/10.1016/j.ijms.2010.09.008

13. Quehenberger O et al (2010) Lipidomics reveals a remarkable diversity of lipids in human plasma. J Lipid Res 51:3299–3305. https://doi.org/10.1194/jlr.M009449

Chapter 10

LC-MS/MS Analysis of the Epoxides and Diols Derived from the Endocannabinoid Arachidonoyl Ethanolamide

Amy A. Rand, Patrick O. Helmer, Bora Inceoglu, Bruce D. Hammock, and Christophe Morisseau

Abstract

Liquid chromatography-tandem mass spectrometry (LC-MS/MS) is a useful tool to characterize the behavior of natural lipids within biological matrices. We report a LC-MS/MS method developed specifically to analyze CYP products of the arachidonoyl ethanolamide (anandamide, AEA), the epoxyeicosatrienoic acid ethanolamides (EET-EAs) and their hydrolyzed metabolites, and the dihydroxyeicosatrienoic acid ethanolamides (DHET-EAs). This method was used to measure EET-EA biotransformation to DHET-EAs by two human epoxide hydrolases: the soluble EH (sEH) and the microsomal EH (mEH). In general, sEH and mEH substrate preference was similar, based on k_{cat}/K_M. The 14,15-EET-EA and 11,12-EET-EA were the most efficiently hydrolyzed, followed by 8,9-EET-EA and 5,6-EET-EA. The method was also used to detect endogenous levels of these lipids in mouse tissues, although levels were below the instrumental detection limit (0.1–3.4 nM). Because both AEA and EETs are biologically active, the method described herein will be invaluable in revealing the role(s) of EET-EAs in vivo.

Key words Liquid chromatography mass spectrometry, Anandamide, CYP450, Epoxide hydrolase, EET-EAs, DHET-EAs, Limit of -detection, Enzyme kinetics, Bioactivity

1 Introduction

The endocannabinoid arachidonoyl ethanolamide (anandamide, AEA) is a natural lipid-signaling molecule found in most tissues that mediates neurological, immune, and cardiovascular functions by primarily targeting cannabinoid receptors [1]. While it is primarily metabolized by fatty acid amide hydrolase (FAAH) to arachidonic acid, it is also metabolized through several cytochrome P450s (CYP450), including CYP3A4, CYP2B6, CYP2D6, CYP4F2, CYP2J2, and CYP4X1 [2–4]. AEA is converted to monooxygenated products in mouse liver and brain microsomes [5]. These products include the epoxyeicosatrienoic acid ethanolamides (EET-EAs), which were detected, upon incubation with AEA, in

Martin Giera (ed.), *Clinical Metabolomics: Methods and Protocols*, Methods in Molecular Biology, vol. 1730,
https://doi.org/10.1007/978-1-4939-7592-1_10, © Springer Science+Business Media, LLC 2018

human liver microsomes and brain mitochondrial fractions [6], as well as in bovine and porcine heart microsomes [4].

While the biological function of EET-EAs needs to be further explored, it is anticipated that they have important physiological roles due to their structural similarity to the epoxyeicosatrienoic acids (EETs) [7]. EETs are formed from CYP oxidation of arachidonic acid and have far-reaching physiological effects involved in pain, inflammation, and kidney and cardiovascular diseases [8, 9]. The 5,6-EET-EA is a potent agonist for the cannabinoid receptor 2 [10], which may mediate brain and liver pathological outcomes such as neurodegeneration and fibrosis, respectively, by triggering anti-inflammatory responses [2]. While endogenous levels of EET-EAs are uncertain and may be produced in an activity-dependent manner, they are transformed by epoxide hydrolases (EHs) to their corresponding dihydroxyeicosatrienoic acid ethanolamides (DHET-EAs) [6, 10]. The biological activity of DHET-EAs has not yet been reported but might be similar to the less bioactive diols from EETs [11]. sEH inhibitors, which enhance levels of EETs and other epoxy fatty acids, have similar activities to cannabinoid receptor 2 agonists [12].

Analysis of EET-EAs and DHET-EAs is important to further our understanding of their biological activity, metabolic fate, and distribution within tissues. Here, we use two sample preparation methods for analysis of EET-EAs and DHET-EAs by LC-MS/MS. We report a sample dilution LC-MS/MS method to examine the hydrolysis of EET-EAs to DHET-EAs by two human epoxide hydrolases: the soluble EH (sEH) and the microsomal EH (mEH). While the EET-EAs had similar turnover with both sEH and mEH (Table 1), enzyme efficiency for each substrate (k_{cat}/K_M) varied; the 14,15-EET-EA and 11,12-EET-EA were the most efficiently hydrolyzed followed by 8,9-EET-EA and 5,6-EET-EA

Table 1
Kinetic constants of recombinant purified human sEH and mEH for the hydrolysis of all the regioisomers of 14,15-EET-EAs

Enzyme	Substrate	K_M (μM)	r^2	k_{cat} (s^{-1})	k_{cat}/K_M (s^{-1} μM^{-1})
Human sEH	14,15-EET-EA	16 ± 1	0.99	3.32 ± 0.02	0.20 ± 0.01
	11,12-EET-EA	19 ± 6	0.92	6.0 ± 0.8	0.31 ± 0.07
	8,9-EET-EA	28 ± 3	0.99	1.9 ± 0.1	0.068 ± 0.009
	5,6-EET-EA	22 ± 4	0.96	0.22 ± 0.02	0.010 ± 0.002
Human mEH	14,15-EET-EA	3.8 ± 0.4	0.99	0.61 ± 0.02	0.16 ± 0.02
	11,12-EET-EA	1.2 ± 0.3	0.99	0.17 ± 0.01	0.15 ± 0.04
	8,9-EET-EA	6.2 ± 0.6	0.99	0.10 ± 0.01	0.016 ± 0.002
	5,6-EET-EA	10.2 ± 1.4	0.99	0.11 ± 0.01	0.010 ± 0.002

Constants (K_M and k_{cat}) were calculated by nonlinear fitting of the Michaelis-Menten equation using the enzyme kinetic module of SigmaPlot version 11. Results are value ± error

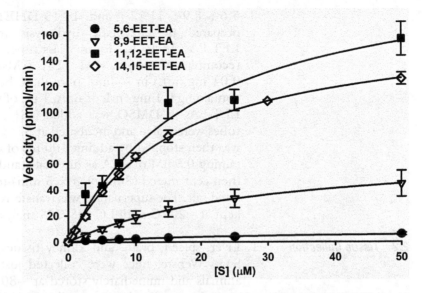

Fig. 1 Determination of the kinetic constants for 5,6-EET-EA; 8,9-EET-EA; 11,12-EET-EA; and 14,15-EET-EA, with the human sEH ([HsEH]$_{final}$ = 4 nM) in NaPO$_4$ buffer (0.1 M pH 7.4 containing 0.1 mg/mL of BSA) at 37 °C. Values were obtained from sextuplicate analysis (average ± standard deviation)

(Fig. 1 and Table 1). We also use solid-phase extraction (SPE), paired with the reported LC-MS/MS method, to detect endogenous levels of EET-EAs and DHET-EAs in mouse liver, kidney, spleen, and brain tissue. Although levels of the EET-EAs and DHET-EAs were below the instrumental limits of detection, the method can be applied to further work using enzymatic inhibitors and/or transgenic mice to raise levels of EET-EAs and DHET-EAs and to understand EET-EA and DHET-EA endogenous distribution and biological function in endocannabinoid signaling.

2   Materials

All solvents used should be Optima grade or 18 MΩ ultrapure water, unless otherwise indicated.

2.1   Standards for Liquid Chromatography-Tandem Mass Spectrometry (LC-MS/MS)

1. 5,6-; 8,9-; 11,12-; and 14,15-EET-EA and standard solutions (Cayman Chemical, Ann Arbor, MI, USA): Make up as 5 μM solution in methanol and store at −20 °C under nitrogen gas until use (*see* **Note 1**).

2. AEA standard solution (Cayman Chemical, Ann Arbor, MI, USA): Make up as 5 μM solution in methanol and store at −20 °C under nitrogen gas until use.

3. 5,6-; 8,9-; 11,12-; and 14,15-DHET-EA standards were prepared by enzymatic hydrolysis from their respective EET-EAs. In 10 × 75 mm glass tube, to 100 μL of purified recombinant mouse sEH [13] ([MsEH]$_{final}$ = 0.6 μM, 0.04 mg/mL) in sodium phosphate buffer (0.1 M pH 7.4) containing 0.1 mg/mL of BSA, 1 μL of solutions of individual EET-EAs in DMSO was added ([S]$_{final}$ = 0–2.5 μM). The tubes were sealed and incubated at 37 °C for 2 h. The reaction was then stopped by adding 100 μL of chilled methanol containing 0.5 μM of AEA as internal standard. The mixture was then centrifuged (3000 × g × 5 min) to remove any precipitated salt. The supernatant was transferred to vials, which were kept at −20 °C until LC-MS/MS analysis.

2.2 Tissue Collection

1. Liver, spleen, brain, and kidney tissues from saline-perfused Swiss Webster mice were collected just after sacrifice of the animals and immediately stored at −80 °C until analysis. All procedures and animal care adhered to the National Institutes of Health Guide for the Care and Use of Laboratory Animals (NIH Publications 8th Edition 2011) and were performed in accordance with the protocols approved by the International Animal Use and Care Committee (IACUC) of the University of California, Davis.

2. The whole liver, spleen, brain, and kidney were used from mice ($n = 3$). Frozen tissues were cut into small pieces with a clean razor blade. Tissues were weighted (100–260 mg) and allocated to 2 mL polypropylene microtubes. Tissues were suspended in 400 μL ice-cold methanol containing 0.1% concentrated acetic acid.

3. Antioxidant solution was prepared by combining 0.6 mg/mL butylated hydroxytoluene (BHT), 0.6 mg/mL triphenylphosphine (TPP), and 0.6 mg/mL ethylenediaminetetraacetic acid (EDTA) in DMSO to give a final concentration of 0.2 mg/mL in methanol. A 10 μL aliquot was added to each tissue sample to prevent autoxidation.

2.3 Solid-Phase Extraction (SPE)

1. Oasis HLB 3 mL 60 mg cartridges (Waters Corporation, Milford, MA, USA).

2. Water/methanol (8:2) with 0.1% acetic acid: Fill a 1 L graduated cylinder with 800 mL pure water, 200 mL methanol, and 1 mL concentrated acetic acid and store in a 1 L reagent bottle.

3. 30% glycerol solution: Measure 6 mL glycerol and 14 mL methanol in a 25 mL graduated cylinder and store in a 20 mL glass screw-cap vial.

4. Methanol: Store methanol in a 500 mL reagent bottle.

5. Ethyl acetate: Store ethyl acetate in a 500 mL reagent bottle.

6. SPE vacuum manifold processing station.

2.4 Liquid Chromatography-Tandem Mass Spectrometry (LC-MS/MS)

1. LC-MS/MS: We used a Waters Acquity Ultra Performance Liquid Chromatograph coupled to a Waters Xevo TQ-S Mass Spectrometer (Waters Corporation, Milford, MA, USA).

2. Liquid chromatography column: Kinetex 1.7 μm XB-C18 (100 Å,150 × 2.1 mm) with a SecurityGuard ULTRA cartridge for C18 UPLC, sub-2 μm with 2.1 mm internal diameters (Phenomenex, Torrance, CA, USA).

3. Water with 0.1% acetic acid and 1 mM ammonium acetate: Fill a 1 L graduated cylinder with 999 mL LC-MS grade water, 1 mL acetic acid, and 77 mg ammonium acetate and store in a 1 L LC glass solvent bottle.

4. Methanol with 0.1% acetic acid and 1 mM ammonium acetate: Fill a 1 L graduated cylinder with 999 mL LC-MS grade methanol, 1 mL acetic acid, and 77 mg ammonium acetate and store in a 1 L LC glass solvent bottle.

5. Glass screw top vials: 2 mL, 32 × 12 mm, and 9 mm Amber Glass Screw Thread Vials are used.

6. Glass inserts for screw top vials: Use a 0.150 mL glass insert with attached plastic spring purchased.

7. Caps for screw top vials: Use 9 mm thread cap with bonded plastic PTFE/silicone septa with pre-slit.

3 Methods

3.1 Processing of Enzyme Kinetic Samples

1. Kinetic parameters for the 5,6-EET-EA; 8,9-EET-EA; 11,12-EET-EA; and 14,15-EET-EA were determined under steady-state conditions using partially purified recombinant human sEH (95% pure) and human mEH (80% pure) [14, 15]. Just before usage, the human sEH was diluted in chilled sodium phosphate buffer (0.1 M pH 7.4) containing 0.1 mg/mL BSA to a final concentration of 4 nM (0.25 μg/mL). Whereas, the human mEH was diluted in chilled Tris/HCl buffer (0.1 M, pH 9.0) containing 0.1 mg/mL of BSA to a final concentration of 200 nM (10 μg/mL). Enzyme solutions were kept on ice until usage.

2. A series of substrate (individual EET-EA) stock solutions were prepared in DMSO. In general, 50–100 μL of eight substrate concentrations ranging from 0 to 5 mM were prepared.

3. In 10 × 75 mm glass tubes, to 100 μL of enzyme solution or buffer for control samples, 1 μL of solutions of individual

EET-EAs in DMSO was added ([S]$_{final}$ = 0–50 µM). The mixture was sealed and incubated at 37 °C for 15–120 min depending on substrate and enzyme. Incubation time was optimized to ensure the enzymatic reaction is in steady-state conditions, with less than 5% of the initial substrate metabolized during the whole course of the hydrolysis. The reaction was then stopped by adding 100 µL of chilled methanol containing 0.5 µM of AEA as internal standard.

4. The mixture was then centrifuged (3000 × g × 5 min) to remove any precipitated salt. The supernatant was transferred to a vial, which was kept at −20 °C until LC-MS/MS analysis.

5. The quantity of product (DHET-EA) formed was determined by LC-MS/MS analysis (Subheading 3.3). The kinetic constants (K$_M$ and k$_{cat}$) were calculated by nonlinear fitting of the Michaelis-Menten equation (v = (k$_{cat}$.[E].[S])/(K$_M$ + [S])) using the enzyme kinetic module of SigmaPlot 11.0. Reactions were done in hexaplicate ($n = 6$).

3.2 Tissue Extraction of EET-EAs and DHET-EAs

1. Between 100 and 260 mg of tissues are weighted in 2 mL plastic vials. Then, 10 µL of the BHT, TPP, and EDTA antioxidant solution was added as well as 400 µL ice-cold methanol containing 0.1% acetic acid. To correct for possible analyte loss throughout the extraction, 5 µL of 1 µM AEA was added as an internal standard (see Note 2) just prior to tissue homogenization.

2. Tissue samples are homogenized by placing two steel beads into each vial and shaken using a bead beater operating at 30 Hz for 10 min (see Note 3). The resulting homogenate solution was centrifuged at 13,200 × g × 10 min.

3. The supernatant is collected into 2 mL polypropylene vials, and the remaining tissue pellet is washed with 100 µL ice-cold methanol containing 0.1% acetic acid by vortexing for 1 min to break up and resuspend the pellet (see Note 4). Samples are centrifuged again at 13, 200 × g × 10 min. This supernatant was combined with the first to give a total volume approximating 0.5 mL. Samples are diluted with 1.5 mL ice-cold water and stored on ice before loading onto the SPE cartridges (see Note 5).

4. Oasis HLB cartridges (3 mL, 60 mg) are mounted on a vacuum manifold processing station under vacuum (~5 bar). Cartridges are preconditioned to remove impurities with 3 mL ethyl acetate and 6 mL methanol [16].

5. Cartridges are equilibrated by loading 6 mL of water/methanol (8:2) with 0.1% acetic acid under atmospheric pressure (see Note 6).

6. Samples are loaded onto the cartridge and left to filter at atmospheric pressure through the cartridge. The approximate flow rate is approximately 1 drop/3 s, monitored by watching the drip rate at the bottom of the cartridge. If the rate is less than this, light vacuum can be applied (*see* **Note 7**).

7. Polar substances from the tissue extracts are eluted through the column using 6 mL of water/methanol (8:2) with 0.1% acetic acid.

8. Apply approximately 10 bar vacuum to the manifold to dry the cartridges (20 min).

9. While cartridges are drying, prepare 2 mL polypropylene Eppendorf vials for the sample eluent, by adding 6 μL 30% glycerol solution in methanol. Label each vial with their respective sample name and place in the vacuum manifold, just below each sample cartridge.

10. EET-EAs, DHET-EAs, and AEA are eluted with 0.5 mL methanol followed by 1.5 mL ethyl acetate into the 2 mL Eppendorf vials at atmospheric pressure. Apply a 10 bar vacuum to elute the remaining solvent from the cartridge.

11. Samples are evaporated under vacuum using a ScanVac Speed-Vac (Neutec Group Inc., Farmingdale, NY) for approximately 2 h (*see* **Note 8**).

12. Samples are reconstituted in 50 μL methanol in a 2 mL polypropylene Eppendorf tube, filtered using Durapore PVDF 0.1 μm filter units (Millipore, Cork, Ireland), and centrifuged at $13,200 \times g \times 10$ min (*see* **Note 9**).

13. Filtrates are transferred to 0.15 mL glass inserts in a 2 mL screw top vial and cap. Samples are stored at -20 °C until LC-MS/MS analysis.

14. Analyte recovery was validated by performing a spike-and-recovery test, the values of which are given in Table 2 (*see* **Note 10**).

3.3 LC-MS/MS Analysis

1. Samples are analyzed on a Waters Acquity Ultra Performance Liquid Chromatograph coupled to a Waters Xevo TQ-S Mass Spectrometer in positive electrospray ionization mode. Samples are injected (5 μL) and separated using a Phenomenex Kinetex column (150 × 2.1 mm; 1.7 μm) at 40 °C using the following mobile phase gradient, consisting of LC-MS grade water (A) and optima LC-MS grade methanol (B) each containing 0.1% acid and 0.1 mM ammonium acetate: initial conditions of 70:30 A/B for 2.9 min ($t = 2.9$ min), changing to 45:55 at 3 min ($t = 3$ min) and decreasing to 35:65 over 5.5 min ($t = 8.5$ min), decreasing to 28:72 over 4 min ($t = 12.5$ min), decreasing to 18:82 over 2.5 min ($t = 15$ min), decreasing to

Table 2
LC-MS/MS parameters, instrumental limits of detection and quantification, and analyte recovery from mouse liver tissue

Analyte	Retention time (min)	MRM transition	Cone voltage (V)	Collision energy (V)	Limit of detection (nM)	Limit of quantification (nM)	Analyte recovery (%)
14,15-DHET-EA	12.56	383.0 > 285.0 (165.0)	16	12 (32)	3.4	11	163 ± 35
11,12-DHET-EA	13.22	383.0 > 285.0 (165.0)	16	12 (32)	1.7	5.7	117 ± 15
8,9-DHET-EA	14.13	383.0 > 285.0 (165.0)	16	12 (32)	2.9	9.8	88 ± 8
5,6-DHET-EA	15.13	383.0 > 285.0 (165.0)	16	12 (32)	5.4	18	167 ± 18
14,15-EET-EA	15.77	346.3 > 62.2 (285.2)	56	16 (16)	0.11	0.38	93 ± 11
11,12-EET-EA	16.17	346.3 > 62.2 (285.2)	56	16 (16)	1.6	5.4	82 ± 11
8,9-EET-EA	16.40	346.3 > 62.2 (285.2)	56	16 (16)	0.42	1.4	77 ± 14
5,6-EET-EA	16.74	346.3 > 62.2 (285.2)	56	16 (16)	0.79	2.6	11 ± 3
AEA	17.79	348.1 > 62.2	15	25			

Analytes measured were all the regioisomers of DHET-EAs and EET-EAs, extracted from tissues using AEA as an internal standard. Mass spectra were collected in multiple reaction monitoring (MRM) mode, with two analyte transitions used for quantification and qualification (shown in brackets). Limits of detection and quantification were obtained from linear regression by calculating the concentration at which the signal is three and ten times greater than the noise, respectively. Recoveries were calculated after spiking a known concentration (25 nM) of analytes into mouse liver tissue and extracted using solid-phase extraction (SPE) prior to LC-MS/MS analysis. Values were obtained from triplicates and are given as average ± standard error

5:95 over 1.5 min ($t = 16.5$ min), holding at 5:95 for 1.5 min ($t = 18$ min), reverting to initial conditions of 70:30 in 0.1 min ($t = 18.1$ min), and re-equilibrating for 2.9 min ($t = 21$ min). Mass spectral analysis was accomplished using a capillary voltage of 3 kV, a desolvation temperature of 200 °C, a desolvation gas (nitrogen 99.99%) flow of 800 L/h, a cone gas flow of 150 L/h, nebulizer pressure of 6 bar, and collision gas flow of 0.15 mL/min. LC-MS/MS parameters MRM transitions for each analyte are shown in Table 2, with analyte dwell times of 25 ms. A representative spectrum of the separated EET-EAs and DHET-EAs is shown in Fig. 2.

2. EET-EAs and DHET-EAs are detected by retention time and mass transitions, values of which are given in Table 2.

3. Analyte concentration was calculated using a matrix-matched and non-matrix-matched standard curve with known concentrations (2.5–60 nM) (*see* **Notes 10** and **11**). The area of all analytes was achieved through peak integration and dividing its value by the area of the AEA internal standard.

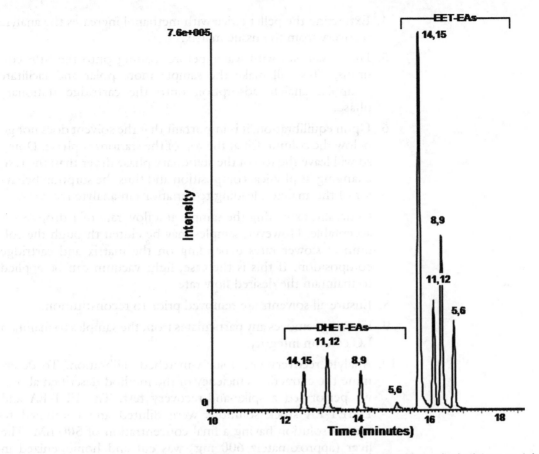

Fig. 2 LC-MS/MS total ion chromatograph of a 60 nM standard mixture containing all the regioisomers of DHET-EAs and EET-EAs

4 Notes

1. Store lipid standards under nitrogen at −20 °C to minimize oxidation.

2. Deuterated standards are not commercially available for EET-EAs and DHET-EAs; therefore, we used AEA (Cayman Chemical, Ann Arbor, MI, USA) as an internal standard for quantifying EET-EAs and DHET-EAs, due to its similar size and structure. We used non-labeled AEA since endogenous levels of AEA were not detected in any tissue. Alternatively, the d4-AEA (Cayman Chemical, Ann Arbor, MI, USA) can be used as an internal standard for extracting EET-EAs and DHET-EAs, especially if there are levels of endogenous AEA in the tissue samples.

3. Ensure vials are tightly capped; otherwise, the sample mixture will be lost during homogenization.

4. Extracting the pellet twice with methanol increases the analyte recovery from the tissue matrix.

5. Dilute samples with water before loading onto the SPE columns. This will make the sample more polar and facilitate nonpolar analyte adsorption onto the cartridge stationary phase.

6. Upon equilibration, it is important that the solvent does not go below the column frit at the top of the stationary phase. Doing so will leave the top of the stationary phase dryer than the rest, changing its physical composition and thus the sorption behavior of the analytes, leading to variations in analyte recovery.

7. Generally, extracting the sample at a flow rate of 1 drop/3 s is acceptable. However, samples may be eluted through the column at slower rates depending on the matrix and cartridge composition. If this is the case, light vacuum can be applied to maintain the desired flow rate.

8. Ensure all solvents are removed prior to reconstitution.

9. Filtering removes any particulates from the samples to maintain LC column integrity.

10. Analyte recovery and matrix-matched calibration: To determine the extraction efficiency of the method described above, we performed a spike-and-recovery test. The EET-EA and DHET-EA stock solutions were diluted and combined to form a solution having a final concentration of 500 nM. The liver (approximately 600 mg) was cut and homogenized in methanol. 100 μL of the homogenate was spiked with 2.5 μL of the 500 nM EET-EA and DHET-EA solution in triplicate. 5 μL of the 500 nM AEA internal standardwas also added to the spiked tissue samples. The rest of the unspiked liver homogenate was divided and extracted to create a matrix-matched calibration curve. All tissue samples (spiked and unspiked) were extracted and analyzed using the SPE-LC-MS/MS- method described above. Samples were reconstituted in 50 μL methanol. Matrix-matched standards were made by adding 0–6 μL of the 500 nM EET-EA and DHET-EA stock solution to generate a concentration range from 0 to 60 nM, as well as 5 μL of the 500 nM AEA. The final volume for each standard was brought to 50 μL with methanol. A non-matrix-matched calibration curve was produced in methanol (rather than reconstituted matrix) using the same concentrations (0–60 nM with 50 nM AEA).

11. Matrix-matched calibration curves are used to correct for matrix effects within the samples. In our case, there was no significant matrix suppression or enhancement when compared to the non-matrix-matched calibration curve; therefore, either curve can be used for quantification.

Acknowledgments

This work received support in part from the National Institute of Environmental Health Sciences R01 ES002710 and NIEHS Superfund Program P42 ES004699. A. Rand acknowledges support from the OSCB training grant NIH/NIEHS T32 CA108459 and from the 2016 AACR Judah Folkman Fellowship for Angiogenesis Research, Grant Number 16-40-18-RAND.

References

1. Mechoulam R, Parker LA (2013) The endocannabinoid system and the brain. Annu Rev Psychol 64:21–47

2. Snider NT, Walker VJ, Hollenberg PF (2010) Oxidation of the endogenous cannabinoid arachidonoyl ethanolamide by the cytochrome P450 monooxygenases: physiological and pharmacological implications. Pharmacol Rev 62:136–154

3. Sridar C, Snider NT, Hollenberg PF (2011) Anandamide oxidation by wild-type and polymorphically expressed CYP2B6 and CYP2D6. Drug Metab Dispos 39:782–788

4. McDougle DR, Kambalyal A, Meling DD, Das A (2014) Endocannabinoids anandamide and 2-arachidonoylglycerol are substrates for human CYP2J2 epoxygenase. J Pharmacol Exp Ther 351:616–627

5. Bornheim LM, Kim KY, Chen B, Correia MA (1995) Microsomal cytochrome P450-mediated liver and brain anandamide metabolism. Biochem Pharmacol 50:677–686

6. Snider NT, Kornilov AM, Kent UM, Hollenberg PF (2007) Anandamide metabolism by human liver and kidney microsomal cytochrome p450 enzymes to form hydroxyeicosatetraenoic and epoxyeicosatrienoic acid ethanolamides. J Pharmacol Exp Ther 321:590–597

7. Zelasko S, Arnold WR, Das A (2015) Endocannabinoid metabolism by cytochrome P450 monooxygenases. Prostaglandins Other Lipid Mediat 116:112–123

8. Imig JD, Hammock BD (2009) Soluble epoxide hydrolase as a therapeutic target for cardiovascular diseases. Nat Rev Drug Discov 8:794–805

9. Morisseau C, Hammock BD (2013) Impact of soluble epoxide hydrolase and epoxyeicosanoids on human health. Annu Rev Pharmacol Toxicol 53:37–58. https://doi.org/10.1146/annurev-pharmtox-011112-140244

10. Snider NT, Nast JA, Tesmer LA, Hollenberg PF (2009) A cytochrome P450-derived epoxygenated metabolite of anandamide is a potent cannabinoid receptor 2-selective agonist. Mol Pharmacol 75:965–972

11. Bellien J, Joannides R (2013) Epoxyeicosatrienoic acid pathway in human health and diseases. J Cardiovasc Pharmacol 61:188–196

12. Wagner K, Inceoglu B, Hammock BD (2011) Soluble epoxide hydrolase inhibition, epoxygenated fatty acids and nociception. Prostaglandins Other Lipid Mediat 96:76–83

13. Argiriadi MA, Morisseau C, Hammock BD, Christianson DW (1999) Detoxification of environmental mutagens and carcinogens: structure, mechanism, and evolution of liver epoxide hydrolase. Proc Natl Acad Sci 96:10637–10642

14. Morisseau C, Bernay M, Escaich A et al (2011) Development of fluorescent substrates for microsomal epoxide hydrolase and application to inhibition studies. Anal Biochem 414:154–162

15. Morisseau C, Wecksler AT, Deng C et al (2014) Effect of soluble epoxide hydrolase polymorphism on substrate and inhibitor selectivity and dimer formation. J Lipid Res 55:1131–1138

16. Yang J, Schmelzer K, Georgi K, Hammock BD (2009) Quantitative profiling method for oxylipin metabolome by liquid chromatography electrospray ionization tandem mass spectrometry. Anal Chem 81:80–85

Chapter 11

Sphingolipid Analysis in Clinical Research

Bo Burla, Sneha Muralidharan, Markus R. Wenk, and Federico Torta

Abstract

Sphingolipids are the most diverse class of lipids due to the numerous variations in their structural components. This diversity is also reflected in their extremely different functions. Sphingolipids are not only constituents of cell membranes but have also emerged as key signaling molecules involved in a variety of cellular functions, such as cell growth and differentiation, proliferation, and apoptotic cell death. Lipidomic analyses in clinical research have identified pathways and products of sphingolipid metabolism that are altered in several human pathologies. In this article, we describe how to properly design a lipidomic experiment in clinical research, how to handle plasma and serum samples for this purpose, and how to measure sphingolipids using liquid chromatography-mass spectrometry.

Key words Sphingolipids, Mass spectrometry, Lipidomics, Sphingolipidomics, Ceramide, Sphingomyelin, Glucosylceramide, Sphingosine-1-phosphate, Clinical mass spectrometry, Quality control

1 Introduction

Sphingolipids (SL) represent a diverse class of molecules that includes lipids with both structural (e.g., sphingomyelins) and signaling functions (e.g., sphingosine-1-phosphate). They can be found in cell membranes and lipoproteins in the blood [1].

The first step in the de novo synthesis of SL takes place in the endoplasmic reticulum, involves the condensation of a serine residue with a fatty acyl-CoA, and is catalyzed by serine palmitoyltransferase (SPT). The product of this reaction is 3-ketodihydrosphingosine. A subsequent enzymatic reaction generates the long-chain base (LCB) dihydrosphingosine that then can be N-acylated with fatty acids of various lengths to give dihydroceramides and, after desaturation, ceramides, the precursors of complex SL (such as sphingomyelins, glycosphingolipids, and ceramide-1-phosphate). The final irreversible step for SL degradation, and recycling to glycerolipids, is catalyzed by the enzyme S1P lyase that degrades sphingosine-1-phosphate (S1P) into a fatty aldehyde (hexadecenal) and phosphoethanolamine [1].

Martin Giera (ed.), *Clinical Metabolomics: Methods and Protocols*, Methods in Molecular Biology, vol. 1730, https://doi.org/10.1007/978-1-4939-7592-1_11, © Springer Science+Business Media, LLC 2018

Since numerous LCBs, with different carbon chain length, position, and number of double bonds and hydroxylations, are acylated with different fatty acids (between C14 and C26), SL include extremely diverse molecules (>300 molecular species identified). This heterogeneity is even increased by the different modifications on the headgroup (glycosylation). SL are lower in abundance (<20%) when compared to their glycerolipid counterparts, but they are very active and play key roles in the regulation of cell and tissue functions.

Many sphingolipid and phospholipid functions have been linked to pathological processes. In particular, genetic disorders linked to the sphingolipid pathway are associated with cardiovascular and metabolic diseases [2]. Advances in mass spectrometry-based lipidomics have greatly expanded our understanding of the extent and complexity of lipid dysregulation in disease conditions. This technology enables the measurement of several hundred individual lipid species in thousands of samples for each study [3, 4]. Lipidomics in population-based studies has helped to clarify that concentrations of specific SL are altered in several metabolic disorders and may serve as prognostic and diagnostic markers [5]. Ceramides have been implicated in mediating or regulating numerous central cellular processes such as proliferation, differentiation and senescence, stress responses, apoptosis, and inflammation [6]. Harmful stimuli can activate these bioactive signaling lipids leading to numerous pathophysiological states, including inflammatory diseases such as obesity [7].

More specifically, four molecular ceramides have been found to be closely linked to cardiovascular conditions [8] and can be used as markers in stratifying a patient's risk for fatal outcome of coronary artery disease (CAD).

Alanine and glycine can also be used, instead of serine, as substrates by SPT and condensed with fatty acyl-CoA to produce deoxysphingolipids. Mutations in SPT can alter its preference for the substrate, and, as a consequence, elevated levels of these lipids can be present in plasma. This is the case of hereditary sensory and autonomic neuropathy type 1, in which increased plasma levels of deoxysphingolipids characterize this rare genetic disease [9]. Deoxysphingolipid levels are also elevated in type 2 diabetes mellitus. Thus targeting deoxysphingolipid synthesis could complement the currently available therapies for this pathology [10].

Altered glycosphingolipid metabolism is associated with common diseases of the kidney, in particular diabetic nephropathy, polycystic kidney disease, and renal cell carcinoma. Patients with these diseases show high levels of glucosylceramide, lactosylceramide, and gangliosides in the kidneys. More importantly from a therapeutic point of view, inhibition of glucosylceramide synthase with the use of small molecules reverses the abnormal renal phenotypes [11]. An important sphingolipid that has been involved in many clinical research studies and functions as a regulator of processes such as immune cell trafficking, angiogenesis, and vascular permeability is S1P [12]. Our group has recently published a new very sensitive method for its detection that expanded the number of S1P molecular species usually detected in biological samples [13].

In MS-based lipidomics, the two most common approaches include analysis of lipids either by direct infusion (shotgun MS) or after chromatographic separation (LC-MS/MS). Both approaches have their limitations and advantages [14]. As the purpose of this manuscript is to guide users in the analysis of specific molecules (sphingolipids) that are present at low levels in plasma and serum, we will describe here an approach based on LC-MS/MS, the most commonly used technique for sensitive targeted studies. Several specific methods for sphingolipid extraction have been published in the last few years and have improved the detection of the low abundant species [15, 16]. In this manuscript, we limit our approach to a simple and fast single-phase extraction protocol that is effective for many lipid classes. We think that when dealing with high number of samples, as is often the case in clinical studies, a fast extraction procedure might be the best choice [17].

Blood serum and plasma are the sample types most commonly used in clinical research studies, and for this reason we will focus our method description on this kind of samples. Please notice that plasma and serum are different matrices that have different sphingolipid concentrations. One should be aware that they are not comparable, and a choice driven by the scope of the study should be made before starting the sample collection. As in any clinical research study, the selection of the subjects should be balanced, knowing that age, gender, and ethnicity have an impact on the SL profiles [18, 19]. The level of specific lipids can also be affected by diet, medications, circadian rhythm, and other factors that should be documented as metadata to improve the final analysis. All these considerations have been described and carefully considered in previous reports, as the use of lipidomics (and metabolomics) in clinical research studies has been the focus of many publications in the last few years [3, 20, 21].We describe here a recommended approach for sphingolipidomics studies in clinical research samples, including suggestions about sampling, storage, analytical design, use of quality controls, and data analysis. Two different analytical methods, one for extraction/analysis of the most common sphingolipid species and one specific for S1P (for which higher sensitivity is required), are described [13].

2   Materials

2.1   Blood Collection

1. Plasma: Blood collection container (e.g., BD Vacutainer and Sarstedt S-Monovette®) with spray-dried K_2EDTA (violet cap) or buffered sodium citrate solution (0.109 M/3.2%, light blue cap). Heparin is not recommended as anticoagulant.

2. Serum: Blood collection container (e.g., BD Vacutainer and Sarstedt S-Monovette®) with or without spray-coated silica as clot activator.

3. Biosafety cabinet.

4. Centrifuges for blood collection tubes.

2.2   Sample Extraction and Derivatization

1. Chemical fume hood.

2. Eppendorf® Safe-Lock 2 mL polypropylene tubes.

3. Ultrasonic bath.

4. Eppendorf ThermoMixer® or equivalent.

5. Microcentrifuge.

2.3   Liquid Chromatography-Mass Spectrometry Systems

1. UHPLC system: Agilent 1290 Infinity LC (Agilent Technologies, Santa Clara, USA).

2. Triple-quadrupole mass spectrometer: Agilent 6490 or 6495 QQQ.

3. Chemically inert autosampler vials (glass or polypropylene) with airtight PTFE-sealed screw septum caps.

4. Agilent ZORBAX RRHD Eclipse Plus C18, 95 Å, 2.1 × 100 mm, 1.8 μm UPLC column.

5. Waters ACQUITY UPLC BEH C18 HILIC 130 Å, 2.1 × 100 mm, 1.7 μm UPLC column.

6. Schott Duran® glass bottles for preparation and storage of mobile phases.

2.4   Reagents and Solutions

All compounds should be of analytical or better grade quality. Solvents (acetonitrile, methanol, 2-propanol, formic acid) must be of LC-MS grade. Ultrapure water should be of LC-MS grade or Milli-Q water (18 MΩ cm).

Internal standards (ISTDs)

1. $S1P$-$^{13}C_2D_2$ solution: Dissolve vial content (1 mg) of D-erythro-Sphingosine-1-phosphate-13C2,D2 (# S681502, Toronto Research Chemicals, Toronto, Canada) in 10 mL methanol by vortexing and sonication until solution is clear. Store at −20°C.

Table 1
Composition of the BUME + ISTD mix

Internal standards (IS)	Monoisotopic mass	Chemical formula	Concentration in BUME + ISTDs (nM)
Sph d17:1	285.2668	$C_{17}H_{35}NO_2$	250
Sph d17:0	287.2824	$C_{17}H_{37}NO_2$	250
S1P d17:1	365.2331	$C_{17}H_{36}NO_5P$	250
S1P d17:0	367.2488	$C_{17}H_{38}NO_5P$	250
LacCer d18:1/12:0	805.5551	$C_{45}H_{79}NO_3$	250
SM d18:1/12:0	646.5050	$C_{35}H_{71}N_2O_6P$	250
GluCer d18:1/12:0	643.5023	$C_{36}H_{69}NO_8$	250
Cer d18:1/12:0	481.4495	$C_{30}H_{59}NO_3$	250
C1P d18:1/12:0	561.4158	$C_{30}H_{60}NO_6P$	250
S1P d18:1 $^{13}C_2D_2$	383.2487	$C_{16}^{13}C_2H_{36}D_2NO_5P$	52.1

2. Cer/Sph mixture II: Ceramide/Sphingoid Internal Standard Mixture II (#LM-6005, Avanti Polar Lipids, USA). Store at −20°C.

Sample processing and lipid extraction

3. BHT solution 100 mM: Dissolve 220.4 mg of 3,5-di-tert-4-butylhydroxytoluene in 10 mL ethanol. Store in aliquots in microtubes at −20°C.

4. BUME: Mix 1 volume 1-butanol (HPLC grade) with 1 volume methanol and add 5 mM (315.3 mg/L) ammonium formate (LC-MS grade). Store at room temperature in a glass bottle.

5. BUME + ISTDs: Add 100 μL Cer/Sph mixture II and 2 μL S1P-$^{13}C_2D_2$ solution to 9.9 mL BUME (*see* Table 1).

6. TMS-diazomethane solution: (Trimethylsilyl)diazomethane, 2 M solution in hexanes (e.g., #AC385330250, ACROS Organics). Store at room temperature in the original bottle in a safety cabinet or ventilated solvent cupboard.

7. Acetic acid, glacial (>99%).

8. Acetic acid 10% in methanol (for inactivation of excess TMS solution).

Mobile phases for sphingolipid LC-MS analyses

9. Mobile phase A for sphingolipid analysis: Mix 600 mL methanol with 400 mL ultrapure water and 2 mL formic acid (98–100%) in a glass bottle, and then add 1.33 mL 7.5 M

ammonium acetate solution (e.g., #A2706, Sigma-Aldrich). Mix well and sonicate for 10 min.

10. Mobile phase B for sphingolipid analysis: Mix 600 mL methanol with 400 mL of 2-propanol and 2 mL formic acid in a glass bottle, and then add 1.33 mL 7.5 M ammonium acetate solution. Mix well and sonicate for 10 min.

Mobile phases for S1P LC-MS analyses

11. Ammonium formate buffer 25 mM pH 4.6: Dissolve 1575 mg ammonium formate (LC-MS grade) in approximately 800 mL ultrapure water. Adjust the pH to 4.6 by adding formic acid 100% (approximately 100 μL). Adjust volume to 1 L with ultrapure water and store in a glass bottle at 4°C in dark.

12. Mobile phase A for S1P analysis: Mix 500 mL acetonitrile and 500 mL of ammonium formate buffer 25 mM pH 4.6 in a glass bottle. Sonicate for 10 min.

13. Mobile phase B for S1P analysis: Mix 950 mL acetonitrile and 50 mL of ammonium formate buffer 25 mM pH 4.6 in a glass bottle. Sonicate for 10 min.

3   Methods

3.1   Experimental Study Design

The study design is a crucial part of the study that affects its outcome and the interpretation of the results. To ensure reliable results, the study must be carefully designed, taking into consideration different aspects, such as composition of the cohorts, sample collection, storage, and analysis order.

1. Age, gender, ethnicity, BMI, diet, medications, and other possible confounding factors should be recorded and considered when designing the analytical workflow and analyzing the final data to reliably establish associations between sphingolipid levels and clinical parameters.

2. A consistent sampling procedure during the whole study is essential since different collecting procedures for plasma or serum have been shown to result in different metabolite isolations (*see* below and notes). For this reason, a strict and well-documented protocol must be followed, and eventual variations should be recorded.

3. Samples should be stored preferably at −80°C to minimize metabolite degradation. Freeze-thaw cycles may affect the concentration of specific molecules so the number of these cycles must be the same for all the samples and well documented.

4. In case of large studies, it is recommended to divide them into smaller batches, to facilitate experimental procedures and allow for instrument cleaning between batches when necessary. Each batch should represent the study population. This is achieved by using a stratified randomization procedure, to exclude the risk of bias or correlation between metabolite levels and analysis order [20].

3.2 Blood Collection and Preparation of Plasma and Serum

The choice between plasma and serum depends on the aims and practical aspects of the study (*see* **Notes 1** and **2**). Blood collection and subsequent processing of drawn blood must be performed uniformly to minimize variability in measured analytes caused by preanalytical effects (*see* **Notes 3–5**) [22]. Blood should be treated as potentially infectious and handled with appropriate safety measures (*see* **Note 1**).

1. Collect fresh blood by venipuncture into collection tubes without (for serum) or with (for plasma) anticoagulants (*see* **Notes 6–8**).

2. For plasma preparation: Place tubes with anticoagulated whole blood on ice immediately after collection (*see* **Note 9**).

3. For serum preparation: Allow collected blood to clot undisturbed at room temperature for 60 min in collection tubes without clot activator and 30 min for tubes with silica clot activator (*see* **Note 10**).

4. Centrifuge as soon as possible at 2000 × g (4 °C) for 15 min (plasma) or 10 min (serum), respectively (*see* **Notes 11** and **12**).

5. Transfer supernatants into new polypropylene tubes without disturbing the cellular fraction.

6. Add 1 μL 100 mM BHT solution in ethanol per 1 mL plasma to prevent oxidation [17], proceed with sample preparation, or store at −80 °C until further processing.

3.3 Preparation of Plasma/Serum and BQC Samples

Preparation of samples in large studies represents a nontrivial challenge. Here below are some suggestions for sample preparation.

1. Thaw plasma/serum samples (study samples) at room temperature, making sure samples are transferred to ice just before they are completely thawed.

2. Vortex all study samples to remove concentration gradients that may have resulted from the thawing process and to ensure sample is completely thawed.

3. Centrifuge study samples at 12,000 × g for 10 min (4 °C) to sediment precipitated material.

4. Stock BQC: Pool aliquots from the supernatant of each study sample (or just a stratified representative subset of all the study

samples) to generate a stock BQC. Prepare a sufficient volume of stock BQC, calculated based on the total estimated number of BQC samples that will be analyzed.

5. Study and BQC samples: Transfer 20 μL supernatant from each centrifuged study sample to a new 2 mL Eppendorf Safe-Lock tubes. Transfer 20 μL supernatant from the stock BQC to a new tube every five or ten samples. Two or more BQCs should be added in the beginning and at the end of each batch.

6. Proceed directly with lipid extraction or store aliquots at −80°C until extraction.

3.4 Lipid Extraction   Once aliquots with the required amount of all the samples have been prepared, two general strategies can be considered: (1) extract lipids from all the samples at the same time, divide them in batches (according to rules mentioned in the previous paragraph), freeze all of them, and analyze each batch separately, i.e., at different days, or (2) divide the samples in several batches (according to rules mentioned in the previous paragraph), and extract lipids from them every day prior to analysis [21].

The lipid extraction procedure described below is based on the method described by Alshehry et al. [17].

1. Study and BQC samples: Add 200 μL BUME + ISTDs to all aliquots (20 μL, see **Notes 13–15**).

2. Blanks: 200 μL BUME + ISTDs (for the blank + ISTD samples) and 200 μL BUME (for blank − ISTD samples) are added into empty tubes of the same type used for the study samples. Blanks can be added to the sample sequence (e.g., one blank + ISTD after each BQC) to monitor contamination from the extraction solvent, ISTD mixes, extraction procedure, and carry-over.

3. Vortex all samples (including blank − ISTD and blank + ISTD) for 30 s, and sonicate them for 30 min ensuring that the temperature of the water bath does not exceed 20°C.

4. After sonication, samples are centrifuged for 10 min at $14,000 \times g$ (20°C) to precipitate proteins and salts, and then the supernatant containing the sphingolipids is transferred into new tubes or autosampler vials.

5. TQC_{SL} samples: Pool aliquots of equal volumes from all extracted BQCs to prepare the stock TQC_{SL}. Estimate the required volume based on injection sequence. Aliquot stock TQC_{SL} into two vials, one used as TQC_{SL} for the LC-MS analysis and the other for the initial conditioning of the LC-MS system.

6. TQC$_{SL}$ dilution series: A serial dilution series, e.g., 1:2, 1:4, and 1:8 with BUME, is prepared from the stock TQC$_{SL}$.

7. Extracts may be directly transferred to autosampler vials for LC-MS analysis or stored at −80°C until needed (*see* **Notes 16** and **17**). The samples can also be dried with nitrogen or in centrifugal concentrator and reconstituted with a desired volume of mobile phase A before analysis.

3.5 Derivatization

This procedure is based on the method described in (Narayanaswamy et al. [13]). Consult safety information detailed in **Notes 18** and **19** before performing this procedure.

1. 90 μL lipid extract from study samples, BQCs, and blanks is transferred to 2 mL Eppendorf Safe-Lock polypropylene tubes in the same sequence as performed for the lipid extraction.

2. 30 μL of TMS-diazomethane 2 M solution in hexanes is added for derivatization of S1P (*see* **Notes 18** and **19**).

3. This mixture is incubated for 20 min at 750 rpm at 22°C using a thermal shaker.

4. The derivatization reaction is stopped after 20 min by adding 1 μL of acetic acid 100% (*see* **Note 19**).

5. Samples are vortexed for 5 s and centrifuged for 10 min at 14,000 × *g* (20 °C) and the supernatant transferred to new vials/tubes.

6. TQC$_{S1P}$ samples: Pool aliquots of equal volume from all derivatized BQCs to prepare the stock TQC$_{S1P}$. Estimate the required volume based on the sample sequence. Aliquot stock TQC$_{S1P}$ into two vials, one used as TQC$_{S1P}$ for the LC-MS analysis and the other one for the initial conditioning of the LC-MS system.

7. TQC$_{S1P}$ dilution series: A serial dilution series, e.g., 1:2, 1:4, and 1:8 with BUME, is prepared from the stock TQC$_{S1P}$.

8. Extracts may be directly transferred to vials for LC-MS analysis or stored at −80°C until needed. The samples can also be dried with nitrogen or in centrifugal concentrator and reconstituted with a desired volume of the starting mobile phase before analysis.

3.6 Liquid Chromatography-Mass Spectrometry Analysis

The protocol described below is based on a modified previously described method by Wang et al. [23].

3.6.1 LC-MS Setup for Sphingolipid Analysis

1. Samples are analyzed by reversed-phase liquid chromatography-electrospray ionization (ESI) MS. Agilent 1290 UHPLC system equipped with an Agilent ZORBAX RRHD Eclipse Plus C18, 95 Å, 2.1 × 100 mm, 1.8 μm, set at 40 °C, is used as a chromatographic system to separate the sphingolipid species.

The flow rate is 0.4 mL/min. Mobile phases A and B are mixed according to the following gradient: 0% B to 10% B from 0 to 3 min, 10 to 40% B from 3 to 5 min, 40 to 55% B from 5 to 5.3 min, 55 to 60% B from 5.3 to 8 min, 60 to 80% B from 8 to 8.5 min, 80% B from 8.5 to 10.5 min, 80 to 90% B from 10.5 to 16 min, 90% B from 10.5 to 19 min, and 90 to 100% B from 19 to 22 min, and re-equilibrate at 0% B from 22.1 to 25.0 min. The total run time is 25.0 min.

2. AJS ESI source parameters: Sheath gas temperature and flow are set to 200°C and 12 L/min, respectively; nebulizer pressure, 25 psi; capillary voltage and nozzle voltage, 3500 V and 500 V, respectively; dry gas temperature and flow, 200°C and 15 L/min, respectively; and the delta EMV, 200 V. Positive high-/low-pressure RF of the iFunnel is set to 210/110.

3. For this analysis, the triple-quadrupole mass spectrometer is operated in positive ionization dynamic multiple reaction monitoring (dMRM) mode. For ceramides (Cer), hexosylceramides (HexCer), and dihexosylceramides (Hex2Cer), the long-chain base product ions generated from intact precursors and from precursors after in-source water loss are monitored. For sphingomyelins (SM) only the product ion of m/z 184 is monitored. MRM transitions, collision energies, and expected dMRM retention time are listed in Table 2. Retention time windows are set to 1.5; cycle time is set to 1000 ms, keeping a minimum dwell time of $\geq$10 ms.

4. Injection volume for all injected samples is 2 µL (*see* **Note 20**).

3.6.2 LC-MS Setup for S1P Analysis

The protocol described below is a modified version of a previously described method by Narayanaswamy et al. [13].

1. S1P analysis is performed using hydrophilic interaction liquid chromatography (HILIC)-ESI MS on Agilent 1290 UHPLC connected to an Agilent 6495 mass spectrometer. A Waters ACQUITY BEH HILIC 130 Å, 1.7 µm, 2.1 × 100 mm, column maintained at 60°C is used for the analysis. The flow rate is 0.4 mL/min. Mobile phases A and B are mixed according to the following gradient: 99.9% B to 40% B from 0 to 5 min, 40 to 10% B from 5 to 5.5 min, 10% B from 5.5 to 6.5 min, and re-equilibrate at 99.9% B from 6.6 to 9.6 min. The total run time is 9.6 min.

2. AJS ESI source parameters: The sheath gas temperature and flow are set at 400°C and 12 L/min, respectively; nebulizer pressure, 25 psi; capillary voltage, 3500 V; nozzle voltage, 500 V; dry gas temperature and flow, 200°C and 12 L/min, respectively; delta EMV, 200; positive high-/low-pressure RF, 200/110.

Table 2
Parameters defined in the MS dMRM method for sphingolipid analysis

Compound name	Precursor ion	Product ion	Ret time (min)	Collision energy
C1P d18:1/12:0 (ISTD)	562.4	264.1	9.8	25
C1P d18:1/12:0 (ISTD)	544.4	264.1	9.8	25
C1P d18:1/12:0 (ISTD)	562.4	464.2	9.8	10
C1P d18:1/16:0	618.5	264.2	11.9	25
C1P d18:1/16:0	600.5	264.2	11.9	25
C1P d18:1/18:0	646.5	264.2	12.9	25
C1P d18:1/18:0	628.5	264.2	12.9	25
C1P d18:1/26:0	758.6	264.2	16.9	25
C1P d18:1/26:0	740.6	264.2	16.9	25
Cer d18:1/12:0 (ISTD)	482.4	264.2	10.6	25
Cer d18:1/12:0 (ISTD)(-H$_2$O)	464.4	264.2	10.6	25
Cer d18:0/14:0	512.5	266.2	13.4	25
Cer d18:0/14:0(-H$_2$O)	494.5	266.2	13.4	25
Cer d18:1/14:0	510.5	264.2	11.8	25
Cer d18:1/14:0(-H$_2$O)	492.5	264.2	11.8	25
Cer d18:1/14:1	508.5	264.2	10.6	25
Cer d18:1/14:1(-H$_2$O)	490.5	264.2	10.6	25
Cer d18:2/14:0	508.5	262.2	10.6	25
Cer d18:2/14:0(-H$_2$O)	490.5	262.2	10.6	25
Cer d18:0/15:0	526.5	266.2	14	25
Cer d18:0/15:0(-H$_2$O)	508.5	266.2	14	25
Cer d18:0/16:0	540.5	266.2	14.6	25
Cer d18:0/16:0(-H$_2$O)	522.5	266.2	14.6	25
Cer d18:1/16:0	538.5	264.2	12.5	25
Cer d18:1/16:0(-H$_2$O)	520.5	264.2	12.5	25
Cer d18:1/16:1	536.5	264.2	11.5	25
Cer d18:1/16:1(-H$_2$O)	518.5	264.2	11.5	25
Cer d18:2/16:0	536.5	262.2	11.8	25
Cer d18:2/16:0(-H$_2$O)	518.5	262.2	11.8	25
Cer d18:0/17:0	554.5	266.2	15.2	25

(continued)

Table 2
(continued)

Compound name	Precursor ion	Product ion	Ret time (min)	Collision energy
Cer d18:0/17:0(-H$_2$O)	536.5	266.2	15.2	25
Cer d18:0/18:0	568.5	266.2	15.8	25
Cer d18:0/18:0(-H$_2$O)	550.5	266.2	15.8	25
Cer d18:0/18:1	566.5	266.2	14.6	25
Cer d18:0/18:1(-H$_2$O)	548.5	266.2	14.6	25
Cer d18:1/18:0	566.5	264.2	14	25
Cer d18:1/18:0(-H$_2$O)	548.5	264.2	14	25
Cer d18:1/18:1	564.5	264.2	12.7	25
Cer d18:1/18:1(-H$_2$O)	546.5	264.2	12.7	25
Cer d18:2/18:0	564.5	262.2	13	25
Cer d18:2/18:0(-H$_2$O)	546.5	262.2	13	25
Cer d18:2/18:1	562.5	262.2	11.8	25
Cer d18:2/18:1(-H$_2$O)	544.5	262.2	11.8	25
Cer d18:0/20:0	596.5	266.2	17	25
Cer d18:0/20:0(-H$_2$O)	578.5	266.2	17	25
Cer d18:1/20:0	594.5	264.2	15.2	25
Cer d18:1/20:0(-H$_2$O)	576.5	264.2	15.2	25
Cer d18:1/20:1	592.5	264.2	14	25
Cer d18:1/20:1(-H$_2$O)	574.5	264.2	14	25
Cer d18:2/20:0	592.5	262.2	14.2	25
Cer d18:2/20:0(-H$_2$O)	574.5	262.2	14.2	25
Cer d18:2/20:1	590.5	262.2	13	25
Cer d18:2/20:1(-H$_2$O)	572.5	262.2	13	25
Cer d18:0/22:0	624.6	266.2	18.2	25
Cer d18:0/22:0(-H$_2$O)	606.6	266.2	18.2	25
Cer d18:0/22:1	622.5	266.2	17	25
Cer d18:0/22:1(-H$_2$O)	604.5	266.2	17	25
Cer d18:1/22:0	622.6	264.2	16.5	25
Cer d18:1/22:0(-H$_2$O)	604.6	264.2	16.5	25
Cer d18:1/22:1	620.6	264.2	15.37	25

(continued)

Table 2
(continued)

Compound name	Precursor ion	Product ion	Ret time (min)	Collision energy
Cer d18:1/22:1(-H$_2$O)	602.6	264.2	15.37	25
Cer d18:2/22:0	620.6	262.2	15.6	25
Cer d18:2/22:0(-H$_2$O)	602.6	262.2	15.6	25
Cer d18:2/22:1	618.6	262.2	14.4	25
Cer d18:2/22:1(-H$_2$O)	600.6	262.2	14.4	25
Cer d18:0/23:0	638.6	266.2	18.8	25
Cer d18:0/23:0(-H$_2$O)	620.6	266.2	18.8	25
Cer d18:0/23:1	636.6	266.2	17.6	25
Cer d18:0/23:1(-H$_2$O)	618.6	266.2	17.6	25
Cer d18:1/23:0	636.6	264.2	17.2	25
Cer d18:1/23:0(-H$_2$O)	618.6	264.2	17.2	25
Cer d18:1/23:1	634.6	264.2	16	25
Cer d18:1/23:1(-H$_2$O)	616.6	264.2	16	25
Cer d18:2/23:0	634.6	262.2	16.2	25
Cer d18:2/23:0(-H$_2$O)	616.6	262.2	16.2	25
Cer d18:2/23:1	632.6	262.2	15	25
Cer d18:2/23:1(-H$_2$O)	614.6	262.2	15	25
Cer d18:0/24:0	652.6	266.2	19.4	25
Cer d18:0/24:0(-H$_2$O)	634.6	266.2	19.4	25
Cer d18:0/24:1	650.6	266.2	18.2	25
Cer d18:0/24:1(-H$_2$O)	632.6	266.2	18.2	25
Cer d18:1/24:0	650.6	264.2	17.9	25
Cer d18:1/24:0(-H$_2$O)	632.6	264.2	17.9	25
Cer d18:1/24:1	648.6	264.2	16.7	25
Cer d18:1/24:1(-H$_2$O)	630.6	264.2	16.7	25
Cer d18:2/24:0	648.6	262.2	16.9	25
Cer d18:2/24:0(-H$_2$O)	630.6	262.2	16.9	25
Cer d18:2/24:1	646.6	262.2	15.6	25
Cer d18:2/24:1(-H$_2$O)	628.6	262.2	15.6	25
Cer d18:0/25:0	666.6	266.2	20	25

(continued)

Table 2
(continued)

Compound name	Precursor ion	Product ion	Ret time (min)	Collision energy
Cer d18:0/25:0(-H$_2$O)	648.6	266.2	20	25
Cer d18:0/25:1	664.6	266.2	18.8	25
Cer d18:0/25:1(-H$_2$O)	646.6	266.2	18.8	25
Cer d18:1/25:0 (ISTD)	646.4	264.2	18.5	25
Cer d18:1/25:0 (ISTD)(-H$_2$O)	664.4	264.2	18.5	25
Cer d18:0/26:0	680.6	266.2	19.4	25
Cer d18:0/26:0(-H$_2$O)	662.6	266.2	19.4	25
Cer d18:0/26:1	678.6	266.2	18.2	25
Cer d18:0/26:1(-H$_2$O)	660.6	266.2	18.2	25
Cer d18:1/26:0	678.6	264.2	19.6	25
Cer d18:1/26:0(-H$_2$O)	660.6	264.2	19.6	25
Cer d18:1/26:1	676.6	264.2	18.9	25
Cer d18:1/26:1(-H$_2$O)	658.6	264.2	18.9	25
Cer d18:2/26:0	676.6	262.2	18.1	25
Cer d18:2/26:0(-H$_2$O)	658.6	262.2	18.1	25
Cer d18:2/26:1	674.6	262.2	16.9	25
Cer d18:2/26:1(-H$_2$O)	656.6	262.2	16.9	25
Hex2Cer d18:1/12:0 (ISTD)	788.5	264.2	9.9	45
Hex2Cer d18:1/12:0 (ISTD)(-H$_2$O)	806.5	264.2	9.9	45
Hex2Cer d18:1/16:0	862.7	264.2	11.2	45
Hex2Cer d18:1/16:0(-H$_2$O)	844.7	264.2	11.2	45
Hex2Cer d18:1/18:0	890.7	264.2	12.4	45
Hex2Cer d18:1/18:0(-H$_2$O)	872.7	264.2	12.4	45
Hex2Cer d18:1/18:1	888.7	264.2	11.4	45
Hex2Cer d18:1/18:1(-H$_2$O)	870.7	264.2	11.4	45
Hex2Cer d18:1/20:0	918.7	264.2	13.6	45
Hex2Cer d18:1/20:0(-H$_2$O)	900.7	264.2	13.6	45
Hex2Cer d18:1/22:0	946.7	264.2	14.6	45
Hex2Cer d18:1/22:0(-H$_2$O)	928.7	264.2	14.6	45
Hex2Cer d18:1/22:1	944.7	264.2	13.5	45

(continued)

Table 2
(continued)

Compound name	Precursor ion	Product ion	Ret time (min)	Collision energy
Hex2Cer d18:1/22:1(-H$_2$O)	926.7	264.2	13.5	45
Hex2Cer d18:1/23:0	960.7	264.2	15.2	45
Hex2Cer d18:1/23:0(-H$_2$O)	942.7	264.2	15.2	45
Hex2Cer d18:1/24:0	974.7	264.2	15.7	45
Hex2Cer d18:1/24:0(-H$_2$O)	956.7	264.2	15.7	45
Hex2Cer d18:1/24:1	972.7	264.2	14.6	45
Hex2Cer d18:1/24:1(-H$_2$O)	954.7	264.2	14.6	45
HexCer 30:1 d18:1/12:0 (ISTD)	644.5	264.2	10	45
HexCer 30:1 d18:1/12:0 (ISTD)(-H$_2$O)	626.5	264.2	10	45
HexCer 32:1 d18:1/14:0	672.6	264.2	10.5	45
HexCer 32:1 d18:1/14:0(-H$_2$O)	654.6	264.2	10.5	45
HexCer d18:0/16:0	702.6	266.2	14	45
HexCer d18:0/16:0(-H$_2$O)	684.6	266.2	14	45
HexCer d18:0/16:1	700.6	266.2	12.9	45
HexCer d18:0/16:1(-H$_2$O)	682.6	266.2	12.9	45
HexCer d18:1/16:0	700.6	264.2	11.52	45
HexCer d18:1/16:0(-H$_2$O)	682.6	264.2	11.52	45
HexCer d18:1/16:1	698.6	264.2	10.5	45
HexCer d18:1/16:1(-H$_2$O)	680.6	264.2	10.5	45
HexCer d18:2/16:0	698.6	262.2	10.5	45
HexCer d18:2/16:0(-H$_2$O)	680.6	262.2	10.5	45
HexCer d18:1/18:0	728.6	264.2	12.9	45
HexCer d18:1/18:0(-H$_2$O)	710.6	264.2	12.9	45
HexCer d18:1/18:1	726.6	264.2	11.7	45
HexCer d18:1/18:1(-H$_2$O)	708.6	264.2	11.7	45
HexCer d18:0/20:0	758.6	266.2	15.1	45
HexCer d18:0/20:0(-H$_2$O)	740.6	266.2	15.1	45
HexCer d18:1/20:0	756.6	264.2	14.1	45
HexCer d18:1/20:0(-H$_2$O)	738.6	264.2	14.1	45
HexCer d18:1/20:1	754.6	264.2	13	45

(continued)

Table 2
(continued)

Compound name	Precursor ion	Product ion	Ret time (min)	Collision energy
HexCer d18:1/20:1(-H$_2$O)	736.6	264.2	13	45
HexCer d18:2/20:0	754.6	262.2	13	45
HexCer d18:2/20:0(-H$_2$O)	736.6	262.2	13	45
HexCer d18:0/22:0	786.6	266.2	17.5	45
HexCer d18:0/22:0	786.6	266.2	17.5	45
HexCer d18:0/22:0(-H$_2$O)	768.6	266.2	17.5	45
HexCer d18:0/22:0(-H$_2$O)	768.6	266.2	17.5	45
HexCer d18:0/22:1	784.6	266.2	16.3	45
HexCer d18:0/22:1(-H$_2$O)	766.6	266.2	16.3	45
HexCer d18:1/22:0	784.6	264.2	15.2	45
HexCer d18:1/22:0(-H$_2$O)	766.6	264.2	15.2	45
HexCer d18:1/22:1	782.6	264.2	14	45
HexCer d18:1/22:1(-H$_2$O)	764.6	264.2	14	45
HexCer d18:2/22:0	782.6	262.2	14	45
HexCer d18:2/22:0(-H$_2$O)	764.6	262.2	14	45
HexCer d18:0/23:0	800.7	266.2	18.1	45
HexCer d18:0/23:0(-H$_2$O)	782.7	266.2	18.1	45
HexCer d18:1/23:0	798.6	264.2	15.9	45
HexCer d18:1/23:0(-H$_2$O)	780.6	264.2	15.9	45
HexCer d18:1/23:1	796.6	264.2	14.7	45
HexCer d18:1/23:1(-H$_2$O)	778.6	264.2	14.7	45
HexCer d18:0/24:0	814.7	266.2	18.6	45
HexCer d18:0/24:0	814.7	266.2	18.6	45
HexCer d18:0/24:0(-H$_2$O)	796.7	266.2	18.6	45
HexCer d18:0/24:0(-H$_2$O)	796.7	266.2	18.6	45
HexCer d18:0/24:1	812.7	266.2	17.5	45
HexCer d18:0/24:1(-H$_2$O)	794.7	266.2	17.5	45
HexCer d18:1/24:0	812.7	264.2	16.5	45
HexCer d18:1/24:0(-H$_2$O)	794.7	264.2	16.5	45
HexCer d18:1/24:1	810.7	264.2	15.3	45

(continued)

Table 2
(continued)

Compound name	Precursor ion	Product ion	Ret time (min)	Collision energy
HexCer d18:1/24:1($-H_2O$)	792.7	264.2	15.3	45
HexCer d18:1/25:0	826.7	264.2	17.1	45
HexCer d18:1/25:0($-H_2O$)	808.7	264.2	17.1	45
HexCer d18:1/25:1	824.7	264.2	16	45
HexCer d18:1/25:1($-H_2O$)	806.7	264.2	16	45
HexCer d18:0/26:0	842.7	266.2	18.7	45
HexCer d18:0/26:0($-H_2O$)	824.7	266.2	18.7	45
HexCer d18:0/26:1	842.7	266.2	17.7	45
HexCer d18:0/26:1($-H_2O$)	824.7	266.2	17.7	45
HexCer d18:1/26:0	840.7	264.2	17.7	45
HexCer d18:1/26:0($-H_2O$)	822.7	264.2	17.7	45
HexCer d18:1/26:1	838.7	264.2	16.5	45
HexCer d18:1/26:1($-H_2O$)	820.7	264.2	16.5	45
SM 32:0	677.53	184.1	11.5	25
SM 30:1 (ISTD)	647.42	184.1	10	25
SM 32:1	675.53	184.1	10.8	25
SM 32:2	673.52	184.1	10.1	25
SM 33:0	691.56	184.1	11.8	25
SM 33:1	689.56	184.1	11.2	25
SM 33:2	687.56	184.1	10.5	25
SM 34:0	705.58	184.1	12	25
SM 34:1	703.58	184.1	11.5	25
SM 34:2	701.56	184.1	10.9	25
SM 35:0	719.58	184.1	12.7	25
SM 35:1	717.59	184.1	12	25
SM 36:0	733.58	184.1	13.3	25
SM 36:1	731.61	184.1	12.6	25
SM 36:2	729.59	184.1	11.8	25
SM 37:0	747.61	184.1	14	25
SM 37:1	745.61	184.1	13.2	25

(continued)

Table 2
(continued)

Compound name	Precursor ion	Product ion	Ret time (min)	Collision energy
SM 38:0	761.6	184.1	14.5	25
SM 38:1	759.64	184.1	13.8	25
SM 38:2	757.61	184.1	13	25
SM 39:1	773.6	184.1	14.4	25
SM 40:0	789.67	184.1	15.7	25
SM 40:1	787.67	184.1	15.1	25
SM 40:2	785.65	184.1	14.2	25
SM 40:3	783.65	184.1	13.3	25
SM 41:1	801.68	184.1	15.8	25
SM 41:2	799.67	184.1	14.8	25
SM 41:3	797.67	184.1	13.8	25
SM 42:0	817.7	184.1	16.9	25
SM 42:1	815.7	184.1	16.5	25
SM 42:2	813.68	184.1	15.5	25
SM 42:3	811.67	184.1	14.5	25
SM 43:1	829.7	184.1	17.1	25
SM 43:2	827.7	184.1	16.1	25
SM 44:1	843.7	184.1	17.9	25
SM 44:2	841.7	184.1	16.9	25
Sph d16:1	272.24	254.2	5.7	5
Sph d16:1	272.24	236.1	5.7	10
Sph d16:1	272.24	224.1	5.7	21
Sph d17:0 (ISTD)	288.32	270.3	6.4	20
Sph d17:0 (ISTD)	288.32	252.2	6.4	20
Sph d17:0 (ISTD)	288.32	240.2	6.4	20
Sph d17:1 (ISTD)	286.32	268.3	6.2	5
Sph d17:1 (ISTD)	286.32	250.2	6.2	17
Sph d17:1 (ISTD)	286.32	238.2	6.2	20
Sph d18:0	302.32	284.2	6.65	20
Sph d18:0	302.32	266.2	6.65	20

(continued)

Table 2
(continued)

Compound name	Precursor ion	Product ion	Ret time (min)	Collision energy
Sph d18:0	302.32	254.2	6.65	21
Sph d18:1	300.3	282.2	6.45	5
Sph d18:1	300.3	264.2	6.45	17
Sph d18:1	300.3	252.2	6.45	21
Sph d18:2	298.25	280.2	6	5
Sph d18:2	298.25	262.2	6	17
Sph d18:2	298.25	250.2	6	21

Table 3
Parameters defined in the MS MRM method for S1P analysis

Compound name	Precursor ion	Product ion
S1P d18:1 $^{13}C_2D_2$ (ISTD)	440.33	113.0
S1P d18:1 $^{13}C_2D_2$ (ISTD)	440.33	60.08
S1P d18:0	438.34	113.0
S1P d18:0	438.34	60.08
S1P d18:1	436.31	113.0
S1P d18:1	436.31	60.08
S1P d18:2	434.29	113.0
S1P d18:2	434.29	60.08
S1P d17:0[a]	424.3	113.0
S1P d17:0[a]	424.3	60.08
S1P d17:1[a]	422.3	113.0
S1P d17:1[a]	422.3	60.08

[a]*see* **Note 13**

3. The triple-quadrupole (QQQ) mass spectrometer is operated in positive ionization multiple reaction monitoring (MRM) mode. Two product ions (from the tetramethylated S1P that is formed after TMS-diazomethane derivatization) are monitored at m/z 60.08 and m/z 113 (*see* Table 3). The fragmentor, collision energy, and cell accelerator voltage are set to 380 V, 29 V, and 3 V, respectively.

4. Injection volume for all samples is 2 μL (*see* **Note 20**).

3.6.3 LC-MS Analyses

1. LC-MS system performances have to meet established quality criteria (*see* **Note 21**).

2. Column is well conditioned before starting the analysis (*see* **Note 22**).

3. The LC-MS system should first be equilibrated by injecting three extracted blanks and then injecting TQC samples until retention times and signals are stable. Ensure that all the peaks are well captured within the dMRM windows for the sphingolipid method (*see* Table 2), or adjust retention time windows if necessary.

4. Injection sequence of samples, BQC, TQC, and extracted blanks should depend on the experimental design. We recommend the following injection sequence:

    (a) TQC dilution series starting with the 1:8 diluted TQC.

    (b) Samples, BQCs, and blanks in the same sequence used for extraction.

    (c) Insert TQCs in the sequence to monitor instrument performance (e.g., after each blank + ISTD).

3.7 Data Analysis

3.7.1 Peak Integration of MRM Chromatograms

1. MS raw data are imported into a MRM data analysis software for small molecules, e.g., Agilent MassHunter Quantitative Analysis, and MRM chromatograms are extracted.

2. Determine peak areas of quantifier transitions by peak integration. Transitions with fragments from intact precursors are used as quantifiers and transitions with fragments from precursors after water loss as qualifiers. For HexCer and Hex2Cer, it is opposite; the transitions with intact precursors are used as qualifiers, and those of precursors with water loss as qualifier. For sphingomyelins (SM) no qualifier transitions were monitored in this protocol (*see* Table 2). For S1P transitions, the product ion at m/z 60.08 is used as a quantifier and the one at m/z 113 as a qualifier. Ensure that the correct peak is picked, that a co-eluting qualifier peak is present, and that integration borders are determined correctly (*see* **Notes 23–25**).

3.7.2 Normalization and Quantification

1. Normalization with internal standards is done by calculating the peak area ratio of the lipid of interest and the corresponding internal standard (ISTD), in a linear interval covered by previously measured calibration curves [13, 16]. Lipids are normalized with the corresponding class-specific compounds: ceramides are normalized with Cer d18:1/12:0; HexCer with GluCer d18:1/12:0; Hex2Cer with LacCer d18:1/12:0; SM with SM d18:1/12:0; C1P with C1P d18:1/12:0; sphingosines d18:1, d18:2, and d16:1 with Sph d17:1; sphingosine d18:0 with Sph d17:0; and all S1Ps with S1P-$^{13}C_2D_2$ d18:1 (*see* **Notes 13** and **26**).

2. Isotope correction of the S1P quantifier transitions (m/z 60.03) is done by using the following formulas:

$$RA_{S1P\ d18:0} = RAraw_{S1P\ d18:0} - RAraw_{S1P\ d18:1} \times 0.03161$$

$$RA_{S1P\ d18:1} = RAraw_{S1P\ d18:1} - RAraw_{S1P\ d18:2} \times 0.03156$$

RA, relative abundance (corrected for isotope interference); RAraw, raw relative abundance (*see* **Note 27**).

Calculation of absolute S1P d18:1 concentrations in the study sample can be done using the following formula:

$$c_{S1Pd18:1} = RA_{S1Pd18:1} \times \left(\frac{\nu_{ISTD}}{\nu_{Sample}} \right) \times c_{ISTD}$$

RAcorr$_{S1P\ d18:1}$, relative abundance of S1P d18:1; ν_{ISTD}, volume of S1P-$^{13}C_2D_2$ ISTD solution (BUME + ISTD); ν_{Sample}, sample volume; c_{ISTD}, molar concentration of S1P d18:1 in the sample; c_{ISTD}, molar concentration of S1P-$^{13}C_2D_2$ in the ISTD (BUME + ISTD) solution (*see* **Note 28** and Table 1).

3. *See* **Note 29** for calculation of concentrations of the other lipids measured with the sphingolipid and S1P methods.

3.7.3 Quality Control

The obtained raw dataset of lipid abundances should be filtered based on quality control (QC) parameters to exclude potential artifacts and noisy signals in downstream data analysis. The criteria for QC filtering and parameters depend on the aims, requirements, and practical aspects of the study.

1. Filtering of lipids exhibiting high variations in QC samples: The coefficient of variation (CoV) in the BQC samples yields information on the variation caused by experimental and analytical effects. We usually tend to exclude lipids with a %CoV of >20–25%, depending on the study. %CoV is calculated using the following formula:

$$\%CoV_{Lipid\ X} = \frac{StdDev\left(RA_{Lipid\ X\ in\ all\ BQCs}\right)}{Mean\left(RA_{Lipid\ X\ in\ all\ BQCs}\right)} \times 100$$

RA, relative abundance.

2. Filtering of lipids that have high background signals. The signal-to-blank peak area ratio of a lipid between sample and corresponding blank yields information on interfering signals from the extraction solvent, ISTD mix, extraction procedure, and carry-over. We usually exclude lipids that have a signal-to-blank ratio <10.

3. Filtering of lipids that do not linearly respond to dilution: Significantly lower or higher response to dilution of extracts may indicate saturation effects, interfering contaminants from

the lipid extraction or LC system, other artifacts from the LC-MS system, or errors in data processing, i.e., peak picking and integration. We usually exclude lipids that respond to dilutions of TQC with a Pearson correlation coefficient < 0.8 (*see* also **Note 20**).

4. The relative abundances of lipid in BQCs can be used to monitor, and to potentially correct, for drifts and batch effects during analyses. We routinely plot relative abundances of each lipid and the principal components 1–3 from a principal component analysis (PCA) of all relative lipid abundances in all BQCs against the analysis order to visualize and detect drifts and batch effects.

4   Notes

1. Personal protection equipment (PPE) should be worn during all experimental steps. Human blood, plasma, and serum samples should be treated as potentially infectious and handled in a biosafety cabinet. All steps involving organic solvents should be performed under a chemical hood, except when organic extraction solvent is added to the plasma samples. Concentrated formic acid should be added into water/solvents and not in reverse order.

2. During blood coagulation, platelets, leukocytes, and erythrocytes release extracellular vesicles and various compounds, including lipids such as sphingosine-1-phosphate, and enzymes that may change lipid profiles ex vivo. Plasma obtained from anticoagulated blood can therefore be considered as the closest to the blood plasma in vivo [22, 24, 25].

3. Blood should always be taken around the same time of the day for all the samples, and fasting status (time between the last meal and blood draw) should also be comparable between the study subjects, to reduce diurnal variations and effect of food intake on plasma lipids [26, 27].

4. Samples from matched reference groups should be taken for each study and at times distributed throughout the study duration to avoid biases, if applicable. Comparisons with samples or with lipidomic results obtained from other studies may be difficult due to differences in the used methodologies that may lead to incorrect conclusions.

5. Collection of capillary blood, e.g., by fingerprick, is not recommended as blood may be contaminated with skin cells, tissue fluids, and disinfectants. These procedures often result in hemolytic samples, especially when the puncture point is

squeezed (*see* **Note 12**), and clotting before mixing with the anticoagulant [28].

6. Proper venipuncture protocols should be applied to prevent hemolysis and clotting [29]. Discarding the first 1–3 mL of blood with a dedicated discard tube may reduce the risk of hemolysis and clots. Special care should be taken when blood is collected via infusion catheters, in which case discarding the first 1–3 mL blood reduces the risk of hemolysis and mixing of infusion solution into the collected blood [30].

7. Heparin has been shown to introduce higher variability in the measurements of certain sphingolipids [27].

8. When using blood collection containers with liquid anticoagulant solution (i.e., citrate tubes), care must be taken not to under- or overfill the tubes to avoid dilution errors and ensure a reproducible final concentration of anticoagulant. The dilution by the anticoagulant solution must be considered in the calculation of the lipid concentrations.

9. When collected whole blood is not kept on ice before centrifugation, specific lipid levels may increase ex vivo, i.e., sphingosine-1-phosphate [22].

10. Clotting time of blood from patients with disorders affecting coagulation and/or who are under anticoagulant therapy may considerably vary. In this case samples from all subjects should be allowed to clot for the same time.

11. Low centrifugation forces or short centrifugation times may lead to residual platelets in the plasma, which may affect levels of lipids abundant in platelets, such as sphingosine-1-phosphate [31]. High centrifugation forces may cause hemolytic serum preparations and alter the lipid levels as well.

12. Hemolytic samples may have altered profiles of specific lipids that are highly abundant in erythrocytes, e.g., sphingosine-1-phosphate, and must therefore be avoided or measured with caution [22].

13. To analyze S1P d17:0 and S1P d17:1, the Cer/Sph mixture II cannot be used, as it contains S1P d17:0 and S1P d17:1 standards, which will interfere with endogenous species. In this case, a separate extraction with BUME containing only S1P-$^{13}C_2D_2$ as internal standard has to be prepared.

14. Before using internal standard solutions stored at −20°C or −80°C, allow solutions to equilibrate to room temperature, sonicate for approximately 10 min, and mix well to ensure compounds are fully dissolved.

15. Polypropylene tubes of some other brands may not be compatible with the butanol/methanol extraction solvent and/or may cause interferences in LC-MS analyses. It is recommended

to test specific tubes by analyzing blanks extracted in these tubes (*see* Subheading 3.4, **step 2**).

16. Note that these extracts can be used for other lipid analyses as well, as, for example, in phospholipid and neutral lipid analyses [17].

17. When extracts that were stored at −80°C will be analyzed, they should be equilibrated at room temperature for 1 h, briefly vortexed, sonicated for 10 min in a sonication bath, and vortexed again, before analysis.

18. (Trimethylsilyl)diazomethane (TMS-diazomethane) solution is highly toxic by inhalation and skin contact, and all experimental steps using this reagent must therefore be handled with appropriate safety measures, i.e., working always under a fume hood and wearing personal protection equipment. People handling this compound should consult the safety data sheet and should be accordingly trained. To minimize risk of larger spills, only the estimated total volume of TMS-diazomethane required for an experiment should be transferred from the stock bottle to a glass or polypropylene tube using a glass syringe. Aliquots are then taken from this tube. Excessive, unused TMS-diazomethane should be neutralized by slowly adding 10% acetic acid in methanol. This reaction is exothermic and produces nitrogen gas; therefore only little amounts of 10% acetic acid should be added at once.

19. After adding TMS-diazomethane, the solution in the tubes turns yellow. After incubation, it may become less colored, and after addition of acetic acid, the solution must become colorless, indicating inactivation of TMS-diazomethane. The amount of acetic acid added is enough to inactivate the TMS-diazomethane in the tube. However, in case the sample is still yellow, do not proceed with analyses, verify your procedure, and add more acetic acid for full inactivation.

20. Injection volumes may change depending on the LC-MS system used. We advise to check if signal intensities of lipid species in the TQC dilution series follow the dilution pattern, especially when using a different LC-MS than the one described in this article. When signals of lipids do not linearly respond to the dilution, a signal saturation effect may be present, in which case lowering the injection volume or diluting the samples may be helpful (*see* Subheading 3.7, **step 10**).

21. LC-MS instrument performances may deteriorate in time, especially during long sample sequences, due to accumulation of contaminants in the system. We suggest using the TQC samples to monitor the system condition and to perform maintenance/cleaning procedures when necessary. At this stage, it will be important to document when these procedures

took place during the analysis, to be able to explain and correct possible batch effects at the end of the analysis (*see* Subheading 3.7, **step 11**).

22. The column should be fully equilibrated prior to the start of the sample analysis. The column can be equilibrated for 30 min using the initial gradient conditions. A minimum of three chromatographic runs using blanks prior to the start of TQC injections is suggested.

23. Interferences from [M + 2] isotopes, in-source fragmentation products, and unknown compounds may be present in the chromatograms [16]. Particularly, in plasma and serum, d18:0 sphingolipids are usually less abundant or not detectable compared to d18:1 sphingolipids, whose M + 2 isotope can cause interferences that may be misinterpreted as d18:0 sphingolipid peaks. For the sphingolipid method data, it is therefore important to ensure that retention times of picked peaks are in accordance with the chain length and desaturation of the lipid species. In reversed-phase separation, lipid species with longer chains elute later, whereas species with higher number of desaturations (double bonds in the acyl and sphingosine chains) elute earlier [16]. In the S1P method all S1P species, including the S1P-^{13}C$_2$D$_2$ d18:1 internal standard, nearly co-elute in overlapping elution profiles due to the HILIC separation [13]. As a consequence, isotope correction has to be performed for S1P d18:0 and d18:1 (*see* **Note 27**). Peaks that do not at least partially co-elute with S1P-^{13}C$_2$D$_2$ d18:1 indicate interferences and must not be considered (*see* **Note 23**) [13].

24. Presence of corresponding co-eluting qualifier peaks should be verified for all the integrated quantifier peaks. Compounds without corresponding co-eluting qualifier may be interferences from unknown compounds and should not be integrated.

25. In-source sugar loss may occur from HexCer and Hex2Cer species and generate ceramide precursors. However, these interferences can be identified by their shorter retention time relative to corresponding "true" ceramide species present in the sample.

26. Internal standards for each lipid class are used to normalize for the differential extraction and ionization efficiencies of different lipid classes.

27. For S1P d18:0 it is necessary to correct for the isotopic interference from M + 2 ions of S1P d18:1, which is usually higher in abundance than S1P d18:0. For S1P d18:1 the isotopic interference from S1P d18:2 M + 2 is usually less than 1% in plasma and could therefore be omitted. No isotope correction

is needed for the lipids in the sphingolipid method, as those lipids whose M + 2 isotopes could interfere are separated by retention time.

28. Accurate quantification can be achieved for S1P d18:1 using the isotope-labeled S1P-$^{13}C_2D_2$ d18:1 internal standard combined with HILIC chromatography. For all other lipid species in the sphingolipids and S1P method matrix, chromatographic and ionization effects may affect calculated concentrations as only one internal standard is used per lipid class. The co-elution of S1P species due to the HILIC separation in the S1P method based may compensate for some of these effects and may therefore allow a more accurate quantification of different S1P species with only one internal standard (*see* **Note 29**).

29. Hypothesizing that the analyte and the selected internal standard have identical response factors, this general equation can be used to calculate the concentration of an analyte in the sample:

$$c_{Analyte} = \frac{Area_{Analyte}}{Area_{ISTD}} \times \frac{\nu_{ISTD}}{\nu_{Sample}} \times c_{ISTD}$$

$Area_{Analyte}$, peak area of analyte; $Area_{Analyte}$, peak area of internal standard (ISTD); V_{ISTD}, volume of ISTD solution; V_{Sample}, sample volume; $c_{Analyte}$, molar concentration of the analyte in the sample; c_{ISTD}, molar concentration of the ISTD in the ISTD solution (*see* Table 1).

References

1. Merrill AH (2011) Sphingolipid and glyco-sphingolipid metabolic pathways in the era of sphingolipidomics. Chem Rev 111 (10):6387–6422. https://doi.org/10.1021/cr2002917

2. Iqbal J, Walsh MT, Hammad SM et al (2017) Sphingolipids and lipoproteins in health and metabolic disorders. Trends Endocrinol Metab. https://doi.org/10.1016/j.tem.2017.03.005

3. Hyötyläinen T, Ahonen L, Poho P et al (2017) Lipidomics in biomedical research-practical considerations. Biochim Biophys Acta. https://doi.org/10.1016/j.bbalip.2017.04.002

4. Zhao Y-Y, Wu S-P, Liu S et al (2014) Ultra-performance liquid chromatography-mass spectrometry as a sensitive and powerful technology in lipidomic applications. Chem Biol Interact 220:181–192. https://doi.org/10.1016/j.cbi.2014.06.029

5. Sillence DJ, Platt FM (2003) Storage diseases: new insights into sphingolipid functions. Trends Cell Biol 13(4):195–203. https://doi.org/10.1016/S0962-8924(03)00033-3

6. Aburasayn H, Batran RA, Ussher JR (2016) Targeting ceramide metabolism in obesity. Am J Physiol Endocrinol Metab 311(2): E423–E435. https://doi.org/10.1152/ajpendo.00133.2016

7. Fucho R, Casals N, Serra D et al (2017) Ceramides and mitochondrial fatty acid oxidation in obesity. FASEB J 31(4):1263–1272. https://doi.org/10.1096/fj.201601156R

8. Laaksonen R, Ekroos K, Sysi-Aho M et al (2016) Plasma ceramides predict cardiovascular death in patients with stable coronary artery disease and acute coronary syndromes beyond LDL-cholesterol. Eur Heart J. https://doi.org/10.1093/eurheartj/ehw148

9. Penno A, Reilly MM, Houlden H et al (2010) Hereditary sensory neuropathy type 1 is caused by the accumulation of two neurotoxic sphingolipids. J Biol Chem 285(15):11178–11187. https://doi.org/10.1074/jbc.M109.092973

10. Zuellig RA, Hornemann T, Othman A et al (2014) Deoxysphingolipids, novel biomarkers for type 2 diabetes, are cytotoxic for insulin-producing cells. Diabetes 63(4):1326–1339. https://doi.org/10.2337/db13-1042

11. Shayman JA (2016) Targeting glycosphingolipid metabolism to treat kidney disease. Nephron 134(1):37–42. https://doi.org/10.1159/000444926

12. Kunkel GT, Maceyka M, Milstien S et al (2013) Targeting the sphingosine-1-phosphate axis in cancer, inflammation and beyond. Nat Rev Drug Discov 12(9):688–702. https://doi.org/10.1038/nrd4099

13. Narayanaswamy P, Shinde SA, Sulc R et al (2014) Lipidomic 'deep profiling': an enhanced workflow to reveal new molecular species of signaling lipids. Anal Chem 86 (6):3043–3047. https://doi.org/10.1021/ac4039652

14. Blanksby SJ, Mitchell TW (2010) Advances in mass spectrometry for lipidomics. Annu Rev Anal Chem (Palo Alto, Calif) 3:433–465. https://doi.org/10.1146/annurev.anchem.111808.073705

15. Sullards MC, Liu Y, Chen Y et al (2011) Analysis of mammalian sphingolipids by liquid chromatography tandem mass spectrometry (LC-MS/MS) and tissue imaging mass spectrometry (TIMS). Biochim Biophys Acta 1811(11):838–853. https://doi.org/10.1016/j.bbalip.2011.06.027

16. Wang J-R, Zhang H, Yau LF et al (2014) Improved sphingolipidomic approach based on ultra-high performance liquid chromatography and multiple mass spectrometries with application to cellular neurotoxicity. Anal Chem 86(12):5688–5696. https://doi.org/10.1021/ac5009964

17. Alshehry ZH, Barlow CK, Weir JM et al (2015) An efficient single phase method for the extraction of plasma lipids. Metabolites 5 (2):389–403. https://doi.org/10.3390/metabo5020389

18. Begum H, Li B, Shui G et al (2016) Discovering and validating between-subject variations in plasma lipids in healthy subjects. Sci Rep 6:19139. https://doi.org/10.1038/srep19139

19. Sales S, Graessler J, Ciucci S et al (2016) Gender, contraceptives and individual metabolic predisposition shape a healthy plasma lipidome. Sci Rep 6:27710. https://doi.org/10.1038/srep27710

20. Dunn WB, Wilson ID, Nicholls AW et al (2012) The importance of experimental design and QC samples in large-scale and MS-driven untargeted metabolomic studies of humans. Bioanalysis 4(18):2249–2264. https://doi.org/10.4155/bio.12.204

21. Kohler I, Verhoeven A, Derks RJ et al (2016) Analytical pitfalls and challenges in clinical metabolomics. Bioanalysis 8(14):1509–1532. https://doi.org/10.4155/bio-2016-0090

22. Yin P, Peter A, Franken H et al (2013) Preanalytical aspects and sample quality assessment in metabolomics studies of human blood. Clin Chem 59(5):833–845. https://doi.org/10.1373/clinchem.2012.199257

23. Wang J-R, Zhang H, Yau LF et al (2014) Improved sphingolipidomic approach based on ultra-high performance liquid chromatography and multiple mass spectrometries with application to cellular neurotoxicity. Anal Chem 86(12):5688–5696. https://doi.org/10.1021/ac5009964

24. Ono Y, Kurano M, Ohkawa R et al (2013) Sphingosine 1-phosphate release from platelets during clot formation: close correlation between platelet count and serum sphingosine 1-phosphate concentration. Lipids Health Dis 12:20. https://doi.org/10.1186/1476-511X-12-20

25. Yu Z, Kastenmuller G, He Y et al (2011) Differences between human plasma and serum metabolite profiles. PLoS One 6(7):e21230. https://doi.org/10.1371/journal.pone.0021230

26. Chua EC-P, Shui G, Lee IT-G et al (2013) Extensive diversity in circadian regulation of plasma lipids and evidence for different circadian metabolic phenotypes in humans. Proc Natl Acad Sci U S A 110(35):14468–14473. https://doi.org/10.1073/pnas.1222647110

27. Hammad SM, Pierce JS, Soodavar F et al (2010) Blood sphingolipidomics in healthy humans: impact of sample collection methodology. J Lipid Res 51(10):3074–3087. https://doi.org/10.1194/jlr.D008532

28. Denery JR, Nunes AAK, Dickerson TJ (2011) Characterization of differences between blood sample matrices in untargeted metabolomics. Anal Chem 83(3):1040–1047. https://doi.org/10.1021/ac102806p

29. Lima-Oliveira G, Volanski W, Lippi G et al (2017) Pre-analytical phase management: a review of the procedures from patient

162   Bo Burla et al.

preparation to laboratory analysis. Scand J Clin Lab Invest 77(3):153–163. https://doi.org/10.1080/00365513.2017.1295317

30. Heiligers-Duckers C, Peters NALR, van Dijck JJP et al (2013) Low vacuum and discard tubes reduce hemolysis in samples drawn from intravenous catheters. Clin Biochem 46 (12):1142–1144. https://doi.org/10.1016/j.clinbiochem.2013.04.005

31. Frej C, Andersson A, Larsson B et al (2015) Quantification of sphingosine 1-phosphate by validated LC-MS/MS method revealing strong correlation with apolipoprotein M in plasma but not in serum due to platelet activation during blood coagulation. Anal Bioanal Chem 407(28):8533–8542. https://doi.org/10.1007/s00216-015-9008-4

## Chapter 12

Shotgun Lipidomics Approach for Clinical Samples

Lars F. Eggers and Dominik Schwudke

Abstract

Shotgun lipidomics offers fast and reproducible identification and quantification of lipids in clinical samples. Lipid extraction procedures based on the methyl *tert*-butyl protocol are well established for performing shotgun lipidomics in biomedical research. Here, we describe a shotgun lipidomics workflow that is well suited for the analysis of clinical samples such as tissue samples, blood plasma, and peripheral blood mononuclear cells.

Key words Shotgun lipidomics, High-resolution mass spectrometry, Clinical samples, Lipid extraction, Lipid identification

Abbreviations

CE	Cholesteryl ester
Cer	Ceramide
DAG	Diacylglycerol
FT-ICR	Fourier transform ion cyclotron resonance
HexCer	Hexosyl ceramide
LPC	Lysophosphatidylcholine
MAG	Monoacylglycerol
MFQL	Molecular fragmentation query language
MS	Mass spectrometry
MS^2	Tandem mass spectrometry
MTBE	Methyl *tert*-butyl ether (UPAC, *tert*-butyl methyl ether)
PC	Phosphatidylcholine
PE	Phosphatidylethanolamine
PG	Phosphatidylglycerol
PI	Phosphatidylinositol
PS	Phosphatidylserine
TAG	Triacylglycerol
TOF	Time of flight

Electronic supplementary material: The online version of this chapter (https://doi.org/10.1007/978-1-4939-7592-1_12) contains supplementary material, which is available to authorized users.

Martin Giera (ed.), *Clinical Metabolomics: Methods and Protocols*, Methods in Molecular Biology, vol. 1730, https://doi.org/10.1007/978-1-4939-7592-1_12, © Springer Science+Business Media, LLC 2018

1   Introduction

Shotgun lipidomics is a well-established approach for lipid analysis, and a number of clinical applications were reported [1–3]. The term "shotgun lipidomics" was coined by Han and Gross [4] and refers to direct analysis of complex lipid extracts by mass spectrometry [5, 6]. The shotgun approach avoids time-consuming chromatographic separations and takes advantage of the superior analytical power of modern tandem mass spectrometers [7]. Application of high-resolving mass analyzers like Orbitrap, FT-ICR, and TOF enables unambiguous quantification of a variety of phospholipids, sphingolipids, and neutral lipids. *Top-down* lipidomics is based on accurate mass determination in MS^1 and allows to determine chemical sum compositions. The heteroatom content is characteristic for many lipid classes, and lipid species can be assigned according to the length of aliphatic chains and number of double bond [6, 8]. The *bottom-up* approach identifies lipids on the basis of specific fragments for the lipid class-specific head groups and fatty acids in MS^2 [9–12]. With fast high-resolving tandem mass spectrometers, a sufficient number of MS^2 spectra can be collected together with MS^1 information. These complex mass spectrometric datasets are analyzed with the dedicated LipidXplorer software [13, 14]. As a result, an improved coverage of the lipidome is achieved because the aliphatic chain compositions can be determined [14, 15].

Here, we describe a shotgun lipidomics workflow suitable for clinical analyses. We describe in detail the sample preparation procedures for tissue homogenization and lipid extraction using the methyl *tert*-butyl ether (MTBE) protocol [16]. We provide detailed information for lipid profiling based on high-resolution MS instrumentation and provide a data analysis strategy for lipid identification and quantitation based on LipidXplorer.

2   Materials

All solvents and chemicals should be of LC-MS grade or otherwise with the highest purity grade available. All glass vials and Pasteur pipets have to be washed three times with each solvent with increasing hydrophobicity, (1) water, (2) methanol, and (3) MTBE, before use (*see* **Note 1**).

2.1 Tissue Homogenization

1. Homogenization buffer: Weigh 1.86 g KCl (74.55 g/mol) in a 500 mL volumetric flask and fill it to the mark with water. Store the solution at 4 °C.

2. Homogenization instrument with sample tubes (20 mL) containing stainless steel beads for tissue amounts below 200 mg; otherwise tubes with rotor-stator element (*see* **Note 2**).

3. Prepare solution of butylhydroxytoluene (BHT, $M = 220.35$ g/mol) in methanol: Solve 10 mg BHT in 10 mL methanol and store the solution at 4 °C (*see* **Note 3**).

2.2 Lipid Extraction

1. Prepare Pasteur pipets, at least two per sample, and Eppendorf tubes (Eppendorf Safe-Lock tubes). Per sample is needed: two tubes, size 2 mL, and three tubes, size 0.5 mL (*see* **Note 4**).

2. Centrifuge and vacuum centrifuge (SpeedVac).

3. Shaker (e.g., Eppendorf MixMate).

4. Methanol solution (containing 3% acetic acid): Mix 20 mL methanol and 600 μL acetic acid in a 50 mL glass bottle. Store the solution at room temperature (*see* **Note 5**).

5. Re-extraction solvent: Mix 20 mL MTBE with 6 mL methanol (with 3% acetic acid) and 5 mL water. From the resulting biphasic system, the upper phase is used for the second extraction step (*see* **Note 6**).

6. Internal standard solution: We use SPLASH LipidoMix mass spectrometry standard from Avanti Polar Lipids (No. 330707) and mix it with ceramide (d18:1/25:0) quantitative mass spectrometry standard (Avanti Polar Lipids No. LM-2225). Purchased vials of SPLASH and ceramide (d18:1/25:0) contain 1 mL standard solution. This solution is quantitatively transferred into a 10 mL volumetric flask and filled with methanol to the ring mark. From this stock, aliquots of 500 μL volume were prepared and stored at −20 °C (*see* **Note 7**) (Table 1).

7. Storage solution: Mix 30 mL of chloroform with 15 mL methanol and 2.25 mL water.

8. For adding internal standard solution, use precision glass capillaries with calibrated volumes (Fig. 1).

2.3 Shotgun Lipidomics

1. 96-well plate or 384-well plate (e.g., Eppendorf Twin Tec).

2. Spray solution: Mix 5 mL chloroform with 10 mL methanol (containing 0.1% w/v = 1 g/L ammonium acetate, $M = 77.08$ g/mol) and 20 mL 2-propanol. The resulting ammonium acetate concentration is 3.7 mM (*see* **Note 8**).

3. Mass spectrometer with high-resolving power ($Rmin@700 = 100,000$); this protocol refers to usage of a Q Exactive Plus (Thermo, Bremen, Germany).

Table 1
Composition of the internal standards. Concentrations are given for this protocol. Note that exact concentrations may vary between batches of SPLASH

Mixture components	Molecular weight (g/mol)	Chemical formula	m/z (positive)	m/z (negative)	Concentration in mixture (pmol/µL)
15:0-18:1(d7) PC	753.11	$C_{41}H_{73}D_7NO_8P$	753.6134 $[M + H]^+$	811.6199 $[M + CH_3COO]^-$	21.34
15:0-18:1(d7) PE	711.03	$C_{38}H_{67}D_7NO_8P$	711.5664 $[M + H]^+$	709.5519 $[M - H]^-$	0.80
15:0-18:1(d7) PS(Na-Salt)	777.02	$C_{39}H_{66}D_7NNaO_{10}P$	–	753.5417 $[M - H]^-$	0.54
15:0-18:1(d7) PG(Na-Salt)	764.02	$C_{39}H_{67}D_7NaO_{10}P$	–	740.5464 $[M - H]^-$	3.81
15:0-18:1(d7) PI(NH$_4$-Salt)	847.13	$C_{42}H_{75}D_7NO_{13}P$	–	828.5625 $[M - H]^-$	1.07
15:0-18:1(d7) PA(Na-Salt)	689.94	$C_{36}H_{61}D_7NaO_8P$	–	666.5097 $[M - H]^-$	1.07
18:1(d7) LPC	528.72	$C_{26}H_{45}D_7NO_7P$	529.3994 $[M + H]^+$	587.4059 $[M + CH_3COO]^-$	4.82
18:1(d7) LPE	486.64	$C_{23}H_{39}D_7NO_7P$	487.3524 $[M + H]^+$	485.3378 $[M - H]^-$	1.08
18:1(d7) Chol ester	658.16	$C_{45}H_{71}D_7O_2$	675.6779 $[M + NH_4]^+$	–	54.11
18:1(d7) MAG	363.59	$C_{21}H_{33}D_7O_4$	381.3704 $[M + NH_4]^+$	–	0.55
15:0-18:1(d7) DAG	587.98	$C_{36}H_{61}D_7O_5$	605.5844 $[M + NH_4]^+$	–	1.60
15:0-18:1(d7)-18:1 TAG	812.37	$C_{51}H_{89}D_7O_6$	829.7985 $[M + NH_4]^+$	–	7.05
18:1(d9) SM	738.12	$C_{41}H_{72}D_9N_2O_6P$	738.6470 $[M + H]^+$	796.6529 $[M + CH_3COO]^-$	4.19
Cholesterol(d7)	393.71	$C_{27}H_{39}D_7O$	411.4326 $[M + NH_4]^+$	–	24.99
Cer(d18:1/25:0)	664.14	$C_{43}H_{85}NO_3$	664.6602 $[M + H]^+$	–	2.44

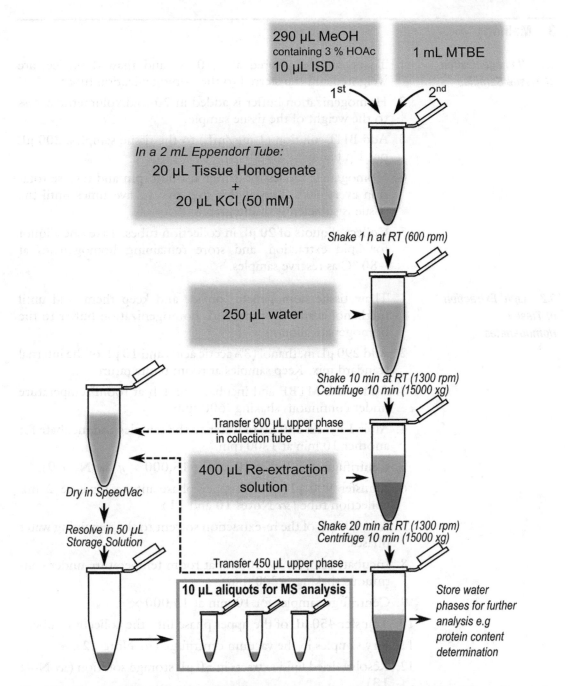

Fig. 1 Workflow of MTBE-based lipid extraction. *MeOH* methanol, *ISD* internal standard, *RT* room temperature

4. Nano-electrospray ion source: This protocol refers to the Tri-Versa NanoMate (Advion, Ithaca, USA). Flow injection strategies are another possibility to automatize shotgun lipidomics [12].

3   Methods

3.1   Homogenization of Tissue Samples

1. Tissue samples, stored at −80 °C and thawed on ice, are weighted and transferred to the homogenization tubes.

2. Homogenization buffer is added in 20-fold volumetric excess to the weight of the tissue sample.

3. Add BHT solution (1 mg/mL) to the tissue samples: 200 µL per 1 g tissue.

4. Homogenize tissues for 2 min at 6000 rpm and reverse rotation every 30 s. Repeat this step two to five times until the tissue is sufficiently disintegrated.

5. Collect aliquots of 20 µL in collection tubes. Take one aliquot for lipid extraction, and store remaining homogenates at −80 °C as reserve samples.

3.2   Lipid Extraction of Tissue Homogenates

1. Thaw tissue homogenates on ice and keep them cold until methanol is added. Add 20 µL homogenization buffer to the homogenate aliquots.

2. Add 290 µL methanol (3% acetic acid) and 10 µL of the internal standard mix. Keep samples at room temperature.

3. Add 1 mL MTBE and incubate for 1 h at room temperature under continuous shaking (600 rpm).

4. After 1 h, add 250 µL of water to the samples and incubate for another 10 min at 1300 rpm.

5. Centrifuge samples for 10 min at $15,000 \times g$ (see **Note 9**).

6. Transfer 900 µL of the upper phase into a separate 2 mL collection tube (see **Notes 10** and **11**).

7. Add 400 µL of the re-extraction solvent to the remaining water phases.

8. Incubate samples for 20 min at room temperature under continuous shaking (1300 rpm).

9. Centrifuge samples for 10 min at $15,000 \times g$.

10. Transfer 450 µL of the upper phase into the collection tube.

11. Dry samples in the vacuum centrifuge (see **Note 12**).

12. Resolve dried lipid extracts in 50 µL storage solution (see **Note 13**).

13. Pipet three to four aliquots with 10 µL in 500 µL sample tubes.

3.3   Lipid Extraction of Blood Plasma

1. Dilute EDTA plasma tenfold with water.

2. 50 µL are transferred to a new tube.

3. 20 µL of the internal standard mix are added and the mixture is well mixed.

4. Add 250 µL methanol (3% acetic acid) and agitate the suspension.

5. Add 1 mL MTBE and incubate the mixture for 1 h with constant stirring.

6. For induction of phase separation, add 250 µL water. Incubate the extraction mixture for another 5 min on a shaker.

7. Afterward, the extraction mixture is centrifuged at 15,000 × *g* for 2 min to improve phase separation.

8. 800 µL of the upper organic phase are collected into a new sample tube. Optional, the remaining phase is extracted a second time as described above (Subheading 3.2). The organic phase should be clear and free of any residual aqueous phase and insoluble material.

9. The organic phase is dried down with the help of a vacuum concentrator. Optional, the aqueous phase can be stored for further determination of the protein content (Subheading 3.4).

10. The dried phase is taken up in 100 µL storage solution and can be stored until mass spectrometric analysis.

3.4 Lipid Extraction of Peripheral Blood Mononuclear Cells (PBMCs)

1. Cell pellets of 1E06 PBMCs are carefully resuspended in 50 µL water.

2. Add 4 µL of the internal standard mix.

3. Add 270 µL methanol (3% acetic acid) and mix the suspension well.

4. Add 1 mL MTBE and incubate the mixture for 1 h with constant stirring.

5. Add 250 µL water to induce phase separation and incubate the suspension for another 5 min.

6. Centrifuge the extraction mixture at 15,000 × *g* to improve phase separation.

7. Collect 800 µL of the upper organic phase and transfer it to a new sample tube. Optional, the remaining phase is extracted a second time as described above (Subheading 3.2). The organic phase should be clear and free of any residual aqueous phase and insoluble material.

8. The combined organic phases are dried down and dissolved in 50 µL storage solution.

3.5 Shotgun Lipidomics

1. Add 190 µL of the spray solution to the 10 µL sample aliquots directly before MS analysis.

2. Vortex each sample and centrifuge for at least 5 min at 15,000 × *g* to prevent that particles clog nanoESI nozzles.

3. Pipet 20 μL of each sample onto a 96-well or 384-well plate. We recommend analyzing each sample twice. Consecutively, two wells are needed per sample (*see* **Note 14**).

4. Set spray voltage to 1.1 kV and back pressure (only for TriVersa NanoMate) at 1.1 psi.

5. Whole analysis duration is 10 min. Ion source switches polarity at 5 min.

6. Instrument acquires MS1 spectra in a range of m/z 350–1000 in both ion modes.

7. MS^2 spectra are acquired in data-independent acquisitions (DIA). An MS^2 spectra are acquired every m/z 1.001 starting at m/z 350.2 in the positive ion mode and at m/z 350.1 in the negative ion mode.

8. When a Q Exactive Plus is utilized, set resolving power of MS^1 scans to 270,000 and of MS^2 scans to 70,000.

3.6 Data Analysis with LipidXplorer

1. Install LipidXplorer Version 1.2.7 according to the documentation available on https://wiki.mpi-cbg.de/lipidx/Main_Page.

2. Convert raw data files into *.mzML file by MS convert from ProteoWizard [17]: http://proteowizard.sourceforge.net/downloads.shtml

3. Data files are imported into LipidXplorer using the import settings *lpdxImportSettings_LE4.ini* as available in the online supplement (*see* **Note 15**).

4. In the online supplement, we provide MFQL (molecular fragmentation query language) scripts for lipid identification based on data acquired on a Q Exactive Plus using both positive and negative ion mode. Detailed information about LipidXplorer and MFQL can be found in the tutorial by Herzog et al. [14] (*see* **Note 16**). Specific MFQLs for SPLASH lipid standard components are also provided.

5. After import of the data and lipid identification with MFQL scripts for each ion mode separately, *.csv files are written providing intensities for each lipid species as well as for the internal standards. In consecutive steps, quantities can be calculated based on this data.

6. We recommend eliminating background ions from the dataset by deleting all identified lipids whose average abundance is lower than ten times their abundance in blank extracts.

4   Notes

1. Industrial cleaned glass surfaces are in our experience not suffi-
    ciently cleaned for lipid analysis. To avoid ionization suppres-
    sion of lipids, we employ for all utilized glassware the described
    cleaning procedure with LC-MS grade solvents.

2. The selection of the homogenization device depends on the
    tissue amounts that should be homogenized. We recommend,
    if available, to use higher tissue amounts about 500 mg to cover
    the inhomogeneity within the tissue. As described in Subhead-
    ing 3, a 20-fold excess of buffer is given to the tissue samples
    during homogenization. When used tissues amount below
    200 mg, the volume of buffer is too low to yield sufficient
    homogenization in rotor-stator system. In this case, use of
    steel beads is beneficial.

3. Always prepare a fresh BHT solution.

4. The polypropylene of Eppendorf Safe-Lock tubes has a well-
    defined chemical background [18]. We recommend to deter-
    mine the chemical background pattern in conjunction of any
    mass spectrometric lipid analysis, especially when changing the
    manufacturer of the extraction tubes. Some of the known
    background signals can be used for mass recalibration.

5. The acidified methanol solution has to be prepared freshly.
    Acetic acid reacts with methanol to form methyl acetate. We
    observed impaired extraction efficiency and reproducibility if
    acidified methanol was not freshly prepared.

6. This biphasic system has an overall composition of 10:3:2.5
    (MTBE, methanol, water; v/v/v), which is exact the same
    composition in the first extraction step. The goal is to keep
    the volumetric ratios of MTBE, water, and methanol constant
    during the second extraction step. If only MTBE is added, the
    volumetric ratios shift and the extraction efficiency could be
    affected.

7. The choice of internal standards has to be customized for the
    applied screening experiment. Lipids with dialkyl-bound side
    chains, for example, are commercially available and do not
    naturally occur in clinical samples. They form the same adducts
    than natural lipids of the same class, although ionization effi-
    ciencies differ from diacyl species. Unfortunately, these species
    undergo different fragmentation mechanisms than diacyl lipids
    which makes them only suitable for *top-down* lipidomics screen.
    Another strategy would be to choose lipids with odd numbers
    of carbon atoms in their fatty acid residues (e.g., 17:0). These
    lipids are not available by mammalian biosynthetic pathways.
    From our experience, "odd-chain" lipid species are detected

with low abundance in human lipidomes. Potentially, these lipids were taken up by nutrition. The SPLASH mix contains lipids with deuterated acyl chains (d7 18:1) making this mixture ideal for *top-down* as well as *bottom-up* strategies. Additional lipid standards might be required depending on the experimental question. Here, we added a ceramide standard.

8. To yield higher sensitivities for anionic phospholipids in negative ion mode, 0.05 mM ammonium chloride in the spray solution can be used.

9. The sample tube contains a biphasic system: The upper phase is composed mainly by MTBE and contains extracted lipids. The lower phase is aqueous and contains small polar molecules (metabolites). On the bottom of the tube, precipitated proteins and biomacromolecules are located.

10. This step has to be carried out with care. Make sure that only the organic phase is collected.

11. To save time to dry down samples, the extracts in the collection tubes can be concentrated in a SpeedVac.

12. This step could be time-consuming. Dry sample at first at room temperature until the pressure notably drops in the SpeedVac (~45 min). Then MTBE is completely evaporated, and the remaining sample contains predominantly water. At this point methanol can be added to the samples to speed up drying. One can increase the temperature to 45 °C. It takes about 1.5 h until complete dryness. As long as there is water in the tube, the samples stay cooled. Make sure that dried samples do not remain an unnecessary long time in the heated vacuum chamber.

13. Care has to be taken when dried samples are resolved in the storage solution. Vortex the samples carefully and extensively.

14. We recommend to keep the temperature of the autosampler between 15 and 18 °C for long sequences.

15. The LipidXplorer import settings were written for data acquired on a Q Exactive Plus instrument and acquisitions as described under Subheading 3.3.

16. The MFQL scripts report quantifier ions in columns with prefixes "INT." The quantifiers were chosen to be specific for each individual lipid species. In negative ion mode, we quantify based on the sum intensity of both fatty acid fragments in MS^2. The bases for quantification are signals of the deuterated fatty acids of the SPLASH internal standard mix. Cardiolipin is quantified as doubly charged precursor ion in negative ion mode based on the first isotope signal as proposed by Han et al. [19]. No internal standard for cardiolipin was utilized; we utilize the response of the PG internal standard as reference.

Quantification of cholesterol esters is based on the fragment ion at m/z 369.35. For ceramides the C18 long-chain base signal at m/z 264.26 is used. For DAG and TAG species, the specific neutral loss of fatty acids is used for quantitation. MAG species are exclusively identified and quantified by their accurate mass in MS^1. SM species are also quantified based on their precursor ion. We recommend to use the negative ion mode for PE and PC quantitation.

All provided MFQL scripts contain empirical determined rules for filtering initial identification results. Intensity ratios of precursor ions and fragment ions were determined from internal standards and are used as quality check to avoid false assignment due to in-source fragmentation and adducts. These ratios are dependent on the instrument platform. MFQL script for all glycerophospholipids contains "(avg(PR. intensity) / (avg(FA1.intensity) + avg(FA2.intensity)) >= X AND avg(PR.intensity) / (avg(FA1.intensity) + avg(FA2. intensity)) <= Y)" where X represent the lower and Y higher limit for this ratio for each class.

Acknowledgments

The research was supported by funds of the EXC 306 Inflammation at Interfaces, German Centre for Infection Research (DZIF TTU-TB), German Centre for Lung Research, Airway Research Center North (ARCN), and Lipidomics Informatics for Life-Sciences (LIFS) of the German Network for Bioinformatics Infrastructure (de.NBI).

References

1. Griese M, Kirmeier HG, Liebisch G, Rauch D, Stückler F, Schmitz G, Zarbock R, ILD-BAL working group of the Kids-Lung-Register (2015) Surfactant lipidomics in healthy children and childhood interstitial lung disease. PLoS One 10(2):e0117985. https://doi.org/10.1371/journal.pone.0117985

2. Graessler J, Schwudke D, Schwarz PEH, Herzog R, Shevchenko A, Bornstein SR (2009) Top-down lipidomics reveals ether lipid deficiency in blood plasma of hypertensive patients. PLoS One 4(7):e6261. https://doi.org/10.1371/journal.pone.0006261

3. Sales S, Graessler J, Ciucci S, Al-Atrib R, Vihervaara T, Schuhmann K, Kauhanen D, Sysi-Aho M, Bornstein SR, Bickle M, Cannistraci CV, Ekroos K, Shevchenko A (2016) Gender, contraceptives and individual metabolic predisposition shape a healthy plasma lipidome. Sci Rep 6:27710. https://doi.org/10.1038/srep27710

4. Han X, Gross RW (2003) Global analyses of cellular lipidomes directly from crude extracts of biological samples by ESI mass spectrometry a bridge to lipidomics. J Lipid Res 44(6):1071–1079

5. Shevchenko A, Simons K (2010) Lipidomics: coming to grips with lipid diversity. Nat Rev Mol Cell Biol 11(8):593–598. https://doi.org/10.1038/nrm2934

6. Schwudke D, Schuhmann K, Herzog R, Bornstein SR, Shevchenko A (2011) Shotgun lipidomics on high resolution mass spectrometers. Cold Spring Harb Perspect Biol 3(9). https://doi.org/10.1101/cshperspect.a004614

7. Heiskanen LA, Suoniemi M, Ta HX, Tarasov K, Ekroos K (2013) Long-term performance and stability of molecular shotgun lipidomic analysis of human plasma samples. Anal Chem 85(18):8757–8763. https://doi.org/10.1021/ac401857a

8. Schwudke D, Hannich JT, Surendranath V, Grimard V, Moehring T, Burton L, Kurzchalia T, Shevchenko A (2007) Top-down lipidomic screens by multivariate analysis of high-resolution survey mass spectra. Anal Chem 79(11):4083–4093. https://doi.org/10.1021/ac062455y

9. Brügger B, Erben G, Sandhoff R, Wieland FT, Lehmann WD (1997) Quantitative analysis of biological membrane lipids at the low picomole level by nano-electrospray ionization tandem mass spectrometry. Proc Natl Acad Sci U S A 94(6):2339–2344

10. Schwudke D, Oegema J, Burton L, Entchev E, Hannich JT, Ejsing CS, Kurzchalia T, Shevchenko A (2006) Lipid profiling by multiple precursor and neutral loss scanning driven by the data-dependent acquisition. Anal Chem 78(2):585–595. https://doi.org/10.1021/ac051605m

11. Schuhmann K, Herzog R, Schwudke D, Metelmann-Strupat W, Bornstein SR, Shevchenko A (2011) Bottom-up shotgun lipidomics by higher energy collisional dissociation on LTQ Orbitrap mass spectrometers. Anal Chem 83(14):5480–5487. https://doi.org/10.1021/ac102505f

12. Leidl K, Liebisch G, Richter D, Schmitz G (2008) Mass spectrometric analysis of lipid species of human circulating blood cells. Biochim Biophys Acta 1781(10):655–664. https://doi.org/10.1016/j.bbalip.2008.07.008

13. Herzog R, Schwudke D, Schuhmann K, Sampaio J, Bornstein S, Schroeder M, Shevchenko A (2011) A novel informatics concept for high-throughput shotgun lipidomics based on the molecular fragmentation query language. Genome Biol 12(1):R8

14. Herzog R, Schwudke D, Shevchenko A (2013) LipidXplorer: software for quantitative shotgun lipidomics compatible with multiple mass spectrometry platforms. Curr Protoc Bioinformatics 43:14 12 11–30. doi:https://doi.org/10.1002/0471250953.bi1412s43

15. Liebisch G, Vizcaino JA, Kofeler H, Trotzmuller M, Griffiths WJ, Schmitz G, Spener F, Wakelam MJ (2013) Shorthand notation for lipid structures derived from mass spectrometry. J Lipid Res 54(6):1523–1530. https://doi.org/10.1194/jlr.M033506

16. Matyash V, Liebisch G, Kurzchalia TV, Shevchenko A, Schwudke D (2008) Lipid extraction by methyl-tert-butyl ether for high-throughput lipidomics. J Lipid Res 49(5):1137–1146. https://doi.org/10.1194/jlr.D700041-JLR200

17. Chambers MC, Maclean B, Burke R, Amodei D, Ruderman DL, Neumann S, Gatto L, Fischer B, Pratt B, Egertson J, Hoff K, Kessner D, Tasman N, Shulman N, Frewen B, Baker TA, Brusniak M-Y, Paulse C, Creasy D, Flashner L, Kani K, Moulding C, Seymour SL, Nuwaysir LM, Lefebvre B, Kuhlmann F, Roark J, Rainer P, Detlev S, Hemenway T, Huhmer A, Langridge J, Connolly B, Chadick T, Holly K, Eckels J, Deutsch EW, Moritz RL, Katz JE, Agus DB, MacCoss M, Tabb DL, Mallick P (2012) A cross-platform toolkit for mass spectrometry and proteomics. Nat Biotechnol 30(10):918–920. https://doi.org/10.1038/nbt.2377. http://www.nature.com/nbt/journal/v30/n10/abs/nbt.2377.html#supplementary-information

18. Schuhmann K, Almeida R, Baumert M, Herzog R, Bornstein SR, Shevchenko A (2012) Shotgun lipidomics on a LTQ Orbitrap mass spectrometer by successive switching between acquisition polarity modes. J Mass Spectrom 47(1):96–104. https://doi.org/10.1002/jms.2031

19. Han X, Yang K, Yang J, Cheng H, Gross RW (2006) Shotgun lipidomics of cardiolipin molecular species in lipid extracts of biological samples. J Lipid Res 47(4):864–879. https://doi.org/10.1194/jlr.D500044-JLR200

Chapter 13

Establishing and Performing Targeted Multi-residue Analysis for Lipid Mediators and Fatty Acids in Small Clinical Plasma Samples

Theresa L. Pedersen and John W. Newman

Abstract

LC-MS/MS- and GC-MS-based targeted metabolomics is typically conducted by analyzing and quantifying a cascade of metabolites with methods specifically developed for the metabolite class. Here we describe an approach for the development of multi-residue analytical profiles, calibration standards, and internal standard solutions in support of a fast, simple, and low-cost plasma sample preparation that captures and quantitates a range of metabolite cascades.

Key words Oxylipins, Endocannabinoids, Bile acids, Nonesterified fatty acids, Targeted metabolomics, LC-MS/MS, Metabolic profiling

1 Introduction

Targeted metabolomics provides sensitive, quantitative data sets that can be used to investigate the metabolic consequences of disease, drug treatment effects, and outcomes of interventions, as well as establishing clinical reference ranges and reference materials for determining accuracy and method validation [1, 2]. The manual optimization of analyte ionization and fragmentation parameters for large multi-residue methods is an overwhelming prospect, especially when transferring a method to a new hardware platform. Over the past 10 years, we have developed an approach that has allowed us to successfully transfer complex multi-residue analyses to laboratories around the world [3–6]. Here we describe a semiautomated alternative chromatographic "gradient optimization" approach to transfer large profiles between LC-MS/MS platforms, where the acquisition method is used to optimize all analytes simultaneously

Electronic supplementary material: The online version of this chapter (https://doi.org/10.1007/978-1-4939-7592-1_13) contains supplementary material, which is available to authorized users.

Martin Giera (ed.), *Clinical Metabolomics: Methods and Protocols*, Methods in Molecular Biology, vol. 1730, https://doi.org/10.1007/978-1-4939-7592-1_13, © Springer Science+Business Media, LLC 2018

and provide a record of the optimization, in a timely manner. This includes establishing global MS parameters, determining analyte retention times, and making a series of acquisition methods to simultaneously optimize analyte-specific parameters under the LC-MS/MS acquisition method conditions.

Additionally, increased dynamic linear ranges of 5–6 orders of magnitude allow for the simultaneous observation and quantification of many metabolites in one analysis, and we have taken advantage of this to fuse the analysis of multiple lipid classes [7, 8]. Here we share an approach for quantifying oxygenated lipids, endocannabinoids and like compounds, bile acids, and free fatty acids, captured in two LC-MS/MS and one GC-MS acquisition methods from a single small plasma extract. Included are detailed instructions for the systematic development of the multi-residue calibration standards for these assays. This includes a worksheet for en masse calculations to make up analyte mixtures and calibration standard solutions, which can be expanded to include additional metabolic cascades of interest.

Classical sample preparation and cleanup methods for multi-residue quantitative analysis are labor intensive and expensive. Increased sensitivity in modern MS/MS allows for substantial reductions in sample volume, eliminating the need for extensive cleanup procedures, reducing matrix effects, and resulting in more robust methods. We present a small-volume protein precipitation with filtration as the plasma sample preparation procedure prior to LC-MS/MS and GC-MS analysis. Included is a summary instruction for quality control procedures to assess method robustness.

2   Materials

The Materials section includes procedures needed to perform the described extraction and quantitative analysis. All LC-MS assay solvents are Fisher LC-MS grade, 0.2 μm filtered. GC-MS assay solvents are Fisher Optima grade or equivalent. All liquid chromatography mobile phases are maintained in glass Pyrex bottles, and the phases and solutions containing organic solvents are stored with Teflon-lined caps. Aqueous phase solutions are stored at 4 °C until use and not longer than 2 weeks (see **Note 1**). Prior to use, prerinse all volumetric flasks, vials, tubes, plates, and glassware with methanol (MeOH), followed by hexane for GC glassware, and allow to dry in a glass beaker loosely covered with foil, in a dust-free chemical exhaust hood. Use analytical volumetric glass barrel syringes and flasks to prepare all solutions containing analytical standards. Glass barrel syringes should be cleaned in MeOH and dried by house vacuum (use a funnel on top of a sidearm flask) between measurements. Standard stock solutions, optimization solutions, and working solutions are stored in amber vials topped with nitrogen gas, with Teflon-lined caps. Use thin ½ in. Teflon tape to seal the outside of caps to prevent solvent evaporation, and mark the meniscus to assess volume stability during storage. It is recommended that all analyte mixtures, calibration standards, surrogate spike solutions, and internal standard sample reconstitution solutions are ampouled under nitrogen gas and stored at −20 °C. Analyte stock solutions are stored at −80 °C. Use a calibrated analytical balance with accuracy to 0.1 mg to weigh neat materials.

2.1 Materials for Calibration Standard Preparation

1. *Volumetric flasks*: A minimum of two 1.0 mL, nine 5 mL, two 10 mL, two 25 mL, and one 100 mL volumetric flasks are needed.

2. *Glass syringes*: Each of the following volumes 10, 50, 100, 250, and 500 μL of gastight syringes with Teflon-tipped plungers and blunt-tipped needles is needed.

3. *Analytes*: Purchase analytes for oxylipin (Table 1), endocannabinoid (Table 2), bile acid (Table 3), and fatty acid (Tables 4 and 5) assays. Be sure to purchase fatty acid methyl esters for GC-MS calibration standard preparation (*see* **Note 2**).

4. *9′ Pasteur pipettes* and bulbs are used for non-volumetric solution and extract transfer.

5. *Calibrated pipettors* are used for sample transfer.

2.2 Liquid Chromatography-Tandem Mass Spectrometry

1. *LC*: Use an ultra-performance liquid chromatograph capable of a running pressure of >8000 psi.

2. *MS/MS*: An atmospheric pressure ionization triple-quadrupole mass spectrometer with an electrospray probe with positive-negative mode switching capabilities and fast scanning is preferred. Otherwise, the profiles can be developed as independent positive and negative mode methods, without ESI switching.

3. *Inserts, caps, and vials*: Waters Corp autosampler-compatible 2 mL amber vials, slit-top caps, and 150 μL glass polyspring autosampler inserts are used for analysis.

2.3 LC-MS/MS of Oxylipin/ Endocannabinoid Profile

1. *Liquid chromatography column*: 150 mm × 2.1 mm i.d., 1.7 μm Acquity C18 BEH column, and a 0.2 μm stainless steel frit as a guard column (Waters Corp, Inc. Milford, MA).

2. *Solvent A (0.1% acetic acid)*: 1 L of water, 1.0 mL of glacial acetic acid. Measure 1 L of 0.2 μm filtered LC-MS grade water in a 1 L graduated cylinder, and add 1 mL of LC-MS grade glacial acetic using a 1 mL pipette. Transfer to a 1 L LC solvent bottle.

3. *Solvent B (10:90 isopropanol/acetonitrile)*: 900 mL of acetonitrile (ACN), 100 mL of LC-MS grade isopropanol (IPA).

Table 1
Oxylipin assay chemical identifiers and quality

#	Abbreviation	InChIKey	Class	Quality
1	CUDA	HPTJABJPZMULFH-UHFFFAOYSA-N	LC ISTD	+++
2	PUHA	VUPFPVOJLPTJQS-UHFFFAOYSA-N	LC ISTD	+++
3	6-Keto PGF1a-d4		OXY SSTD	+++
4	TXB2-d4		OXY SSTD	+++
5	PGF2a-d4		OXY SSTD	+++
6	PGD2-d4		OXY SSTD	+++
7	LTB4-d4		OXY SSTD	+++
8	14,15-DiHETrE-d11		OXY SSTD	+++
9	9-HODE-d4		OXY SSTD	+++
10	20-HETE-d6		OXY SSTD	+++
11	12-HETE-d8		OXY SSTD	+++
12	5-HETE-d8		OXY SSTD	+++
13	12(13)-EpOME-d4		OXY SSTD	+++
14	10-Nitrooleate-d17		OXY SSTD	++
15	AA-d8		OXY SSTD	+++
16	6-keto-PGF1a	KFGOFTHODYBSGM-ZUNNJUQCSA-N	PG	+++
17	PGF3a	SAKGBZWJAIABSY-SAMSIYEGSA-N	PG	+++
18	PGE3	CBOMORHDRONZRN-QLOYDKTKSA-N	PG	+++
19	TXB2	XNRNNGPBEPRNAR-JQBLCGNGSA-N	TX	+++
20	9,12,13-TriHOME	MDIUMSLCYIJBQC-MVFSOIOZSA-N	Triol	+++
21	PGF2a	PXGPLTODNUVGFL-UAAPODJFSA-N	PG	+++
22	PGE2	XEYBRNLFEZDVAW-ARSRFYASSA-N	PG	+++
23	PGE1	GMVPRGQOIOIIMI-DWKJAMRDSA-N	PG	+++
24	PGD2	BHMBVRSPMRCCGG-OUTUXVNYSA-N	PG	+++
25	15-keto-PGE2	YRTJDWROBKPZNV-KMXMBPPJSA-N	PG	+++
26	RvD1	OIWTWACQMDFHJG-NJIQAZPPSA-N	Triol	+++
27	Lipoxin A4	IXAQOQZEOGMIQS-SSQFXEBMSA-N	Diol	+++
28	LTB5	BISQPGCQOHLHQK-HDNPQISLSA-N	LT	+++
29	15,16-DiHODE	LKLLJYJTYPVCID-OHPMOLHNSA-N	Diol	+++
30	8,15-DiHETE	NNPWRKSGORGTIM-RCDCWWQHSA-N	Diol	+++
31	12,13-DiHODE	RGRKFKRAFZJQMS-OOHFSOINSA-N	Diol	+++

(continued)

Table 1
(continued)

#	Abbreviation	InChIKey	Class	Quality
32	9,10-DiHODE	QRHSEDZBZMZPOA-ZJSQCTGTSA-N	Diol	+++
33	17,18-DiHETE	XYDVGNAQQFWZEF-JPURVOHMSA-N	Diol	+++
34	5,15-DiHETE	UXGXCGPWGSUMNI-BVHTXILBSA-N	Diol	+++
35	6 trans LTB4	VNYSSYRCGWBHLG-UKNWISKWSA-N	Diol	+++
36	14,15-DiHETE	BLWCDFIELVFRJY-QXBXTPPVSA-N	Diol	+++
37	LTB4	KFGOFTHODYBSGM-ZUNNJUQCSA-N	Diol	+++
38	12,13-DiHOME	CQSLTKIXAJTQGA-FLIBITNWSA-N	Diol	+++
39	9,10-DiHOME	XEBKSQSGNGRGDW-YFHOEESVSA-N	Diol	+++
40	19,20-DiHDoPE	FFXKPSNQCPNORO-MBYQGORISA-N	Diol	+++
41	14,15-DiHETrE	SYAWGTIVOGUZMM-ILYOTBPNSA-N	Diol	+++
42	11,12-DiHETrE	LRPPQRCHCPFBPE-KROJNAHFSA-N	Diol	+++
43	9,10-e-DiHO	VACHUYIREGFMSP-SJORKVTESA-N	Diol	+++
44	9-HOTE	YUPHIKSLGBATJK-OBKPXJAFSA-N	R-OH	+++
45	12(13)Ep-9-KODE	RCMABBHQYMBYKV-BUHFOSPRSA-N	Epox,R = O	++
46	13-HOTE	KLLGGGQNRTVBSU-JDTPQGGVSA-N	R-OH	+++
47	8,9-DiHETrE	DCJBINATHQHPKO-TYAUOURKSA-N	Diol	+++
48	15-Deoxy PGJ2	VHRUMKCAEVRUBK-GODQJPCRSA-N	PG	+++
49	15-HEPE	UDXLGBLAJBYLSZ-XBCQTNLFSA-N	R-OH	+++
50	20-HETE	NNDIXBJHNLFJJP-DTLRTWKJSA-N	R-OH	+++
51	12-HEPE	MCRJLMXYVFDXLS-QGQBRVLBSA-N	R-OH	+++
52	5,6-DiHETrE	GFNYAPAJUNPMGH-QNEBEIHSSA-N	Diol	+++
53	9-HEPE	OXOPDAZWPWFJEW-FPRWAWDYSA-N	R-OH	+++
54	13-HODE	HNICUWMFWZBIFP-IRQZEAMPSA-N	R-OH	+++
55	5-HEPE	FTAGQROYQYQRHF-FCWZHQICSA-N	R-OH	+++
56	9-HODE	NPDSHTNEKLQQIJ-SIGMCMEVSA-N	R-OH	+++
57	15(16)-EpODE	HKSDVVJONLXYKL-OHPMOLHNSA-N	Epox	+++
58	17(18)-EpETE	GPQVVJQEBXAKBJ-JPURVOHMSA-N	Epox	+++
59	15-HETE	JSFATNQSLKRBCI-VAEKSGALSA-N	R-OH	+++
60	13-KODE	JHXAZBBVQSRKJR-BSZOFBHHSA-N	R = O	++
61	9(10)-EpODE	JTEGNNHWOIJBJZ-ZJSQCTGTSA-N	Epox	+++
62	17-HDoHE	SWTYBBUBEPPYCX-VIIQGJSXSA-N	R-OH	+++

(continued)

Table 1
(continued)

#	Abbreviation	InChIKey	Class	Quality
63	12(13)-EpODE	BKKGUKSHPCTUGE-OOHFSOINSA-N	Epox	+++
64	15-KETE	YGJTUEISKATQSM-USWFWKISSA-N	R = O	++
65	14-HDoHE	ZNEBXONKCYFJAF-BGKMTWLOSA-N	R-OH	+++
66	11-HETE	GCZRCCHPLVMMJE-RSPKXIRXSA-N	R-OH	+++
67	14(15)-EpETE	RGZIXZYRGZWDMI-QXBXTPPVSA-N	Epox	+++
68	9-KODE	LUZSWWYKKLTDHU-ZJHFMPGASA-N	R = O	++
69	11(12)-EpETE	QHOKDYBJJBDJGY-BVILWSOJSA-N	Epox	+++
70	12-HETE	ZNHVWPKMFKADKW-FYMOKONMSA-N	R-OH	+++
71	8-HETE	NLUNAYAEIJYXRB-VYOQERLCSA-N	R-OH	+++
72	9-HETE	KATOYYZUTNAWSA-DLJQHUEDSA-N	R-OH	+++
73	19(20)-EpDoPE	OSXOPUBJJDUAOJ-MBYQGORISA-N	Epox	+++
74	5-HETE	KGIJOOYOSFUGPC-JGKLHWIESA-N	R-OH	+++
75	12(13)-EpOME	CCPPLLJZDQAOHD-FLIBITNWSA-N	Epox	+++
76	14(15)-EpETrE	WLMZMBKVRPUYIG-LTCHCNGXSA-N	Epox	+++
77	4-HDoHE	IFRKCNPQVIJFAQ-JGDWKEERSA-N	R-OH	+++
78	16(17)-EpDoPE	BCTXZWCPBLWCRV-ZYADFMMDSA-N	Epox	+++
79	9(10)-EpOME	XEBKSQSGNGRGDW-YFHOEESVSA-N	Epox	+++
80	5-KETE	MEASLHGILYBXFO-XTDASVJISA-N	R = O	++
81	11(12)-EpETrE	DXOYQVHGIODESM-IQCOFVSKSA-N	Epox	+++
82	8(9)-EpETrE	DBWQSCSXHFNTMO-TYAUOURKSA-N	Epox	+++
83	10-Nitrolinoleate	LELVHAQTWXTCLY-XYWKCAQWSA-N	Nitrolipid	++
84	9(10)-EpO	IMYZYCNQZDBZBQ-UHFFFAOYSA-N	Epox	+++
85	10-Nitrooleate	WRADPCFZZWXOTI-UHFFFAOYSA-N	Nitrolipid	++
86	9-Nitrooleate	CQOAKBVRRVHWKV-UHFFFAOYSA-M	Nitrolipid	++

Quality: poor (−); fair (+); good (++); excellent (+++)

Measure 900 mL of 0.2 μm filtered LC-MS grade ACN in a 1 L graduated cylinder, and bring to a 1 L final volume with 100 mL of IPA. Transfer to a 1 L LC solvent bottle with a Teflon-lined cap.

2.4 LC-MS/MS of Bile Acid Profile

1. *Liquid chromatography column*: 100 mm × 2.1 mm i.d., 1.7 μm Acquity C18 BEH column, and a 0.2 μm stainless steel frit as a guard column (Waters Corp, Inc., Milford, MA).

Table 2
Endocannabinoid target identifiers and assay quality

#	Abbreviation	InChIKey	Class	Quality
1	CUDA	HPTJABJPZMULFH-UHFFFAOYSA-N	LC ISTD	+++
2	PUHA	VUPFPVOJLPTJQS-UHFFFAOYSA-N	LC ISTD	+++
3	NA-Gly-d8		Endo SSTD	+++
4	2-AG-d5		Endo SSTD	++
5	PGF2a EA-d4		Endo SSTD	+++
6	P-EA-d4		Endo SSTD	+++
7	A-EA-d8		Endo SSTD	+++
8	PGF2a-EA	XCVCLIRZZCGEMU-FPLRWIMGSA-N	Acyl-amides	+++
9	PGE2-EA	GKKWUSPPIQURFM-IGDGGSTLSA-N	Acyl-amides	+++
10	PGD2-EA	KEYDJKSQFDUAGF-YIRKRNQHSA-N	Acyl-amides	+++
11	15-HETE-EA	XZQKRCUYLKDPEK-BPVVGZHASA-N	Acyl-amides	+++
12	11(12)-EET	TYRRSRADDAROSO-KROJNAHFSA-N	Acyl-amides	+++
13	aL-EA	HBJXRRXWHSHZPU-PDBXOOCHSA-N	Acyl-amides	+++
14	DH-EA	CXWASNUDKUTFPQ-KUBAVDMBSA-N	Acyl-amides	+++
15	A-EA	LGEQQWMQCRIYKG-DOFZRALJSA-N	Acyl-amides	+++
16	L-EA	KQXDGUVSAAQARU-HZJYTTRNSA-N	Acyl-amides	+++
17	NA-Gly	YLEARPUNMCCKMP-DOFZRALJSA-N	Acyl-amides	+++
18	DGL-EA	ULQWKETUACYZLI-QNEBEIHSSA-N	Acyl-amides	+++
19	P-EA	HXYVTAGFYLMHSO-UHFFFAOYSA-N	Acyl-amides	+++
20	NO-Gly	HPFXACZRFJDURI-KTKRTIGZSA-N	Acyl-amides	+++
21	O-EA	BOWVQLFMWHZBEF-KTKRTIGZSA-N	Acyl-amides	+++
22	D-EA	FMVHVRYFQIXOAF-DOFZRALJSA-N	Acyl-amides	+++
23	S-EA	OTGQIQQTPXJQRG-UHFFFAOYSA-N	Acyl-amides	+++
24	PGF2a-1G	NWKPOVHSHWJQNI-OMVDPNNKSA-N	Glyceryl ester	++
25	PGE2-1G	RJXVYMMSQBYEHN-SDTVLRMPSA-N	Glyceryl ester	++
26	2-AG	RCRCTBLIHCHWDZ-DOFZRALJSA-N	Glyceryl ester	++
27	2-LG	IEPGNWMPIFDNSD-HZJYTTRNSA-N	Glyceryl ester	++
28	1-AG	DCPCOKIYJYGMDN-HUDVFFLJSA-N	Glyceryl ester	++
29	1-LG	WECGLUPZRHILCT-GSNKCQISSA-N	Glyceryl ester	++
31	2-OG	UPWGQKDVAURUGE-KTKRTIGZSA-N	Glyceryl ester	++
32	1-OG	RZRNAYUHWVFMIP-QJRAZLAKSA-N	Glyceryl ester	++

Quality: poor (−); fair (+); good (++); excellent (+++)

Table 3
Bile acid target identifiers and assay quality

#	Abbreviation	InChIKey	Class	Quality
1	CUDA	HPTJABJPZMULFH-UHFFFAOYSA-N	LC ISTD	+++
2	PUHA	VUPFPVOJLPTJQS-UHFFFAOYSA-N	LC ISTD	+++
3	CA-d6		BA SSTD	+++
4	CDCA-d4		BA SSTD	+++
5	DCA-d4		BA SSTD	+++
6	GCA-d4		BA SSTD	+++
7	GCDCA-d4		BA SSTD	+++
8	LCA-d4		BA SSTD	+++
9	TCDCA-d4		BA SSTD	+++
10	TDHCA	UBDJSBRKNHQFPD-PYGYYAGESA-N	2° BA Conj	+++
11	T-ω-MCA		m-BA Conj	+++
12	T-α-MCA	XSOLDPYUICCHJX-QQXJNSDFSA-N	m-BA Conj	+++
13	T-β-MCA	XSOLDPYUICCHJX-UZUDEGBHSA-N	m-BA Conj	+++
14	TUDCA	HMXPOCDLAFAFNT-BHYUGXBJSA-N	2° BA Conj	+++
15	TCA	WBWWGRHZICKQGZ-HZAMXZRMSA-N	1° BA Conj	+++
16	ω-MCA	DKPMWHFRUGMUKF-NTPBNISXSA-N	m-BA	+++
17	GUDCA	GHCZAUBVMUEKKP-XROMFQGDSA-N	2° BA Conj	+++
18	α-MCA	DKPMWHFRUGMUKF-JDDNAIEOSA-N	m-BA	+++
19	GHDCA	SPOIYSFQOFYOFZ-BRDORRHWSA-N	2° BA Conj	+++
20	GCA	RFDAIACWWDREDC-FRVQLJSFSA-N	1° BA Conj	+++
21	β-MCA	DKPMWHFRUGMUKF-CRKPLTDNSA-N	m-BA	+++
22	TCDCA	BHTRKEVKTKCXOH-BJLOMENOSA-N	1° BA Conj	+++
23	TDCA	AWDRATDZQPNJFN-VAYUFCLWSA-N	2° BA Conj	+++
24	CA	BHQCQFFYRZLCQQ-OELDTZBJSA-N	1° BA	+++
25	UDCA	RUDATBOHQWOJDD-UZVSRGJWSA-N	2° BA	+++
26	GCDCA	GHCZAUBVMUEKKP-GYPHWSFCSA-N	1° BA Conj	+++
27	GDCA	WVULKSPCQVQLCU-BUXLTGKBSA-N	2° BA Conj	+++
28	TLCA	QBYUNVOYXHFVKC-GBURMNQMSA-N	2° BA Conj	+++
29	CDCA	RUDATBOHQWOJDD-BSWAIDMHSA-N	1° BA	+++
30	DCA	KXGVEGMKQFWNSR-LLQZFEROSA-N	2° BA	+++
31	GLCA	XBSQTYHEGZTYJE-OETIFKLTSA-N	2° BA Conj	+++
32	LCA	SMEROWZSTRWXGI-HVATVPOCSA-N	2° BA	++

Quality: poor (−); fair (+); good (++); excellent (+++)

Table 4
Fatty acid target identifiers and assay quality

#	Abbreviation	InChIKey (fatty acids)	Class	Quality
1	C23:0-me	VORKGRIRMPBCCZ-UHFFFAOYSA-N	GC ISTD	+++
2	C6:0-me	FUZZWVVXGSFPDMH-UHFFFAOYSA-N	SFA	–
3	C8:0-me	WWZKQHOCKIZLMA-UHFFFAOYSA-N	SFA	–
4	C10:0-me	GHVNFZFCNZKVNT-UHFFFAOYSA-N	SFA	+
5	C12:0-me	POULHZVOKOAJMA-UHFFFAOYSA-N	SFA	++
6	C14:0-me	TUNFSRHWOTWDNC-UHFFFAOYSA-N	SFA	+++
7	C14:1n5-me	YWWVWXASSLXJHU-WAYWQWQTSA-N	MUFA	+
8	C15:0-me	WQEPLUUGTLDZJY-UHFFFAOYSA-N	SFA	+++
9	C15:1n5-me		ISTD	+++
10	C16:0-d31-me	MQOCIYICOGDBSG-HXKBIXQXSA-M	SSTD	+++
11	C16:0-me	IPCSVZSSVZVIGE-UHFFFAOYSA-N	SFA	+++
12	C16:1n7t-me	SECPZKHBENQXJG-BQYQJAHWSA-N	Trans-fat	+++
13	C16:1n7-me	SECPZKHBENQXJG-FPLPWBNLSA-N	MUFA	+++
14	C17:0-me	KEMQGTRYUADPNZ-UHFFFAOYSA-N	SFA	+++
15	C17:1n7-me		MUFA	+++
16	C18:0-d35-me	QIQXTHQIDYTFRH-KNAXIHRDSA-N	SSTD	+++
17	C18:0-me	QIQXTHQIDYTFRH-UHFFFAOYSA-N	SFA	+++
18	C18:1n9-me	ZQPPMHVWECSIRJ-KTKRTIGZSA-N	MUFA	+++
19	C18:1n7-me	UWHZIFQPPBDJPM-FPLPWBNLSA-N	MUFA	+++
20	C18:2n6-me	OYHQOLUKZRVURQ-HZJYTTRNSA-N	PUFA-n6	+++
21	C18:3n6-me	VZCCETWTMQHEPK-QNEBEIHSSA-N	PUFA-n6	+++
22	C18:3n3-me	DTOSIQBPPRVQHS-PDBXOOCHSA-N	PUFA-n3	+++
23	C18:4n3-me	JIWBIWFOSCKQMA-LTKCOYKYSA-N	PUFA-n3	+++
24	C19:0-me	ISYWECDDZWTKFF-UHFFFAOYSA-N	SFA	+++
25	C19:1n9-me	BBOWBNGUEWHNQZ-KTKRTIGZSA-N	MUFA	+++
26	9c,11 t–CLA-me	JBYXPOFIGCOSSB-GOJKSUSPSA-N	Trans-fat	+++
27	10 t,12c–CLA-me	GKJZMAHZJGSBKD-BLHCBFLLSA-N	Trans-fat	+++
28	C20:0-me	VKOBVWXKNCXXDE-UHFFFAOYSA-N	SFA	+++
29	C20:1n9-me	BITHHVVYSMSWAG-KTKRTIGZSA-N	MUFA	+++
30	C20:2n6-me	XSXIVVZCUAHUJO-HZJYTTRNSA-N	PUFA-n6	+++
31	C20:3n6-me	HOBAELRKJCKHQD-QNEBEIHSSA-N	PUFA-n6	+++

(continued)

Table 4
(continued)

#	Abbreviation	InChIKey (fatty acids)	Class	Quality
32	C20:4n6-me	YZXBAPSDXZZRGB-DOFZRALJSA-N	PUFA-n6	+++
33	C20:3n3-me	AHANXAKGNAKFSK-PDBXOOCHSA-N	PUFA-n3	+++
34	C21:0-me	CKDDRHZIAZRDBW-UHFFFAOYSA-N	SFA	+++
35	C20:4n3-me	YNVYKJQCWARJFA-ZKWNWVNESA-N	PUFA-n3	+++
36	C20:5n3-me	JAZBEHYOTPTENJ-JLNKQSITSA-N	PUFA-n3	+++
37	C22:0-me	UKMSUNONTOPOIO-UHFFFAOYSA-N	SFA	+++
38	C22:1n9-me	DPUOLQHDNGRHBS-KTKRTIGZSA-M	SSTD	+++
39	C22:2n6-me	HVGRZDASOHMCSK-HZJYTTRNSA-N	PUFA-n6	+++
40	C22:4n6-me	TWSWSIQAPQLDBP-DOFZRALJSA-N	PUFA-n6	+++
41	C22:5n6-me	AVKOENOBFIYBSA-WMPRHZDHSA-N	PUFA-n6	+++
42	C22:3n3-me	WBBQTNCISCKUMU-PDBXOOCHSA-N	SSTD	+++
43	C22:5n3-me	YUFFSWGQGVEMMI-RCHUDCCISA-N	PUFA-n3	+++
44	C22:6n3-me	MBMBGCFOFBJSGT-KUBAVDMBSA-N	PUFA-n3	+++
45	C24:0-me	QZZGJDVWLFXDLK-UHFFFAOYSA-N	SFA	+++
46	C24:1n9-me	GWHCXVQVJPWHRF-KTKRTIGZSA-M	MUFA	+++

Quality: poor (−); fair (+); good (++); excellent (+++)

Table 5
Internal standards, lipid class surrogates, and FAME derivatization controls

#	Abbreviation	InChIKey	Class	Use
1	TAG-tri(16:0)-d93	PVNIQBQSYATKKL-JGBASCRPSA-N	TG	TG SSTD
2	PC-di(18:0)-d70)	NRJAVPSFFCBXDT-MNVUKWGGSA-N	PL	PL SSTD
3	CE-C22:1n9	SQHUGNAFKZZXOT-QXAJUEOOSA-N	CE	CE SSTD
4	C22:3n3	WBBQTNCISCKUMU-PDBXOOCHSA-N	NEFA	FAME SSTD
5	C15:1n5		NEFA	FAME ISTD

2. *Solvent A (0.1% formic acid)*: 1 L of water, 1.0 mL of formic acid. Measure 1 L of water in a 1 L graduated cylinder, and add 1 mL of LC-MS grade formic acid using a 1 mL pipette. Transfer to a 1 L LC solvent bottle.

3. *Solvent B (0.1% formic acid in ACN)*: 1 L of ACN, 1 mL formic acid. Measure 1 L of ACN in a 1 L graduated cylinder, and add 1 mL of LC-MS grade formic acid using a 1 mL pipette. Transfer to a 1 L LC solvent bottle with a Teflon-lined cap.

2.5 Preparing LC-MS/MS Analyte Stocks and Optimization Solutions

Prepare the individual analyte optimization solutions and the "all analyte" optimization solution described below at the same time while the stock solutions are being handled. Clean one 2 mL amber vial for each analyte optimization solution and one vial for the all analyte optimization solution being prepared for each assay.

1. *Vials*: Per assay, ~200 2 mL and ~50 5 mL amber borosilicate vials with solid Teflon-lined caps for solution storage.

2. *LC-MS analyte stock solutions*: 0.10–10.0 mg/mL of analyte in MeOH. One mg/mL is ideal for the volumetric transfers used to make the LC-MS/MS calibration solutions. Stock solutions of each analyte are prepared or purchased in a concentration range from 0.10 to 10.0 mg/mL. For neat compounds, weigh materials, record weight, transfer to an appropriate volumetric flask, and solubilize with nitrogen or argon purged MeOH. If necessary, dilute concentrated solutions by transferring the appropriate volume with an analytical syringe to an appropriate volumetric flask and bring to volume in purged MeOH. Vortex each new solution 2 s to mix and transfer to a glass vial with a Pasteur pipette. Top the analyte stock solutions with nitrogen gas, cap, seal, mark the meniscus, and label vial. Store analyte stocks at −80 °C.

3. *LC-MS analyte optimization solutions*: 0.10–10.0 μg/mL of analyte in MeOH. Using one vial per analyte, transfer 1.0 μL of each analyte stock solution into a 2 mL amber glass vial containing 1 mL purged 1:1 MeOH/ACN (v/v). Label the vial with the analyte name, seal, and store at −20 °C.

4. *LC-MS all analyte optimization solution*: 0.10–10.0 μg/mL of all analytes in MeOH. Transfer 1 μL of each analyte stock solution into a 2 mL amber vial containing 1 mL purged 1:1 MeOH/ACN (v/v). Label as "All Analyte Opt" and store at −20 °C.

2.6 Calibration Standard Preparation Worksheet

Generating a spreadsheet containing target weights with all volumetric deliveries (and volumetric constraints) provides an excellent format for both projecting approximate target concentrations, recording actual weights and volumes, and developing the volumetrics for making a complex suite of analytes (*see* **Note 3**). Supplementary File 1 provides an organized approach for making the complex mixtures needed for this protocol. Simple mathematical functions are embedded in the sheet, where the adjustment of delivery and final volumes will change the solution and standard concentrations, en masse. This worksheet can be expanded with the positive mode endocannabinoids (Table 6) and/or used to build other multi-residue profiles, such as the bile acids (Table 7) and fatty acids (Table 8). If your mass spectrometer software quantifies internal standards on a curve, you will need to adjust your internal

Table 6
Representative oxy/endo fusion calibrations for the 6500 QTRAP (nM)

#	Mixture	Compound	Mix target (μM)	Range (nM)
1	IS	PHAU	50	100
2	IS	CUDA	50	100
3	Oxy SSTD	14,15-DiHETrE-d11	13	25
4	Oxy SSTD	6-keto PGF1a-d4	12	25
5	Oxy SSTD	TXB2-d4	12	25
6	Oxy SSTD	PGD2-d4	13	25
7	Oxy SSTD	LTB4-d4	13	25
8	Oxy SSTD	12-HETE-d8	12	25
9	Oxy SSTD	9-HODE-d4	12	25
10	Oxy SSTD	5-HETE-d8	13	25
11	Oxy SSTD	20-HETE-d6	12	25
12	Oxy SSTD	AA-d8	13	25
13	Oxy SSTD	10-Nitrooleate-d17	13	25
14	Oxy SSTD	PGF2a-d4	13	25
15	Oxy SSTD	12(13)-EpOME-d4	12	25
16	Endo SSTD	NA-Gly-d8	6.2	22
17	Endo SSTD	2-AG-d5	6.3	22
18	Endo SSTD	PGF2a EA-d4	6.2	22
19	Endo SSTD	P-EA-d4	6.3	22
20	Endo SSTD	A-EA-d8	6.2	22
21	Oxy mix 1	9,12,13-TriHOME	15	[0.0116–1210]
22	Oxy mix 1	12,13-DiHOME	16	[0.0122–1270]
23	Oxy mix 1	9,10-DiHOME	16	[0.0122–1270]
24	Oxy mix 1	EKODE	16	[0.0123–1280]
25	Oxy mix 1	13-HODE	16	[0.0124–1300]
26	Oxy mix 1	9-HODE	16	[0.0124–1300]
27	Oxy mix 1	13 KODE	17	[0.013–1360]
28	Oxy mix 1	9-KODE	17	[0.013–1360]
29	Oxy mix 1	12(13)-EpOME	17	[0.013–1350]
30	Oxy mix 1	9(10)-EpOME	17	[0.013–1350]
31	Oxy mix 1	9,10-DiHODE	16	[0.0123–1280]

(continued)

Table 6
(continued)

#	Mixture	Compound	Mix target (μM)	Range (nM)
32	Oxy mix 1	12,13-DiHODE	15	[0.0118–1230]
33	Oxy mix 1	15,16-DiHODE	16	[0.0123–1280]
34	Oxy mix 1	13-HOTE	16	[0.0125–1300]
35	Oxy mix 1	9-HOTE	16	[0.0125–1300]
36	Oxy mix 1	9(10)-EpODE	16	[0.0123–1280]
37	Oxy mix 1	12(13)-EpODE	16	[0.0125–1300]
38	Oxy mix 1	15(16)-EpODE	16	[0.0123–1280]
39	Oxy mix 1	6-keto-PGF1a	15	[0.0116–1210]
40	Oxy mix 1	TXB2	15	[0.0116–1210]
41	Oxy mix 1	PGF2a	16	[0.0121–1260]
42	Oxy mix 1	PGE1	16	[0.0121–1260]
43	Oxy mix 1	PGE2	16	[0.0122–1270]
44	Oxy mix 1	PGE3	16	[0.0123–1280]
45	Oxy mix 1	PGD2	16	[0.0122–1270]
46	Oxy mix 1	PGJ2	15	[0.0115–1200]
47	Oxy mix 1	15-deoxy PGJ2	16	[0.0121–1260]
48	Oxy mix 1	LTB5	6.0	[0.00482–502]
49	Oxy mix 1	6-trans-LTB4	15	[0.0114–1190]
50	Oxy mix 1	LTB4	15	[0.0114–1190]
51	Oxy mix 1	Lipoxin a4	14	[0.0109–1130]
52	Oxy mix 1	LTE4	11	[0.00874–910]
53	Oxy mix 2	14,15-DiHETrE	15	[0.0113–1180]
54	Oxy mix 2	11,12-DiHETrE	15	[0.0113–1180]
55	Oxy mix 2	8,9-DiHETrE	15	[0.0113–1180]
56	Oxy mix 2	5,6-DiHETrE	15	[0.0113–1180]
57	Oxy mix 2	20-HETE	16	[0.012–1250]
58	Oxy mix 2	15-HETE	16	[0.012–1250]
59	Oxy mix 2	12-HETE	16	[0.012–1250]
60	Oxy mix 2	11-HETE	16	[0.012–1250]
61	Oxy mix 2	9-HETE	16	[0.0121–1260]
62	Oxy mix 2	8-HETE	16	[0.012–1250]

(continued)

Table 6
(continued)

#	Mixture	Compound	Mix target (μM)	Range (nM)
63	Oxy mix 2	5-HETE	16	[0.012–1250]
64	Oxy mix 2	15-KETE	16	[0.0121–1260]
65	Oxy mix 2	5-KETE	16	[0.0121–1260]
66	Oxy mix 2	14(15)-EpETrE	16	[0.012–1250]
67	Oxy mix 2	11(12)-EpETrE	16	[0.012–1250]
68	Oxy mix 2	8(9)-EpETrE	16	[0.012–1250]
69	Oxy mix 2	5(6)-EpETrE	16	[0.012–1250]
70	Oxy mix 3	14,15-DiHETE	15	[0.0114–1190]
71	Oxy mix 3	17,18-DiHETE	15	[0.0114–1190]
72	Oxy mix 3	5,15-DiHETE	15	[0.0114–1190]
69	Oxy mix 3	8,15-DiHETE	15	[0.0114–1190]
73	Oxy mix 3	15(S)-HEPE	16	[0.0121–1260]
74	Oxy mix 3	12(S)-HEPE	16	[0.0121–1260]
75	Oxy mix 3	5(S)-HEPE	16	[0.0121–1260]
76	Oxy mix 3	17(R)-HDoHE	15	[0.0111–1160]
77	Oxy mix 3	14(15)-EpETE	16	[0.0121–1260]
78	Oxy mix 3	17(18)-EpETE	16	[0.0121–1260]
79	Oxy mix 3	16(17)-EpDoPE	15	[0.0111–1160]
80	Oxy mix 3	19(20)-EpDoPE	15	[0.0111–1160]
81	Oxy mix 3	19,20-DiHDoPE	14	[0.0106–1100]
82	Oxy mix 3	Resolvin D1	13	[0.0102–1060]
83	Oxy mix 3	11(12)-EpETE	16	[0.0121–1260]
84	Oxy mix 3	9-HEPE	16	[0.0121–1260]
85	Oxy mix 3	4-HDoHE	15	[0.0111–1160]
86	Oxy mix 3	14-HDoHE	15	[0.0111–1160]
87	Oxy mix 3	PGF3a	14	[0.0109–1130]
88	Oxy mix 3	10-Nitrolinoleate	15	[0.0118–1230]
89	Oxy mix 3	10-Nitrooleate	15	[0.0117–1220]
90	Oxy mix 3	9-Nitrooleate	15	[0.0117–1220]
91	Oxy mix 3	9,10-EpO	13	[0.0103–1070]
92	Oxy mix 3	9,10-e-DiHO	15	[0.0116–1200]

(continued)

Table 6
(continued)

#	Mixture	Compound	Mix target (μM)	Range (nM)
98	Oxy mix 3	9,10-t-DiHHex	17	[0.0128–1330]
99	Oxy mix 3	15-keto PGE2	16	[0.0123–1280]
100	Endo mix	P-EA	13	[0.0128–1340]
101	Endo mix	S-EA	12	[0.0117–1220]
102	Endo mix	O-EA	15	[0.0147–1540]
103	Endo mix	L-EA	15	[0.0148–1550]
104	Endo mix	aL-EA	16	[0.0149–1560]
105	Endo mix	DGL- EA	14	[0.0137–1430]
106	Endo mix	A-EA	14	[0.0138–1440]
107	Endo mix	D-EA	13	[0.0128–1330]
108	Endo mix	DH-EA	13	[0.0129–1350]
109	Endo mix	NA-Gly	14	[0.0133–1380]
110	Endo mix	NO-Gly	12	[0.0113–1180]
111	Endo mix	1-LG	23	[0.0217–2260]
112	Endo mix	2-LG	42	[0.0406–4230]
113	Endo mix	1-AG	21	[0.0203–2110]
114	Endo mix	2-AG	21	[0.0203–2110]
115	Endo mix	1-OG	42	[0.0404–4210]
113	Endo mix	2-OG	42	[0.0404–4210]
114	Endo mix	15-HETE-EA	3.0	[0.00264–275]
115	Endo mix	PGE2-EA	10	[0.00971–1010]
116	Endo mix	PGD2-EA	13	[0.0121–1260]
117	Endo mix	PGF2a-EA	13	[0.0121–1260]
118	Endo mix	PGE2 1G	12	[0.0113–1170]
119	Endo mix	PGF2a 1G	12	[0.0112–1170]
120	Endo mix	11(12)-EpETrE-EA	3.0	[0.00264–275]

standard volumetric deliveries accordingly to create appropriate concentrations for your standards. The worksheet will guide the following steps: (1) record analyte stock information, (2) establish volumetric transfers, (3) prepare internal standard (ISTD) and surrogate (SSTD) reconstitution solution used to make calibration standards, (4) prepare calibration solutions, (5) prepare sample ISTD reconstitution solution, and (6) prepare the surrogate spike solution.

Table 7
Representative bile acid calibration ranges (nM) for the 4000 QTRAP

#	Mixture	Compound	Mix target (μM)	Range (nM)
1	ISTD	PUHA	10	100
2	ISTD	CUDA	10	100
3	BA-SSTD	CA-d6	10	100
4	BA-SSTD	CDCA-d4	10	100
5	BA-SSTD	DCA-d4	10	100
6	BA-SSTD	GCA-d4	10	100
7	BA-SSTD	GCDCA-d4	10	100
8	BA-SSTD	LCA-d4	10	100
9	BA-SSTD	TCDCA-d4	10	100
10	BA mix 1	CA	10	[0.245–5510]
11	BA mix 1	CDCA	10	[0.201–4520]
12	BA mix 1	DCA	10	[0.183–4120]
13	BA mix 1	GCA	10	[0.214–4810]
14	BA mix 1	GCDCA	10	[0.212–4770]
15	BA mix 1	GUDCA	10	[0.149–3360]
16	BA mix 1	TCA	10	[0.209–4710]
17	BA mix 2	GDCA	10	[0.176–1320]
18	BA mix 2	GHDCA	10	[0.159–1190]
19	BA mix 2	GLCA	10	[0.188–1410]
20	BA mix 2	HDCA	10	[0.168–1260]
21	BA mix 2	LCA	10	[0.18–1350]
22	BA mix 2	TCDCA	10	[0.194–1460]
23	BA mix 2	TDCA	10	[0.176–1320]
24	BA mix 2	TDHCA	10	[0.188–1410]
25	BA mix 2	TLCA	10	[0.165–1240]
26	BA mix 2	TUDCA	10	[0.229–1720]
27	BA mix 2	UDCA	10	[0.298–2240]
28	BA mix 3	α-MCA	10	[0.245–1840]
29	BA mix 3	β-MCA	10	[0.245–1840]
30	BA mix 3	ω-MCA	10	[0.196–1470]
31	BA mix 3	T-α-MCA	10	[0.186–1390]
32	BA mix 3	T-β-MCA	10	[0.186–1390]
33	BA mix 3	T-ω-MCA	10	[0.186–1390]

2.7 Gas Chromatography Mass Spectrometry

The gas chromatography mass spectrometry hardware specification described below is for the analysis of fatty acid methyl esters (FAMEs) on an Agilent 6890 with the second electronic-controlled inlet configured as a backflush unit to facilitate the removal of high-boiling components (e.g., cholesterol) and shorten run times from 46 to 25 min. A modern GC with factory-implemented backflush capabilities can replace this configuration.

1. *GC-MS*: Agilent 6890 Gas Chromatograph with dual split/splitless injection ports under electronic pressure control, coupled to an Agilent 5975N Mass Selective Detector with simultaneous selected ion monitoring/full-scan spectral acquisition capabilities; Agilent 7683B Series Injector Autosampler (Agilent Technologies, Santa Clara, CA, USA).

2. *Gas chromatography column*: 30 m × 0.25 mm, 0.25 μm DB-225 ms (Agilent Technologies, Santa Clara, CA, USA).

3. *Pre-column, backflush column, and post-column transfer line*: 5 m × 0.25 mm deactivated fused silica (Agilent Technologies, Santa Clara, CA, USA).

4. *Column connections*: Columns are connected using straight (pre-column) or T- (post-column and backflush) unions.

5. *Inserts, caps, and vials*: Agilent autosampler-compatible 2 mL vials, caps, and 150 μL glass polyspring autosampler inserts are used.

6. *Carrier gas*: High purity 99.9990% helium is used as the chromatography carrier gas.

7. *Syringe rinse solvents*: Optima grade n-hexane and isooctane (*see* **Note 4**).

2.8 Gas Chromatography Stock Mixtures

1. *FAME stocks and mixes*: Using a standard preparation worksheet as described in Subheading 2.7, prepare the fatty acid methyl ester (FAME) mixes defined in Table 8. Weigh each of the materials into the appropriate volumetric flask and dissolve in Fisher Optima grade or equivalent n-hexane to achieve the final solutions (*see* **Note 5**).

2. *FAME ISTD stock*: Target concentration ~1.5 mM C23:0 methyl ester in hexane.

3. *FAME ISTD reconstitution solution*: Target concentration 4 μM C23:0 methyl ester in hexane.

4. *100 μM FAME SSTD spike solution*: 100 μM C22:3n3, 100 μM CE-22:1n9, 100 μM PC-18:0-d70, TAG-16:0-d93, in 1:1 MeOH/toluene. Using the standard preparation worksheet described in Subheading 2.7, prepare a 100 μM solution in 1:1 MeOH/toluene (v/v) of deuterated tri-palmitoyl-

Table 8
Representative FAME calibration concentrations (μM)

#	Mixture	Compound	Mix target (μM)	Range (μM)
1	FAME ISTD stock	C23:0-me	1400	4
2	FAME mix 1	C6:0-me	200	[0.0154–51.2]
3	FAME mix 1	C8:0-me	300	[0.0253–84.3]
4	FAME mix 1	C10:0-me	400	[0.0429–143]
5	FAME mix 1	C12:0-me	400	[0.0373–124]
6	FAME mix 1	C14:0-me	500	[0.0495–165]
7	FAME mix 1	C14:1n5-me	80	[0.00832–27.7]
8	FAME mix 1	C15:0-me	80	[0.00781–26]
9	FAME mix 1	C16:0-me	600	[0.0592–197]
10	FAME mix 1	C16:1n7-me	70	[0.00745–24.8]
11	FAME mix 1	C17:0-me	70	[0.00703–23.4]
12	FAME mix 1	C17:1n7-me	70	[0.00708–23.6]
13	FAME mix 1	C18:0-me	400	[0.0402–134]
14	FAME mix 1	C18:1n9-me	300	[0.027–89.9]
15	FAME mix 1	C18:2n6-me	200	[0.0204–67.9]
16	FAME mix 1	C18:3n3-me	100	[0.0109–36.2]
17	FAME mix 1	C19:0-me	60	[0.0064–21.3]
18	FAME mix 1	C20:0-me	60	[0.00613–20.4]
19	FAME mix 1	C20:1n9-me	100	[0.0123–41.1]
20	FAME mix 1	C22:0-me	60	[0.00564–18.8]
21	FAME mix 1	C24:0-me	50	[0.00523–17.4]
22	FAME mix 2	C18:4n3-me	10	[0.00124–4.12]
23	FAME mix 2	C20:3n6-me	40	[0.00359–12]
24	FAME mix 2	C20:4n6-me	300	[0.0344–115]
25	FAME mix 2	C20:4n3-me	30	[0.00299–9.95]
26	FAME mix 2	C22:2n6-me	70	[0.00698–23.3]
27	FAME mix 2	C22:5n6-me	80	[0.0078–26]
28	FAME mix 3	C16:1n7t-me	80	[0.00837–27.9]
29	FAME mix 3	C19:1n9-me	90	[0.00941–31.4]
30	FAME mix 3	9c,11 t–CLA-me	70	[0.00735–24.5]
31	FAME mix 3	C20:2n6-me	70	[0.00703–23.4]

(continued)

Table 8
(continued)

#	Mixture	Compound	Mix target (μM)	Range (μM)
32	FAME mix 3	C21:0-me	30	[0.00309–10.3]
33	FAME mix 3	C20:3n3-me	60	[0.0055–18.3]
34	FAME mix 3	C20:5n3-me	50	[0.0051–17]
35	FAME mix 3	C22:4n6 me	80	[0.0081–27]
36	FAME mix 3	C22:6n3-me	200	[0.0188–62.7]
37	FAME mix 4	C18:1n7-me	30	[0.00295–9.85]
38	FAME mix 4	C18:3n6-me	40	[0.00403–13.4]
39	FAME mix 4	10 t,12c–CLA-me	30	[0.00336–11.2]
40	FAME mix 4	C22:5n3-me	70	[0.00681–22.7]
41	FAME mix 4	C24:1n9-me	80	[0.00836–27.9]
42	FAME mix 5	C15:1n5-me	70	[0.00659–22]
43	FAME mix 5	C16:0-d31-me	100	[0.0108–36]
44	FAME mix 5	C18:0-d35-me	100	[0.00227–7.56]
45	FAME mix 5	C22:1n9-me	80	[0.00753–25.1]
46	FAME mix 5	C22:3n3-me	40	[0.00397–13.2]

glycerides (TAG-tri(16:0)-d93), deuterated di-stearoyl-phosphatidylcholine (PC-di(18:0)-d70), the rare cholesteryl ester (CE-22:1n9), and the rare NEFA (C22:3n3) (*see* **Note 6**).

2.9 Sample Preparation Materials

1. *Tubes (or plates)*: 2.0 mL Eppendorf Safe-Lock micro-centrifuge tube, natural, round bottom.

2. *Plates*: Thermo Fisher Scientific Nunc™ 96-well, 500 μL U-bottom plate, natural, non-treated, mfr. No. 267245.

3. *Non-slit 96-well cap mats*: Thermo Fisher Scientific Nunc™ 96-well non-slit cap mats, natural, mfr. No. 276002.

4. *Spin filters*: Ultrafree-MC VV Centrifugal Filters Durapore (R) PVDF 0.1 μm EMD Millipore Corporation # UFC30VV00.

5. *Extraction antioxidant solution*: 0.2 mg/mL butylated hydroxytoluene, 0.2 mg/mL EDTA in MeOH. Dissolve 10 mg butylated hydroxytoluene (BHT) in 10 mL of Optima grade MeOH. Dissolve 10 mg ethylenediaminetetraacetic acid (EDTA) in 10 mL of LC-MS grade water. Mix solutions 1:1 to prepare a 0.2 mg/mL BHT/EDTA solution. Sub-aliquot 1 mL proportions into clean vials, seal, and store at −20 °C.

6. *Centrifugal evaporator*: Genevac EZ-2 (SP Scientific, Warminster, PA, USA).

7. *Centrifuge*: A refrigerated centrifuge capable of holding 2 mL Eppendorf tubes and maintaining 6 °C (e.g., Eppendorf 5430 R, model # 5428).

8. *Plasma reference material*: Prepare or purchase a large homogeneous volume of plasma, and sub-aliquot into 110 μL volumes in polypropylene Eppendorf tubes and store at −80 °C (*see* **Note 7**).

9. NEFA extraction solvents: LC-MS grade water, Optima grade cyclohexane, and 1 M ultrahigh purity ammonium acetate.

2.10 FAME Derivatization Reagents

Free fatty acids are derivatized to produce methyl ester for GC-MS analysis. To control for derivatization efficiency, a fatty acid derivatization surrogate control is introduced after sample extraction and prior to derivatization.

1. *Methylation regent*: (Trimethylsilyl)diazomethane solution 2.0 M in hexane.

2. *Derivatization SSTD*: 62.5 μM pentadecenoic acid (15:1n5) in MeOH. In a 10 mL volumetric flask, dissolve 15 mg (15:1n5 Nu-Chek Prep #U-38-A) in MeOH to yield a 6.25 mM solution. In a second 10 mL volumetric flask, dilute 100 μL of the 6.25 mM solution with 10 mL MeOH to the derivatization SSTD.

3 Methods

The Methods section contains instruction for building the LC-MS/MS and GC-MS instrument acquisition methods for analysis of plasma sample extracts, developing the multi-residue calibration standards, the plasma sample preparation protocol, and a review of post-acquisition QC.

3.1 Building the Multi-residue LC-MS/MS Instrument Acquisition Method Using the Gradient Optimization Approach

1. *Establishing global MS/MS parameters*: The analyte optimization solutions (*see* Subheading 2.6) are required to build the LC-MS/MS acquisition method(s). To optimize the −ESI oxylipin MRM global ion source parameters and collision gas pressures (Table 9), infuse the 14,15-DiHETrE-d11 optimization solution while collecting data for its defined precursor to product ion (i.e., $Q1 > Q3$) mass transition listed in Table 10 (*see* **Note 8**). Transfer the optimized global parameters into the table of global parameters (Table 9). Establish the +ESI endocannabinoid MRM experiment by repeating the global optimization steps by infusing the A-EA-d8 optimization solution, while collecting data for its defined $Q1 > Q3$ mass transition listed in Table 11. The +ESI and −ESI source and collision gas

Table 9
Sciex QTRAP global MS parameters for oxylipin, endocannabinoid, and bile acid assays

Assay	Mode	IS (V)	TEM (°C)	CUR (L/min)	GS1 (L/min)	GS2 (L/min)	CAD (pressure)
Oxylipin	−ESI	−4500	525	35	60	50	Med
Endo	+ESI	5500	525	35	60	50	Med
Bile acid	−ESI	−4500	600	35	60	50	Med

CUR curtain gas flow, *CAD* collision gas pressure setting, *IS* ion source voltage, *TEM* source temperature, *GS1* nebulizer gas, *GS2* heater gas

Table 10
Oxy/endo −ESI analyte-specific parameters for Sciex 6500 QTRAP

#	Analyte	t_R	Q1	Q3	DCP	EP	CE	CXP	SSTD
1	PUHA −esi	3.06	249	130	−30	−10	−18	−13	−
2	6-Keto PGF1α-d4	3.54	373	167	−70	−10	−36	−13	6-Keto PGF1α-d4
3	6-Keto PGF1α	3.51	369	163	−70	−10	−33	−13	6-Keto PGF1α-d4
4	PGF3α	4.12	351	307	−60	−10	−24	−13	PGF2α-d4
5	PGE3	4.25	349	269	−35	−10	−21	−13	PGD2-d4
6	TXB2-d4	4.55	373	173	−50	−10	−45	−13	PUHA −esi
7	TXB2	4.61	369	169	−50	−10	−21	−13	TXB2-d4
8	9_12_13-TriHOME	4.82	329	211	−60	−10	−30	−13	PGF2α-d4
9	PGF2α-d4	4.87	357	197	−70	−10	−33	−13	CUDA −esi
10	PGF2α	4.89	353	193	−70	−10	−33	−13	PGF2α-d4
11	PGE2	4.99	351	271	−40	−10	−21	−13	PGD2-d4
12	PGE1	5.15	353	317	−40	−10	−18	−13	PGD2-d4
13	PGD2-d4	5.16	355	275	−40	−10	−27	−13	CUDA −esi
14	PGD2	5.26	351	271	−40	−10	−24	−13	PGD2-d4
15	15-Keto PGE2	5.3	349	331	−50	−10	−15	−13	PGD2-d4
16	Resolvin D1	5.78	375	121	−60	−10	−39	−13	PGD2-d4
17	Lipoxin A4	5.89	351	217	−45	−10	−27	−13	PGD2-d4
18	LTB5	6.99	333	195	−50	−10	−21	−13	LTB4-d4
19	15_16-DiHODE	7.49	311	235	−55	−10	−27	−13	14_15-DiHETrE-d11
20	8_15-DiHETE	7.58	335	235	−65	−10	−21	−13	14_15-DiHETrE-d11
21	12_13-DiHODE	7.59	311	183	−60	−10	−27	−13	14_15-DiHETrE-d11
22	9_10-DiHODE	7.64	311	201	−65	−10	−27	−13	14_15-DiHETrE-d11

(continued)

Table 10
(continued)

#	Analyte	t_R	Q1	Q3	DCP	EP	CE	CXP	SSTD
23	17_18-DiHETE	7.8	335	247	−70	−10	−24	−13	14_15-DiHETrE-d11
24	5_15-DiHETE	7.81	335	173	−40	−10	−18	−13	14_15-DiHETrE-d11
25	6-trans-LTB4	7.86	335	195	−85	−10	−21	−13	LTB4-d4
26	CUDA −esi	8.7	339	214	−70	−10	−30	−13	–
27	LTB4-d4	8.48	339	163	−70	−10	−33	−13	CUDA −esi
28	14_15-DiHETE	8.51	335	207	−40	−10	−24	−13	14_15-DiHETrE-d11
29	LTB4	8.58	335	195	−50	−10	−24	−13	LTB4-d4
30	12_13-DiHOME	8.9	313	183	−70	−10	−30	−13	14_15-DiHETrE-d11
31	9_10-DiHOME	9.34	313	201	−65	−10	−27	−13	14_15-DiHETrE-d11
32	14_15-DiHETrE-d11	9.5	348	207	−65	−10	−24	−13	CUDA −esi
33	19_20-DiHDoPE	9.55	361	273	−80	−10	−24	−13	14_15-DiHETrE-d11
34	14_15-DiHETrE	9.6	337	207	−60	−10	−24	−13	14_15-DiHETrE-d11
35	11_12-DiHETrE	10.2	337	167	−55	−10	−27	−13	14_15-DiHETrE-d11
36	9_10-e-DiHO	10.4	315	297	−110	−10	−30	−13	14_15-DiHETrE-d11
37	12(13)-Ep-9-KODE	10.4	309	291	−90	−10	−21	−13	9-HODE-d4
38	13-HOTrE	10.6	293	195	−70	−10	−24	−13	9-HODE-d4
39	8_9-DiHETrE	10.8	337	127	−65	−10	−27	−13	14_15-DiHETrE-d11
40	15-Deoxy PGJ2	10.8	315	271	−50	−10	−21	−13	PGD2-d4
41	20-HETE-d6	11	325	281	−80	−10	−21	−13	CUDA −esi
42	15-HEPE	11	317	219	−55	−10	−18	−13	12-HETE-d8
43	20-HETE	11	319	275	−95	−10	−24	−13	20-HETE-d6
44	12-HEPE	11.4	317	179	−45	−10	−18	−13	12-HETE-d8
45	5,6-DiHETrE	11.5	337	145	−55	−10	−24	−13	14_15-DiHETrE-d11
46	9-HEPE	11.6	317	167	−45	−10	−18	−13	12-HETE-d8
47	13-HODE	11.9	295	195	−90	−10	−24	−13	9-HODE-d4
48	5-HEPE	12	317	115	−40	−10	−21	−13	5—HETE-d8
49	9-HODE-d4	12	299	172	−85	−10	−27	−13	CUDA −esi
50	9-HODE	12.1	295	171	−70	−10	−24	−13	9-HODE-d4
51	15(16)-EpODE	12.2	293	275	−75	−10	−18	−13	12(13)-EpOME-d4
52	17(18)-EpETE	12.3	317	259	−55	−10	−15	−13	12(13)-EpOME-d4
53	15-HETE	12.3	319	219	−55	−10	−18	−13	12-HETE-d8

(continued)

Table 10
(continued)

#	Analyte	t_R	Q1	Q3	DCP	EP	CE	CXP	SSTD
54	13-KODE	12.3	293	179	−80	−10	−27	−13	9-HODE-d4
55	9(10)-EpODE	12.4	293	275	−65	−10	−18	−13	12(13)-EpOME-d4
56	9-HOTrE	10.4	293	171	−60	−10	−21	−13	9-HODE-d4
57	17-HDoHE	12.4	343	281	−55	−10	−18	−13	12-HETE-d8
58	12(13)-EpODE	12.5	293	183	−50	−10	−24	−13	12(13)-EpOME-d4
59	15-KETE	12.6	317	273	−60	−10	−18	−13	12-HETE-d8
60	14-HDoHE	12.7	343	281	−60	−10	−18	−13	12-HETE-d8
61	11-HETE	12.7	319	167	−45	−10	−21	−13	12-HETE-d8
62	14(15)-EpETE	12.7	317	247	−35	−10	−18	−13	12(13)-EpOME-d4
63	9-KODE	12.8	293	185	−100	−10	−27	−13	9-HODE-d4
64	12-HETE-d8	12.8	327	184	−60	−10	−21	−13	CUDA −esi
65	11(12)-EpETE	12.8	317	167	−40	−10	−18	−13	12(13)-EpOME-d4
66	12-HETE	12.9	319	179	−60	−10	−21	−13	12-HETE-d8
67	8-HETE	13.1	319	155	−45	−10	−21	−13	12-HETE-d8
68	9-HETE	13.2	319	167	−60	−10	−18	−13	12-HETE-d8
69	5-HETE-d8	13.5	327	116	−70	−10	−18	−13	CUDA −esi
70	19(20)-EpDoPE	13.5	343	281	−60	−10	−18	−13	12(13)-EpOME-d4
71	5-HETE	13.5	319	115	−50	−10	−18	−13	5-HETE-d8
72	12(13)-EpOME-d4	13.6	299	198	−90	−10	−24	−13	CUDA −esi
73	12(13)-EpOME	13.7	295	195	−85	−10	−21	−13	12(13)-EpOME-d4
74	14(15)-EpETrE	13.8	319	219	−50	−10	−18	−13	12(13)-EpOME-d4
75	4-HDoHE	13.8	343	281	−60	−10	−18	−13	12-HETE-d8
76	16(17)-EpDoPE	13.9	343	274	−45	−10	−15	−13	12(13)-EpOME-d4
77	9(10)-EpOME	13.9	295	171	−75	−10	−21	−13	12(13)-EpOME-d4
78	5-KETE	14.2	317	203	−75	−10	−27	−13	5-HETE-d8
79	11(12)-EpETrE	14.2	319	167	−40	−10	−21	−13	12(13)-EpOME-d4
80	8(9)-EpETrE	14.4	319	155	−40	−10	−18	−13	12(13)-EpOME-d4
81	10-Nitrolinoleate	14.6	324	277	−40	−10	−18	−13	10-Nitrooleate-d17
82	9_10-EpO	15.1	297	279	−105	−10	−27	−13	12(13)-EpOME-d4
83	10-Nitrooleate-d17	15.3	343	308	−65	−10	−18	−13	CUDA −esi
84	10-Nitrooleate	15.4	326	280	−40	−10	−24	−13	10-Nitrooleate-d17

(continued)

Table 10
(continued)

#	Analyte	t_R	Q1	Q3	DCP	EP	CE	CXP	SSTD
85	9-Nitrooleate	15.5	326	308	−50	−10	−18	−13	10-Nitrooleate-d17
86	C20:5n3	15.5	301	257	−60	−10	−15	−13	C20:4n6-d8
87	C18:3n3	15.6	277	259	−115	−10	−24	−13	C20:4n6-d8
88	C22:6n3	16.2	327	283	−45	−10	−15	−13	C20:4n6-d8
89	C20:4n6-d8	16.5	311	267	−60	−10	−18	−13	C20:4n6-d8
90	C20:4n6	16.6	303	259	−40	−10	−18	−13	C20:4n6-d8
91	C18:2n6	16.9	279	261	−185	−10	−38	−13	C20:4n6-d8

CUR curtain gas flow, *CAD* collision gas pressure setting, *IS* ion source voltage, *TEM* source temperature, *GS1* nebulizer gas, *GS2* heater gas

Table 11
Oxy/endo +ESI analyte-specific parameters for Sciex 6500 QTRAP

#	Analyte	t_R	Q1	Q3	DCP	EP	CE	CXP	SSTD
1	PUHA +esi	3.06	251.2	114.1	65	10	21	12	–
2	PGF2α-EA-d4	3.62	384.3	62.1	50	10	42	12	PUHA +esi
3	PGF2a-EA	3.63	380.3	62.1	45	10	39	12	PGF2α-EA-d4
4	PGE2-EA	3.65	378.301	62.1	65	10	39	12	PGF2α-EA-d4
5	PGD2-EA	3.92	378.302	62.1	65	10	42	12	PGF2α-EA-d4
6	PGF2a 1G	4.29	411.3	301.2	40	10	21	12	PGF2α-EA-d4
7	PGE2 1G	4.33	409.3	317.2	75	10	21	12	PGF2α-EA-d4
8	CUDA +esi	8.7	341.3	216.2	50	10	24	12	–
9	15-HETE-EA	9.62	346.3	62.1	75	10	21	12	A-EA-d8
10	11(12)-EpETre-EA	11.73	364.3	62.1	75	10	45	12	A-EA-d8
11	aL-EA	13.46	322.2	62.1	60	10	21	12	A-EA-d8
12	DH-EA	14.61	372.3	62.1	55	10	45	12	A-EA-d8
13	A-EA-d8	14.72	356.3	63.1	50	10	45	12	CUDA +esi
14	A-EA	14.78	348.3	62.1	70	10	39	12	A-EA-d8
15	L-EA	14.81	324.2	62.1	70	10	21	12	A-EA-d8
16	NA-Gly-d8	14.86	370.3	76.1	45	10	21	12	CUDA +esi
17	NA-Gly	14.91	362.3	76.1	65	10	21	12	NA-Gly-d8
18	2-AG-d5	15.41	384.3	287.2	60	10	21	12	CUDA +esi

(continued)

Table 11
(continued)

#	Analyte	t_R	Q1	Q3	DCP	EP	CE	CXP	SSTD
19	2-AG	15.43	379.3	287.2	110	10	24	12	2-AG-d5
20	DGL-EA	15.45	350.3	62.1	35	10	36	12	A-EA-d8
21	2-LG	15.57	355.3	263.2	25	10	12	12	2-AG-d5
22	P-EA-d4	15.62	304.2	62.1	90	10	18	12	CUDA +esi
23	1-AG	15.64	379.3	287.2	95	10	21	12	2-AG-d5
24	P-EA	15.64	300.2	62.1	70	10	18	12	P-EA-d4
25	1-LG	15.82	355.3	263.2	40	10	12	12	2-AG-d5
26	O-EA	16.1	326.2	62.1	105	10	21	12	A-EA-d8
27	D-EA	16.12	376.3	62.1	105	10	45	12	A-EA-d8
28	NO-Gly	16	340.2	76.2	60	10	21	12	NA-Gly-d8
29	2-OG	16.97	357.3	265.2	50	10	15	12	2-AG-d5
30	1-OG	17.31	357.3	265.2	55	10	15	12	2-AG-d5
31	S-EA	17.79	328.2	62.1	100	10	21	12	P-EA-d4

CUR curtain gas flow, *CAD* collision gas pressure setting, *IS* ion source voltage, *TEM* source temperature, *GS1* nebulizer gas, *GS2* heater gas

flows and temperatures must be kept constant due to the inability to adjust with the speed of +ESI/−ESI switching (*see* **Note 9**).

2. *Creation of single MRM methods to establish retention times*: Using the following instructions, build a series of methods, each with ~20 analytes. (1) Create a method for the first 20 analytes by importing, pasting, or manually programing the analyte names and values for precursor (*Q1*) and product (*Q3*) ions from Table 10; (2) set analyte source voltage and collision energy (CE) at the midrange, based on your mass spectrometer user manual; (3) enter the −ESI global parameters from the infusions in Subheading 3.1, **step 1**; (4) program the liquid chromatography parameters from Table 12; (5) name and save the method.

Repeat these five steps for all oxylipin analytes, being sure to save as new files for each set of ~20 analytes. Repeat for the +ESI endocannabinoid targets. For bile and acids, refer to Tables 13 and 14.

3. *Establishing retention times and creating the "all analyte" method*: Inject 5 μL of the "all analyte" solution for each of the methods to establish analyte retention times for each of the groups of 20 analytes (*see* **Notes 10** and **11**). Update Table 10

Table 12
Oxy/endo UPLC solvent gradient

Time	% B
0.01	25
1.00	40
2.50	42
4.50	50
10.50	65
12.50	75
13.25	80
17.25	85
18.25	95
18.75	100
19.00	100
19.10	25
20.00	Stop

Solvent A: 0.1% acetic acid in water
Solvent B: 10:90 isopropanol/acetonitrile
Flow: 250 μL/min

Table 13
Bile acid positive mode analyte-specific parameters for Sciex 4000 QTRAP

#	Analyte	t_R	Q1	Q3	DCP	EP	CE	CXP	SSTD
1	PHAU	2.6	249.2	130	−65	−10	−20	−5	–
2	TDHCA	2.82	508.2	80	−155	−10	−110	−7	TCDCA-d4
3	T-ω-MCA	3.33	514.3	80	−155	−10	−110	−2	TCDCA-d4
4	T-α-MCA	3.47	514.3	80	−155	−10	−110	−4	TCDCA-d4
5	T-β-MCA	3.56	514.3	80	−155	−10	−110	−4	TCDCA-d4
6	TUDCA	5.48	498.3	80	−155	−10	−110	−9	TCDCA-d4
7	TCA	5.91	514.3	80	−185	−10	−115	−9	TCDCA-d4
8	ω-MCA	6.96	407.3	407	−115	−10	−30	−10	CA-d4
9	GUDCA	7.12	448.3	74	−115	−10	−70	−4	GCA-d4
10	α-MCA	7.3	407.3	407	−115	−10	−30	−4	CA-d4
11	GHDCA	7.32	448.3	74	−120	−10	−70	−4	GCA-d4
12	GCA-d4	7.4	468.3	74	−125	−10	−70	−4	PHAU

(continued)

Table 13
(continued)

#	Analyte	t_R	Q1	Q3	DCP	EP	CE	CXP	SSTD
13	GCA	7.41	464.3	74	−125	−10	−70	−4	GCA-d4
14	β-MCA	7.71	407.3	407	−115	−10	−30	−9	CA-d4
15	TCDCA-d4	8.63	502.3	80	−175	−10	−110	−9	PHAU
16	TCDCA	8.66	498.3	80	−145	10	−110	−9	TCDCA-d4
17	TDCA	9.4	498.3	80	−140	−10	−110	−9	TCDCA-d4
18	CA-d4	10.1	411.3	411	−120	−10	−30	−4	CUDA
19	CA	10.2	407.3	407	−125	−10	−30	−4	CA-d4
20	UDCA	10.3	391.3	391	−125	−10	−30	−4	DCA-d4
21	GCDCA-d4	10.8	452.3	74	−120	−10	−65	−9	CUDA
22	GCDCA	10.8	448.3	74	−125	−10	−65	−4	GCDCA-d4
23	GDCA	11.5	448.3	74	−125	−10	−65	−3	GCDCA-d4
24	CUDA	12.1	339.3	214	−65	−10	−35	−9	−
25	TLCA	12.3	482.3	80	−150	−10	−110	−9	LCA-d4
26	CDCA-d4	12.4	395.3	395	−125	−10	−25	−9	CUDA
27	CDCA	12.4	391.3	391	−130	−10	−30	−9	CDCA-d4
28	DCA-d4	12.5	395.3	395	−125	−10	−30	−4	CUDA
29	DCA	12.5	391.3	391	−130	−10	−30	−4	DCA-d4
30	GLCA	12.6	432.3	74	−120	−10	−65	−10	GCDCA-d4
31	LCA-d5	13	380.3	380	−135	−10	−30	−8	CUDA
32	LCA	13	375.3	375	−130	−10	−35	−10	LCA-d5

CUR curtain gas flow, *CAD* collision gas pressure setting, *IS* ion source voltage, *TEM* source temperature, *GS1* nebulizer gas, *GS2* heater gas

with your retention times and import or paste into your MS method, as a scheduled MRM method or as defined MRM periods. Scheduled MRM allows for more scans across peaks and higher sensitivity. For a scheduled MRM method, use a 60-s retention time window for all analytes, or define in advanced settings. For defined MRM period methods, set dwell times at ~10–20 ms as appropriate for your instrument. Import the updated −ESI and +ESI MRM experiments into a single method, and save method as the "all analyte Tr" method (*see* **Note 12**).

4. *Regio-isomeric purity check for analyte solutions*: Before conducting analyte-specific source and collision energy

Table 14
Bile acid UPLC solvent gradient

Time	% B
1.01	10
0.50	10
1.00	25
11.00	40
12.50	95
14.00	95
14.50	10
16.00	Stop

Solvent A: 0.1% formic acid in water
Solvent B: 0.1% formic acid in acetonitrile
Flow: 250 μL/min

optimizations or making the calibration solutions, each analyte optimization solution should be analyzed by retention time for regio-isomeric purity to allow for adjusting calibration concentrations. Inject each individual analyte optimization solution using your all analyte Tr method. Export area responses to check for substantial contamination and lack of analyte signal. Occasionally (sometimes often), the contents of manufacturer stock can contain substantial amounts of other regio-isomers or analytes. We have observed up to 99% impurity (or mislabeled manufacturers stock)! In this case the analyte should be returned to the manufacturer and purchased from an alternative source, if possible.

5. *"Gradient optimization" of analyte-specific source and collision energies*: Check the user manual for your instrument's range of source voltages and collision energies. In your all analyte Tr method, adjust the source voltage in both +ESI and −ESI MRM to the lowest voltage in the range (e.g., −15/15). This can be set as a global parameter or analyte-specific parameter. Table 10 can be updated and imported or pasted into the MS method, for this purpose. Save this method (e.g., Oxy SV15). Now adjust the source voltage to −20/20 and save as Oxy SV20. Repeat for the entire range in increments of 5. Collect data in a series of 5 μL injections of your all analyte optimization solution (or the dilution) using each method. Export integrated area responses to Excel, and using the MAX function, determine the voltage producing the maximum response for each analyte (*see* **Note 13**). Replace the estimated source voltages with these optimized values in Table 10. Import the

updated Table 10 into the acquisition method and save (e.g., Oxy SV-OPT). Using the Oxy SV-Opt method, repeat this process for collision energies in increments of 3. You now have an optimized method! Fine-tuning of global mass spectrometer parameters can be done to boost specific analyte responses.

3.2 Preparing Calibration Standards, Surrogate Spike, and Internal Standard Solutions for the Analytical Assay

Before making the analyte standard mixtures described below, build your acquisition method, run each optimization solution for isomeric purity, and note concentration adjustments to your stock solutions. Refer to the worksheet in Supplementary File 1. The procedures described below can be followed to prepare all solutions and calibration standards for the assay, using your analyte stock mixture concentrations and the concentration range defined in Tables 6–8. These concentration ranges are for general guidance and can be updated to suit the linear range of your instrument.

1. *In the worksheet, record analyte stock information* (refer to worksheet, **step 1**): Use the Name Box pulldown in the top left corner of the Excel worksheet to locate each of the following steps. Record the amount (mg) of analyte and the µL volume of solution in your analyte stock solutions.

2. *Establish desired calibration standard range* (refer to **step 2**): In the sheet, there are four adjustments that can be made to obtain your desired Cal Standard concentration values: the volume of analyte stock solution delivered to make the mixtures, the final volume of the mixtures, the volume of mixture used to make the Cal Standard, and the final volume of the Cal Standard (*see* **Note 14**). Adjust these values to assure reasonable volumetric deliveries of stock solutions and mixtures to reach your targets. For example, as shown in the Supplementary File 1 worksheet, our internal standard mixture (ISTD Mix) contains 50 µM of each of the two internal standards, with a target of 100 nM in the calibration standards. To make this mixture, use a glass barrel analytical syringe to deliver the following volumes into a 10 mL volumetric flask: 170 µL of a 1 mg/mL solution of cyclohexyl ureido dodecanoic acid (CUDA) and 25 µL of 5 mg/mL phenyl ureido hexanoic acid (PUHA). Bring to volume in purged MeOH. Make the remaining mixtures with the analyte stock solutions in your sheet.

3. *Prepare an IS SSTD dilution solution for Cal Standards that are serial diluted* (refer to **step 3**): Make sure the concentration of your IS SSTD dilution solution matches the concentrations of the calculated values for the higher-concentration Cal Standards that are not serial diluted; see Cal 5–9 in the oxy sheet. *Special Instructions*: If you quantify surrogates (SSTDs) on a

standard curve (e.g., using Micromass MassLynx or Agilent MassHunter software), calculate their delivery volumes like the analytes, and do not include SSTD in the dilution solution! Instead, use purged MeOH to bring all Cal Standards to final volume. Adjust the worksheet accordingly.

4. *Prepare calibration standards* (refer to **step 4**): Make calibration standards (5–9) by delivering volumes of the analyte mixtures to the indicated Cal Standard volumetric flasks. Assure you do not add more than the volume of the flask! Bring to final volume in purged MeOH. If you are maintaining constant internal standard and surrogate concentrations across your calibration standards, make Cal Standards (1–4) as a serial dilution, by delivering the indicated volume of Cal Standard 6 to the Cal 1–4 volumetric flasks, and bring to final volume with the IS SSTD dilution solution for Cal Standards. Cal Standard 0 contains no analytes, is comprised of the remaining IS SSTD dilution solution for Cal Standards, and allows real-time corrections for the addition of non-isotopically labeled impurities into samples at extraction.

5. *Prepare LC ISTD sample reconstitution solution* (refer to **step 5**): Refer to worksheet, for example, deliver 200 μL of the LC ISTD mixture into a 100 mL volumetric flask. Bring to volume in purged 1:1 MeOH/ACN for a final concentration of 100 nM. If you are maintaining constant concentration of LC-ISTD across all Cal Standards, the concentration of this solution should be the same concentration of your Cal Standards.

6. *Prepare surrogate spike solution* (refer to **step 6**): As shown on the sheet, we make surrogate spike solution by delivering 2 mL of the SSTD mixture into a 25 mL volumetric flask and bring to a final volume in purged MeOH. Use the sample spike calculator to assure that the final concentration of SSTD spike solution in samples corresponds to the final concentration in Cal Standards.

3.3 The Shake and Shoot Sample Prep Protocol

This protocol captures the free oxylipins, endocannabinoids, bile acids, and free fatty acids listed in Tables 6–8.

1. *Sample extraction*: Label prerinsed 2 mL Eppendorf tubes or map out a rinsed 1 mL polypropylene plate. Thaw plasma samples on wet ice (*see* **Note 15**). Per ~44 samples, include a sample replicate, a plasma reference material, and a blank (50 μL LC-MS grade water). If you do not have enough sample to make a replicate, then use two aliquots of the plasma reference material. For small studies include one replicate and one blank. Centrifuge plasma samples at 15,000 rcf (relative centrifugal force, g force), at 4 °C for 5 min to concentrate solids on the bottom of the tube. Spike newly labeled clean tubes or

wells with 5 μL of extraction antioxidant solution, 5 μL of 1000 nM oxy/endo surrogate spike solution (SSTD), 5 μL of 1000 nM bile acid SSTD, and 5 μL 100 μM FAME SSTD (*see* **Note 16**). Transfer a 50 μL aliquot of either plasma supernatant or water blank into the labeled tubes/plate, cap, and gently vortex 2 s to mix. Add 200 μL LC internal standard solution (LC ISTD). The total sample volume at this stage equals 270 μL of an 81% alcohol solution. Cap and vortex at medium speed for 30 s to mix-denature-extract the samples. Centrifuge plasma extracts at 15,000 rcf, at 4 °C for 5 min to concentrate solids on the bottom of the tube.

2. *Preparation for NEFA analysis*: Aliquot 150 μL of sample extract to a 2 mL amber vial, add 289 μL of isopropanol to bring the total alcohol amount in the extract to 410 μL, and mix for 30 s. Add 520 μL cyclohexane and mix for 2 min. Add 484 μL of LC-MS grade water followed by 57 μL of ultrahigh purity 1 M ammonium acetate, and mix for 2 min (*see* **Note 17**). The total water content of the extract, including water from the sample itself, must equal a total of 570 μL at this point. Centrifuge for 5 min at 15,000 rcf to break emulsion, and transfer approximately 400 μL of upper organic phase to a second labeled 2 mL glass vial being careful not to collect any of the aqueous phase. Add 520 μL of cyclohexane to the remaining aqueous sample, vortex 2 min to mix, and transfer ~600 μL of the upper phase and combine with the first extract. Again, it is critical not to collect water at this step! Centrifuge extracts for 5 min at 15,000 rcf. Inspect for the appearance of any water droplets, and if observed, remove with a pasture pipette. Remove solvent by centrifugal evaporation in a speed vac. If using a Genevac system, use the low boiling point (i.e., low BP) setting with no heat for ~30 min. It is critical that the dry extract is absolutely free from water to prevent hydrolysis of esterified fatty acids during derivatization. Confirm complete solvent removal and reconstitute dry residues in 100 μL toluene and vortex 30 s to dissolve residues. Create a derivatization blank by placing 100 μL of toluene into a separate vial. Dilute samples and blanks with 100 μL of MeOH and 20 μL of derivatization SSTD (62.5 μM 15:1n5). Add 45 μL of (trimethylsilyl)diazomethane, cap, and vortex to mix for 10 s. React for 30 min at room temperature, and then, evaporate to dryness on a centrifugal vacuum evaporator, at low BP setting. Reconstitute the sample in 100 μL of FAME ISTD reconstitution solution. Vortex 10 s to mix. Store at −20 °C until GC-MS analysis.

3. *Preparation for oxy/endo and bile acid analysis*: Maintain sample extract on wet ice for 15 min. In batches of 8, aliquot 100 μL of extract to a Millipore Durapore PVDF 0.1 μm

centrifugal filter, and centrifuge at 7400 rcf for 3 min, at 4 °C (*see* **Note 18**). Transfer filtrate with a Pasteur pipette to an autosampler insert, cap, and vortex 2 (s) to remove bubbles. Store at −20 °C until analysis. Use slit-top caps for analysis, and analyze 5 μL injections by LC-MS/MS. Build your sample list starting with two MeOH shots, followed by a set of concentration standards, two MeOH (*see* **Note 19**), 20–30 samples, a methanol, three calibration standards, a methanol, 20–30 samples, etc. End your acquisition with a set of calibration standards (*see* **Note 20**).

3.4 Gas Chromatography Mass Spectrometry Analysis of Fatty Acid Methyl Esters

Fatty acid methyl esters are quantified in 1 μL splitless injections against an eight-point calibration curve of authentic standards.

1. *FAME GC instrument configuration*: The injector is configured for fast injection of 1 μL aliquots after three sample pumps, and post-injection syringe cleaning is conducted with three rinses of hexane and isooctane. The front injection port, which receives the sample, has a temperature of 230 °C and is operated in splitless mode with an initial pressure of 14.75 psi and a 1-min purge delay and 18.8 mL total flow. The rear injection port has a 2 psi initial pressure, 0 min purge delay, and a 35 mL/min total flow. The initial column temperature of 64 °C is held for 1 min, then increased at 32 °C/min to 224 °C, held for 8 min, increased at 16 °C/min to 240 °C, and held for 8 min. The chromatographic column carrier gas is regulated in a ramped flow mode. Initial flow is 1.1 mL/min held for 12 min, increased at 0.2 mL/min/min to 2.1 mL/min, held for 4 min, ramped at 2 mL/min to 0.1 mL/min, and held for 8.00 min, with outlet pressure 2 psi. In contrast, the rear backflush column is operated in a ramped pressure mode. Initial pressure is 2 psi, held for 21 min, ramped at 30 psi/min to 32 psi, held for 8.00 min, and then ramped at 30 mL/min to 2.0 psi, outlet pressure ambient (*see* **Note 21**).

2. *FAME MS instrument configuration*: FAMEs are detected with a 5973N mass spectral detector with simultaneous selected ion monitoring/full-scan (SIM/SCAN) spectral acquisition capabilities. Zone temperatures are transfer line: 240 °C, source: 230 °C, quadrupole: 180 °C. Mass spectrometer parameters: filament delay of 3.00 min; scan parameters (50–400 *m/z*); selected ion monitoring parameters are *m/z* 55.10, 67.10, 69.10, 74.10, 77.10, 79.10, and 368.40. Analytes are quantified using the SIM ion data and identified based on retention time and characteristic mass fragmentation as described in Table 15.

3.5 Quantitation

Use internal standard methodologies with ratio response to quantitate surrogate recoveries and analyte concentrations. With this in mind, organize your quantitation method to fit the constraints of

Table 15
FAME GC-MS quantitation and identification parameters

Analyte	t_R (min)	Target ion (m/z)	Q1 ion (m/z)	Q2 ion (m/z)
C23:0	17.55	368.4	74.1	79.1
C6:0	4.30	74.1	55.1	
C8:0	4.70	74.1	55.1	
C10:0	5.20	74.1	55.1	
C12:0	6.06	74.1	55.1	
C14:0	7.12	74.1	55.1	
C14:1n5	7.33	74.1	67.1	55.1
C15:0	7.79	74.1	55.1	
C15:1n5	8.05	74.1	67.1	55.1
C16:0-d31	8.32	77.1	74.1	79.1
C16:0	8.59	74.1	55.1	
C16:1n7t	8.74	74.1	67.1	55.1
C16:1n7	8.80	74.1	67.1	55.1
C17:0	9.51	74.1	55.1	
C17:1n7	9.76	74.1	67.1	55.1
C18:0-d35	10.17	77.1	74.1	79.1
C18:0	10.58	74.1	55.1	
C18:1n9	10.79	74.1	67.1	55.1
C18:1n7	10.86	74.1	67.1	55.1
C18:2n6	11.22	67.1	55.1	79.1
C18:3n6	11.45	79.1	67.1	55.1
C18:3n3	11.76	79.1	67.1	55.1
C19:0	11.81	74.1	55.1	
C19:1n9	11.99	74.1	55.1	55.1
C18:4n3	12.05	79.1	67.1	55.1
9c,11t–CLA	12.22	67.1	55.1	79.1
10 t,12c–CLA	12.36	67.1	55.1	79.1
C20:0	13.02	74.1	55.1	
C20:1n9	13.26	74.1	67.1	55.1
C20:2n6	13.76	67.1	55.1	79.1
C20:3n6	14.02	79.1	67.1	55.1

(continued)

Table 15
(continued)

Analyte	t_R (min)	Target ion (*m/z*)	Q1 ion (*m/z*)	Q2 ion (*m/z*)
C20:4n6	14.15	79.1	67.1	55.1
C21:0	14.33	74.1	55.1	
C20:3n3	14.43	79.1	67.1	55.1
C20:4n3	14.70	79.1	67.1	55.1
C20:5n3	14.84	79.1	67.1	55.1
C22:0	15.82	74.1	55.1	
C22:1n9	16.14	74.1	67.1	55.1
C22:2n6	16.82	67.1	55.1	79.1
C22:4n6	17.44	79.1	67.1	55.1
C22:5n6	17.59	79.1	67.1	55.1
C22:3n3	17.74	108.1	67.1	79.1
C22:5n3	18.43	79.1	67.1	55.1
C22:6n3	18.61	79.1	67.1	55.1
C24:0	19.74	74.1	55.1	
C24:1n9	20.26	74.1	67.1	55.1

your quantitation software. Use the ratio of ISTD/SSTD to determine surrogate recoveries. If possible use the SSTDs as internal standards for the analytes to correct for losses directly in your quantitation software. Otherwise, treat your analytical surrogates as analytes, determine the percentage recovered, and apply these corrections to your measured target analyte concentrations. Surrogate to analyte assignments are indicated for LC-MS analyses in the analyte-specific parameter tables (Tables 10, 11, and 13). Calibration curves should be weighted $1/x$, and linear by default, but should be inspected, and modest quadratic fits can be used (*see* **Note 22**). Review all integrations, modify integration parameters to maximize integration quality, and manually adjust integrations if needed.

3.6 Assessing Quality Controls

There are five basic quality controls used for these quantitative assays.

1. *Internal standard response (ISTD)*: The internal standard area response can be graphed chronologically, shot to shot, over the entire acquisition. This display can be used to determine if any injections were not successful, as shown by a severe lack of signal. Common causes are from not removing air bubbles

from inserts prior to injection (LC-MS) or a lack of extract in the insert due to evaporation or multiple injections (LC-MS and GC-MS). The sample should be reshot as soon as possible. The ISTD response will also show if there has been substantial instrumental drift, displayed as an increase or decrease in signal over time. This is typically due to progressive ion source contamination (LC-MS and GC-MS) or post-cleaning system stabilization (GC-MS), but increases can also be due to sample concentration in the vial. Replace slit-top caps between assays. The CUDA and PUHA LC ISTDs are observed in positive and negative mode (*see* **Note 23**).

2. *Relative retention time*: Plot analyte/ISTD relative retention times, by analyte over the course of the run to assure correct peak picking and integration.

3. *Surrogate recoveries*: The surrogates can be compared as a ratio response against theoretical and measured values to determine performance (% recovery) of the sample preparation.

4. *Replicate analysis*: Replicates can be compared to determine overall method variance for the analytes. Variance (%RSD) can be plotted against sample concentration. Variance will increase as concentration decreases, and this plot provides a simple means to determine the concentration at which your variance is no longer constant. For ESI-LC-MS applications, replicate precision of analytes above the concentration providing constant variance is generally considered acceptable if less than 30%. For GC-MS applications, this cutoff is generally less than 20%.

5. *Method blank*: Use to determine if the reagents are contaminated (*see* **Note 24**).

4 Notes

1. There is a risk of bacterial growth in solutions with <10% organic, and these solutions should be discarded regularly. Bacterial overgrowth in an idle LC can cause serious problems with the hardware. Maintain idle solvent lines in organic solvent to avoid growth.

2. We have successfully obtained these compounds using a variety of vendors including Avanti Polar Lipids (Birmingham, AL, USA), Cayman Chemical (Ann Arbor Michigan, USA), Larodan Fine Chemicals (Malmo, Sweden), and Sigma-Aldrich (Saint Louis, MO, USA). Due to inconsistent naming of chemical structures, the compound International Chemical Identifier (InChI) keys are provided to aid in compound searches. However, a complete set of InChIKeys for the fatty acid methyl esters (FAMEs) were not available while preparing this

protocol, and we provide the free fatty acid structures to locate these methyl esters, use the InChIKey to locate the compound in PubChem, find a chemical name for the acid, append this name with methyl ester, and search the worldwide web.

3. We have found this worksheet a necessary tool to assure reasonable volumetric deliveries and final volumes for the concentration targets of the analytes in the assay. It also functions as a complete record of the analyte stock solutions, mixtures, calibration standards, spike solutions, and the internal standard reconstitution solutions. See Supplementary File 1 for a working example of this spreadsheet as a Microsoft Excel workbook.

4. High-boiling solvent as second rinse prevents syringe from drying while over hot GC injection port.

5. Preparing separate mixtures allows for the confirmation of spectral purity for chromatographically co-eluting analytes. When weighing viscous oils into a mixture, we have found it easiest to deliver materials sequentially to volumetric flask using a pasture pipette, depositing the material below the volumetric demarcation line, and accepting measured weights within 20% of the targeted weight. To accomplish this, it is critical that you confirm and document that your analytical balance can make accurate measures of 1 mg in the 10 mg range. To test this, place a volumetric flask of appropriate size on the balance, tare, and measure the accuracy of weighing a certified 10 mg weight and a 10 mg + 1 mg weight. Repeat this measure no less than three times and allow time for drift between measures of at least 5 min.

6. Prior to extraction, samples are spiked with the FAME SSTD which contains a suite of complex lipid esters. If we observe the fatty acid products from the SSTD triglycerides, glycerophospholipid, and cholesteryl ester, it indicates hydrolysis during the NEFA derivatization, likely due to water, and the assay has failed. In addition, with the increased availability of isotopically labeled fatty acids, the authors would encourage replacement of the rare compounds used here with deuterated compounds. In particular, the C22:3n3 can have problematic resolution from the C22:5n6 on some DB-225ms and replacement with d8 C20:4n6 should enhance method robustness.

7. We have found that Cryovials designed for −80 °C storage often have high fatty acid backgrounds and should not be used.

8. If you have an AB Sciex 6500 QTRAP, the DCP and CE voltages of the oxylipin/endocannabinoid analytes from Table 1 and the global parameters from Table 2 can be used to build the acquisition method. These values also approximate the AB Sciex 4000 and can be used to capture signal for determining analyte retention time prior to analyte-specific optimization; *see* Subheading 3.2, **step 5**. Bile acid tables are for the Sciex 4000 QTRAP.

9. After the method is built, choose the parameter values that have the least effect on overall signal.

10. The analytes in Table 10 are in retention time order for the Waters Corp. 150 mm C_{18} BEH Acquity column, as described in Subheading 2. Some analytes are regio-isomers and have identical precursor and product ions, so the retention time order is very important to maintain, so as to not switch their identities.

11. If an analyte signal is not observed, a manual optimization by infusion of the analyte's optimization solution should be conducted, according to the MS user manual. Optimize source voltage by MS scan and collision energy by MS/MS or MRM scan. The optimized analyte source voltage and collision energy parameters can now be entered into the appropriate 20 analyte method. Rerun the method with the "all analyte" solution to determine the analyte retention times.

12. If any of your peaks are above 1e6, dilute a portion of the all analyte optimization solution, inject, and check peak. Assure other analyte peaks are observed, if not consider conducting a manual optimization to observe.

13. Use the value at the highest source voltage setting, where there is typically less background noise.

14. In this sheet, the Cal Standard range reflects in general, the dynamic linear range of our instrument.

15. Transfer samples from −80 to −20 °C overnight to reduce time to thaw.

16. Please use the spike calculator from **step 6** to develop your spiking solution.

17. Developing this protocol around a 1 M ammonium acetate addition with a separate addition of water allows flexibility to increase or decrease the amount of sample delivered to the assay, such that the final molar concentration in the extract can be maintained at 0.1 M. If such adjustments are made, be sure to maintain the alcohol/cyclohexane/0.1 M ammonium acetate ratio at 8:10:11 v/v/v.

18. If anybody can find a solvent-compatible hydrophilic polypropylene filter plate, at 0.2 μm, please contact the corresponding author:)

19. Determine if there is carryover from your high standards to a methanol injection.

20. It is critical to track the response of the ISTD over the run to assure there is no substantial loss of signal.

21. For calibration solutions, methods should be created that stop rather than execute the backflush routine, thus reducing the time required to calibrate the instrument.

22. Level zero should only be used in the calibration curves when a significant background is detected to avoid inaccuracies in the curves.

23. The chromatographic behavior of both CUDA and PUHA is sensitive to pH, and therefore shifts in this behavior are good indications of changes in mobile phase pH. These can be included in a broad array of assays as a measure to check instrument sensitivity.

24. Surrogate behavior in method blanks generally does not track those observed in biological samples, particularly in LC-MS applications. Therefore, the quantitated values are often inappropriate to directly correct for any observed contamination and should only be used to qualify data that may be affected. When present, contamination associated with derivatization steps in the GC-MS applications should be removed from reported values.

Acknowledgment

The authors would like to thank Ira J. Gray, Michael R. La Frano, and William R. Keyes for their technical support on the development and implementation of these assays over the past 10 years. The methods reported in this chapter were developed as part of research projects funded by the United States Department of Agriculture (5306-51530-016-00D, 2032-51530-019-00-D, 2032-51530-022-00-D) and DK/NIDDK NIH HHS Grant U24 DK097154/United States

References

1. Griffiths WJ, Koal T, Wang Y, Kohl M, Enot DP, Deigner HP (2010) Targeted metabolomics for biomarker discovery. Angew Chem Int Ed Engl 49(32):5426–5445

2. Su LJ, Fiehn O, Maruvada P, Moore SC, O'Keefe SJ, Wishart DS, Zanetti KA (2014) The use of metabolomics in population-based research. Adv Nutr 5(6):785–788

3. Strassburg K, Huijbrechts AM, Kortekaas KA, Lindeman JH, Pedersen TL, Dane A, Berger R, Brenkman A, Hankemeier T, van Duynhoven J, Kalkhoven E, Newman JW, Vreeken RJ (2012) Quantitative profiling of oxylipins through comprehensive LC-MS/MS analysis: application in cardiac surgery. Anal Bioanal Chem 404 (5):1413–1426

4. Yang J, Schmelzer K, Georgi K, Hammock BD (2009) Quantitative profiling method for oxylipin metabolome by liquid chromatography electrospray ionization tandem mass spectrometry. Anal Chem 81(19):8085–8093

5. Lundstrom SL, Levanen B, Nording M, Klepczynska-Nystrom A, Skold M, Haeggstrom JZ, Grunewald J, Svartengren M, Hammock BD, Larsson BM, Eklund A, Wheelock AM, Wheelock CE (2011) Asthmatics exhibit altered oxylipin profiles compared to healthy individuals after subway air exposure. PLoS One 6(8):e23864

6. Sun Y, Koh HW, Choi H, Koh WP, Yuan JM, Newman JW, Su J, Fang J, Ong CN, van Dam RM (2016) Plasma fatty acids, oxylipins, and risk of myocardial infarction: the Singapore Chinese Health Study. J Lipid Res 57(7):1300–1307

7. Grapov D, Adams SH, Pedersen TL, Garvey WT, Newman JW (2012) Type 2 diabetes associated changes in the plasma non-esterified fatty acids, oxylipins and endocannabinoids. PLoS One 7(11):e48852

8. Agrawal K, Hassoun LA, Foolad N, Pedersen TL, Sivamani RK, Newman JW (2017) Sweat lipid mediator profiling: a noninvasive approach for cutaneous research. J Lipid Res 58 (1):188–195

Chapter 14

Chemical Isotope Labeling LC-MS for Human Blood Metabolome Analysis

Wei Han and Liang Li

Abstract

Blood is a widely used biofluid in discovery metabolomic research to search for clinical metabolite biomarkers of diseases. Analyzing the entire human blood metabolome is a major analytical challenge, as blood, after being processed into serum or plasma, contains thousands of metabolites with diverse chemical and physical properties as well as a wide range of concentrations. We describe an enabling method based on high-performance chemical isotope labeling (CIL) liquid chromatography-mass spectrometry (LC-MS) for in-depth quantification of the metabolomic differences in comparative blood samples with high accuracy and precision.

Key words Chemical isotope labeling, Dansylation, DmPA, LC-MS, Metabolomics, Blood

1 Introduction

Blood samples are being extensively used for discovery metabolomics with the goal of finding sensitive and specific biomarkers that are indicative of healthy and diseased states. Despite great advances in blood metabolomic profiling techniques in the past decades, in-depth quantitative analysis of the blood metabolome is still a major analytical challenge. Because of diverse chemical and physical properties of blood metabolites, a conventional strategy of increasing the metabolomic coverage is to use several analytical tools with different metabolite detectability to analyze the same sample. For example, a widely used metabolomic profiling platform includes the use of reversed-phase (RP) liquid chromatography-mass spectrometry (LC-MS) to analyze the hydrophobic metabolites and hydrophilic interaction (HILIC) LC-MS to analyze the relatively polar metabolites. For each separation, positive and negative ion modes of MS detection are separately carried out in order to ionize as many metabolites as possible. This platform is relatively easy to implement. However, the overall metabolomic coverage from the combined LC-MS analyses is still limited. In addition, untargeted

Martin Giera (ed.), *Clinical Metabolomics: Methods and Protocols*, Methods in Molecular Biology, vol. 1730,
https://doi.org/10.1007/978-1-4939-7592-1_14, © Springer Science+Business Media, LLC 2018

metabolomic analysis using LC-MS without standards provides limited quantification accuracy and precision due to matrix and ion suppression effects.

An alternative strategy of performing metabolomic analysis is to classify the metabolites into several subgroups based on the presence of common functional moieties and then perform in-depth analysis of the individual chemical-group-submetabolomes [1]. The combined data from the submetabolomes would allow the analysis of the entire metabolome with high coverage. This divide-and-conquer strategy does not require many analytical measurements of the same sample. For example, when we examined the chemical structures of the endogenous human metabolites in the Human Metabolome Database (HMDB), we found that more than 95% of these metabolites contain one or more of the four functional groups: amine, carboxyl, hydroxyl, and carbonyl. Thus, in principle, if we could analyze all the metabolites within these four submetabolomes, a near-complete metabolomic profile could be generated.

Our laboratory has been developing a high-performance chemical isotope labeling (CIL) LC-MS approach to analyze the individual chemical-group-submetabolomes with very high coverage [1–3]. The idea is to use a rationally designed chemical labeling reagent to react with a common functional group within a submetabolome (e.g., amines) to form metabolite derivatives, followed by MS detection. The properties of the labeled metabolites are altered, compared to the original unlabeled metabolites, to such an extent that all the labeled metabolites can be retained on RPLC for efficient separation without the need of switching to another mode of separation column, ionized with high efficiency to improve overall detection sensitivity, and detected in positive ion mode only to avoid the need of running negative ion mode. For accurate relative quantification, a differential isotope labeling strategy using heavy and light isotope reagents is used. To this end, we have reported the use of ^{12}C-/^{13}C-dansyl chloride (DnsCl) for profiling the amine/phenol submetabolome [1], ^{12}C-/^{13}C-dimethylaminophenacyl (DmPA) bromide for profiling the carboxylic acid submetabolome [2], base-activated ^{12}C-/^{13}C-dansyl chloride (b-DnsCl) labeling for profiling the hydroxyl submetabolome [3], and ^{12}C-/^{13}C-dansyl hydrazine labeling for profiling the ketone/aldehyde submetabolome including many sugars [4]. Each labeling method generates a submetabolome profile with very high metabolic coverage. For example, dansyl labeling of the amine/phenol submetabolome detects over 2000 metabolites routinely from a plasma sample.

Figure 1 shows the high-performance CIL LC-MS workflow for quantitative metabolomics. A control sample is prepared by mixing small aliquots of individual samples to form a pool, followed by heavy (e.g., ^{13}C-dansyl) labeling. This ^{13}C-labeled control is

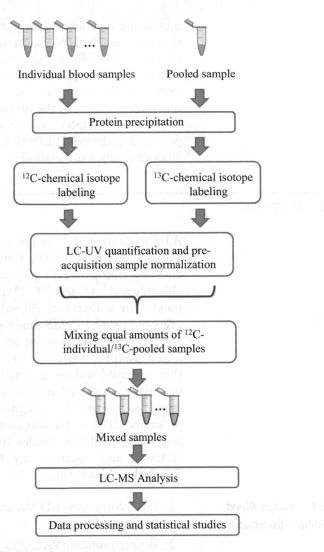

Fig. 1 CIL LC-MS workflow

spiked into all the light or ^{12}C-labeled individual samples and thus serves as a global internal standard [5]. The ^{12}C-labeled metabolite and its corresponding ^{13}C-labeled counterpart in a ^{12}C-/^{13}C-mixture are detected as a peak pair in MS, and their peak area ratio reflects the concentration difference of the metabolite in an individual sample vs. the control. Since the same ^{13}C-labeled control is used for preparing all individual ^{12}C-/^{13}C-mixtures, the peak ratio values can be used for relative quantification of individual metabolites in different samples. We have developed the required data processing software for analyzing the CIL LC-MS data [6]. In addition, over 650 metabolite standards have been individually labeled with the proper labeling reagents for positive metabolite identification [7]. In the library of standards, each labeled metabolite contains triplet parameters: molecular mass, LC retention time, and MS/MS spectrum. Library search program has been developed for fully automated search for rapid metabolite identification [7]. For putative metabolite identification or structure match, accurate mass search is used to match the detected labeled metabolite masses to the metabolites in metabolome databases.

In this chapter, we describe the updated protocols for performing dansyl and DmPA labeling of metabolites and LC-MS analysis of labeled metabolites for human blood metabolomics.

2 Materials

All chemicals and reagents were purchased from Sigma-Aldrich Canada (Markham, ON, Canada) unless otherwise noted. The chemical isotope labeling reagents, including ^{12}C- and ^{13}C-dansyl chloride and ^{12}C- and ^{13}C-DmPA bromide, are available on http://mcid.chem.ualberta.ca. All solutions and LC mobile phases were prepared using LC-MS grade solvents (*see* **Note 1**), and the solutions were stored at room temperature. Blood samples were collected from participants with standard operating procedures after they reviewed and signed the informed consents. All human samples were processed in a Containment Level II laboratory and stored in a Level I or regular chemistry laboratory after being processed. The waste was disposed following the waste disposal regulations. All the studies involving human subject have been reviewed and approved by the University of Alberta Health Research Ethics Board.

2.1 Human Blood Sample Collection

1. We recommend BD Vacutainer 10 mL serum collection tubes (*see* **Note 2**).

2. We recommend VACUETTE 5 mL plasma collection tubes.

2.2 Dansylation Labeling

1. 250 mM NaHCO$_3$/Na$_2$CO$_3$ buffer solution: Weigh 26.5 g of anhydrous Na$_2$CO$_3$ and 21.0 g of anhydrous NaHCO$_3$ in a clean 1 L glass bottle. Measure 1 L of water in a 1 L volumetric flask, and transfer the water to the glass bottle to dissolve the solid (*see* **Note 3**).

2. 250 mM NaOH solution: Weigh 1.0 g of NaOH and transfer it to a Nalgene lab quality bottle. Measure 100 mL of water in 100 mL volumetric flask, and transfer the water to the plastic bottle. Dissolve the NaOH.

3. 425 mM formic acid solution: Dilute 1.60 mL of formic acid to 100 mL with (50/50, v/v) acetonitrile/water.

4. 20 mg/mL dansyl chloride solution: Before the chemical isotope labeling experiment, 20 mg/mL ^{12}C-dansyl chloride

(^{12}C-DnsCl) solution is prepared by dissolving 20 mg of ^{12}C-DnsCl in 1 mL of acetonitrile, and 20 mg/mL ^{13}C-dansyl chloride (^{13}C-DnsCl) solution is prepared by dissolving 20 mg of ^{13}C-DnsCl in 1 mL of acetonitrile.

2.3 DmPA Labeling

1. 0.5 M triethanolamine solution: Dissolve 7.46 g of triethanolamine in 100 mL of acetonitrile.

2. 10 mg/mL DmPA bromide solution: Before the chemical isotope labeling experiment, 10 mg/mL ^{12}C-DmPA bromide (^{12}C-DmPABr) solution is prepared by dissolving 10 mg of ^{12}C-DmPABr in 1 mL of acetonitrile, and 10 mg/mL ^{13}C-DmPA bromide (^{13}C-DmPABr) solution is prepared by dissolving 10 mg of ^{13}C-DmPABr in 1 mL of acetonitrile.

2.4 LC-MS Analysis

1. Mobile phase A: 1 mL of formic acid and 50 mL of acetonitrile are transferred into a 1 L volumetric flask and then diluted to 1 L with water.

2. Mobile phase B: 1 mL of formic acid is transferred into a 1 L volumetric flask and then diluted to 1 L with acetonitrile.

3. HPLC column: Agilent reversed-phase Eclipse plus C18 column (2.1 mm × 100 mm, 1.8 μm particle size, 95 Å pore size) for LC-MS analysis.

3 Methods

3.1 Human Serum Samples

Ten milliliters of venipuncture blood are collected into a BD Vacutainer 10 mL serum collection tube. An individual sample is allowed to clot spontaneously at room temperature for 1 h and then centrifuged at 1,500 × *g* for 15 min to separate serum and cells. The supernatant (serum) is divided into multiple 250 μL aliquots in 1.5 mL microcentrifuge tubes for analysis or storage in a −80 °C freezer. Also, a pooled serum sample is prepared by mixing equal volumes of individual serum samples. This pooled sample is processed with ^{13}C-labeling reagents to serve as the internal standard during the quantitative analyses (*see* **Note 4**).

3.2 Human Plasma Samples

Blood should be collected by venipuncture into a VACUETTE 5 mL plasma collection tube with lithium heparin (*see* **Note 5**). The tube is then inverted for ten times to ensure blood mixing with the inner tube coating material. The sample is left on the bench for 30 min and then centrifuged at 2,100 × *g* for 30 min at 4 °C. The plasma sample is divided into multiple 250 μL aliquots in 1.5 mL microcentrifuge tubes for analysis or storage in a −80 °C freezer. Also, a pooled plasma sample is prepared by mixing equal volumes of individual plasma samples.

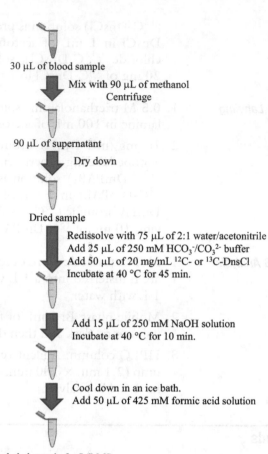

30 μL of blood sample

Mix with 90 μL of methanol
Centrifuge

90 μL of supernatant

Dry down

Dried sample

Redissolve with 75 μL of 2:1 water/acetonitrile
Add 25 μL of 250 mM HCO$_3^-$/CO$_3^{2-}$ buffer
Add 50 μL of 20 mg/mL ^{12}C- or ^{13}C-DnsCl
Incubate at 40 °C for 45 min.

Add 15 μL of 250 mM NaOH solution
Incubate at 40 °C for 10 min.

Cool down in an ice bath.
Add 50 μL of 425 mM formic acid solution

Labeled sample for LC-MS
analysis

Fig. 2 Dansylation labeling scheme

3.3 Dansylation Labeling Reaction

The workflow of dansylation labeling is shown in Fig. 2.

1. Frozen serum or plasma samples are thawed (*see* **Note 6**) in an ice bath and then centrifuged at $15,000 \times g$ for 15 min.

2. Thirty microliters of supernatant are transferred into a microcentrifuge tube and mixed with 90 μL of methanol.

3. Store the mixture at −20 °C for 2 h to precipitate the proteins. After this, the mixture is centrifuged at $15,000 \times g$ for 15 min.

4. Take 90 μL of supernatant and dry using a Speed-Vac centrifugal evaporator.

5. Redissolve to 75 μL with 2:1 water/ACN. Then add 25 μL of 250 mM sodium carbonate/sodium bicarbonate buffer to the sample to generate a basic environment for the dansylation reaction.

6. Vortex, spin down, and mix with 50 μL of freshly prepared ^{12}C-DnsCl solution (20 mg/mL) (for light labeling) or ^{13}C-DnsCl solution (20 mg/mL) (for heavy labeling).

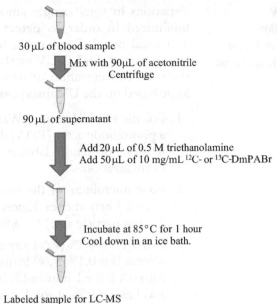

30 μL of blood sample

Mix with 90μL of acetonitrile
Centrifuge

90 μL of supernatant

Add 20 μL of 0.5 M triethanolamine
Add 50 μL of 10 mg/mL ^{12}C- or ^{13}C-DmPABr

Incubate at 85°C for 1 hour
Cool down in an ice bath.

Labeled sample for LC-MS
analysis

Fig. 3 DmPA labeling scheme

7. After 45 min incubation at 40 °C, 10 μL of 250 mM NaOH are added to the reaction mixture to quench the excess dansyl chloride. The solution is then incubated at 40 °C for another 10 min.

8. Finally, add formic acid (425 mM) in 1:1 ACN/H$_2$O to consume excess NaOH and to make the solution acidic. The labeled samples are now ready for LC-MS analysis or stored at a –80 °C freezer (*see* **Note 7**).

3.4 DmPA Labeling Reaction

The workflow is shown in Fig. 3.

1. Frozen serum or plasma samples are thawed in an ice bath and then centrifuged at 15,000 × *g* for 15 min.

2. Transfer 30 μL of supernatant into a microcentrifuge tube and mix with 90 μL of acetonitrile.

3. Store the mixture at –20 °C for 2 h to precipitate the proteins. After this, the mixture is centrifuged at 15,000 × *g* for 15 min.

4. Take 90 μL of supernatant, and mix with 20 μL of 0.5 M triethanolamine and 50 μL of freshly prepared ^{12}C-DmPA bromide solution (10 mg/mL) (for light labeling) or ^{13}C-DmPA bromide solution (10 mg/mL) (for heavy labeling).

5. Incubate the mixture at 85 °C for 1 h (*see* **Note 8**).

6. Finally, the samples are cooled down in an ice bath and are ready for LC-MS analysis. The labeled samples can also be stored in a –80 °C freezer for future use.

3.5 LC-UV Quantification and Pre-acquisition Sample Normalization

Variations in total sample amount in different samples must be minimized in order to detect the concentration differences of individual metabolites caused by the biological or clinical factors being studied. An LC-UV method [8] can be applied to determine the total concentration of dansylated amine/phenol submetabolome based on the UV absorption of the dansyl group (*see* **Note 9**).

1. For the LC-UV setup, a Waters ACQUITY UPLC system with a photodiode array (PDA) detector can be used for the quantification of dansyl-labeled metabolites for sample amount normalization.

2. Four microliters of the labeled serum or plasma are injected onto a Phenomenex Kinetex C18 column (2.1 mm × 5 cm, 1.7 μm particle size) for a fast step-gradient run.

3. Solvent A is 0.1% (v/v) formic acid in 5% (v/v) ACN/H_2O, and solvent B is 0.1% (v/v) formic acid in ACN. The gradient starts with 0% B for 1 min and is increased to 95% B within 0.01 min and held at 95% B for 1 min to ensure complete elution of all labeled metabolites. The flow rate used is 0.45 mL/min.

4. The peak area, which can represent the total labeled metabolite concentration in the sample, is integrated using the Empower software (6.00.2154.003).

5. Based on the quantification results, the ^{12}C- and ^{13}C-labeled samples are mixed in equal mole amounts for LC-MS analysis. When a ^{12}C-labeled individual sample is mixed with the ^{13}C-labeled pooled sample, each peak pair ratio in the LC-MS result represents the relative concentration of one specific metabolite in the individual sample, with respect to the average concentration of the sample set.

3.6 LC-FTICR-MS

1. The LC-Fourier transform ion cyclotron resonance (FTICR)-MS analysis is performed using an Agilent 1100 series binary system (Agilent, Palo Alto, CA) connected to a 9.4 T Apex-Qe FTICR-MS (Bruker, Billerica, MA). The MS data are acquired in the positive ion mode with an electrospray ionization (ESI) source. An Agilent reversed-phase Eclipse plus C18 column (2.1 mm × 100 mm, 1.8 μm particle size, 95 Å pore size) is used for chromatographic separation (*see* **Note 10**). The mobile phase A is 0.1% formic acid in ACN/H_2O (5/95, v/v), and the mobile phase B is 0.1% formic acid in ACN.

2. All dansyl-labeled samples are analyzed with a 32-min gradient: 0 min (20% B), 0–3.5 min (20–35% B), 3.5–18 min (45–65% B), 18–21 min (65–95% B), 21–24 min (95–99% B), and 24–32 min (99% B). The column is re-equilibrated with the initial mobile phase conditions for 15 min before injecting the next sample. The flow rate is 180 μL/min, and the injection volume is 5.0 μL.

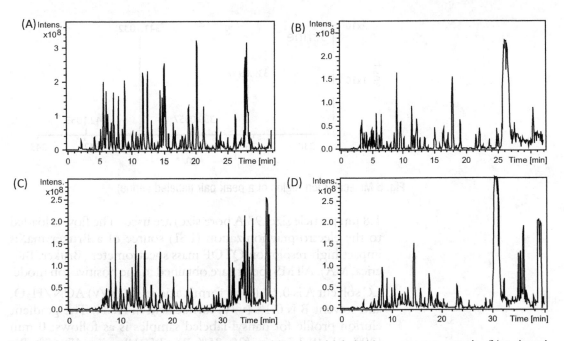

Fig. 4 Representative LC-MS chromatograms from **(a)** a dansyl-labeled human serum sample, **(b)** a dansyl-labeled human plasma sample, **(c)** a DmPA-labeled human serum sample, and **(d)** a DmPA-labeled human plasma sample (acquired by LC-FTICR-MS)

3. The 40-min gradient for all DmPA-labeled samples is 0 min (20% B), 0–9 min (20–50% B), 9–22 min (50–65% B), 22–26 min (65–80% B), 26–28 min (80–98% B), and 28–40 min (98% B). The column is re-equilibrated with the initial mobile phase conditions for 15 min before injecting the next sample. The flow rate is 180 μL/min, and the injection volume is 2.5 μL.

4. The MS conditions used for FTICR-MS are as follows: nitrogen nebulizer gas, 2.3 L/min; dry gas flow, 7.0 L/min; dry temperature, 190 °C; capillary voltage, 4,200 V; spray shield, 3,700 V; acquisition size, 256 k; scan range, 200–1100; and ion accumulation time, 1 s.

5. Figure 4 shows the representative base peak chromatograms of (A) a dansyl-labeled serum, (B) a dansyl-labeled plasma, (C) a DmPA-labeled serum, and (D) a DmPA-labeled plasma. In Fig. 5, the peak pair of dansyl-serine is given as an example for a representative mass spectrum. The peak with m/z of 339.0957 is the ^{12}C-dansyl-labeled serine from the individual sample, and the peak with m/z of 341.1032 is the ^{13}C-dansyl-labeled serine from the pooled sample.

3.7 LC-QTOF-MS

1. For LC-quadrupole time-of-flight (QTOF)-MS, an Agilent 1100 series binary system (Agilent, Palo Alto, CA) and an Agilent reversed-phase Eclipse plus C18 column (2.1 mm × 100 mm,

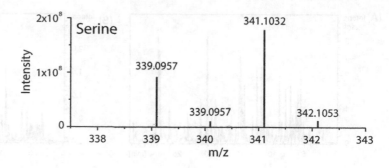

Fig. 5 Molecular ion region of a peak pair (labeled serine)

1.8 μm particle size, 95 Å pore size) are used. The flow is loaded to the electrospray ionization (ESI) source of a Bruker maXis impact high-resolution QTOF mass spectrometer (Bruker, Billerica, MA). All MS spectra are obtained in the positive ion mode.

2. LC solvent A is 0.1% (v/v) formic acid in 5% (v/v) ACN/H_2O, and solvent B is 0.1% (v/v) formic acid in ACN. The gradient elution profile for dansyl-labeled samples is as follows: 0 min (20% B), 0–3.5 min (20–35% B), 3.5–18 min (45–65% B), 18–21 min (65–95% B), 21–24 min (95–99% B), and 24–32 min (99% B). The column is re-equilibrated with the initial mobile phase conditions for 15 min before injecting the next sample. The flow rate is 180 μL/min and the injection volume is 10.0 μL.

3. The 40-min gradient for running the DmPA-labeled samples is: 0 min (20% B), 0–9 min (20–50% B), 9–22 min (50–65% B), 22–26 min (65–80% B), 26–28 min (80–98% B), and 28–40 min (98% B). The column is re-equilibrated with the initial mobile phase conditions for 15 min before injecting the next sample. The flow rate is 180 μL/min and the injection volume is 5.0 μL.

4. The MS conditions used for QTOF-MS are as follows: end plate offset, 500 V; capillary voltage, 4,500 V; nebulizer, 1.0 bar; dry gas, 8.0 L/min; dry temperature, 230 °C; transfer time, 40 μs; and prepulse storage, 10 μs.

3.8 Data Processing and Statistical Analysis

The data processing workflow is shown in Fig. 6 (*see* **Note 11**). A software tool, IsoMS, is used to process the raw data generated from multiple LC-MS runs by peak picking, peak pairing, peak-pair filtering, and peak-pair intensity ratio calculation [6].

1. Align peak pairs detected from multiple samples using IsoMS-Align.

2. Fill missing ratio values using the zero-fill program [9].

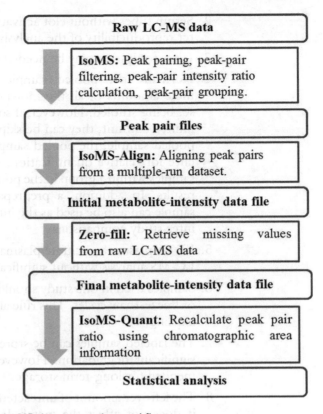

Fig. 6 IsoMS data processing workflow

3. Finally, use IsoMS-Quant to determine the chromatography-peak-intensity ratio of a ^{12}C-/^{13}C-pair [10]. The final metabolite relative concentration file can be exported to SIMCA-P+ 12.0 software (Umetrics, Umeå, Sweden) for multivariate statistical analysis (*see* **Note 12**).

3.9 Metabolite Identification

Positive metabolite identification is performed based on mass and retention time matching to a labeled standard library [3, 7]. Putative identification is done based on accurate mass matches to the metabolites in the Human Metabolome Database (HMDB) (8,021 known human endogenous metabolites) and in the evidence-based metabolome library (EML) (375,809 predicted human metabolites with 1 reaction) using MyCompoundID (www. MyCompoundID.org) [11]. The mass accuracy tolerance window is set at 10 ppm for database search.

4   Notes

1. HPLC grade organic solvents or purified deionized water are found to be adequate for preparing solutions for dansylation.

2. Serum tubes without clot activators are recommended for better reproducibility of the analysis results.

3. Ultra-sonication may be needed to dissolve the solid.

4. Normally the pooled sample is made by mixing equal mole amounts of all the individual samples from the sample set being studied. However, if some of the samples are in very limited amount, they can be skipped from contributing to the pooled sample; the pooled sample works as an internal reference for relative quantification, and thus not including a few samples within a group in the pooled sample does not affect the results. In addition, a preprepared universal serum/plasma sample can also be used as the internal standard, which enables inter-study comparisons.

5. EDTA plasma and citrate plasma can also be used for the CIL LC-MS analysis without significant matrix effect.

6. Samples used in one study should experience the same number of freeze-thaw cycles. This rule also applies to the samples after labeling.

7. The labeled samples can be stored at 4 °C for 1 week without significant degradation. However, −80 °C freezer is recommended for long-term storage.

8. The leftover amount of unreacted DmPA is very small, and thus it does not affect the metabolomic profiling of blood. If it interferes with the detection of a specific compound with similar retention time, triphenylacetic acid can be used to quench the excess amount of DmPA.

9. To reduce the effect of real samples on the lifetime of a C18 column, all samples should be centrifuged at $15,000 \times g$ for 10 min before analysis. A pre-column filter should also be helpful, particularly for running blood samples.

10. Generally, the standard deviation of the distribution of total amine-/phenol-containing metabolite amounts among human blood samples is less than 15%. If a variance of as large as 50% is allowed in a specific study, post-acquisition normalization methods can be used, instead of the pre-acquisition normalization. Nevertheless, the LC-UV quantification is recommended for profiling blood samples for increased accuracy.

11. The data processing packages, as well as metabolite identification libraries, are available on the MyCompoundID website: http://www.mycompoundid.org.

12. Other metabolomic statistical tools, such as MetaboAnalyst, can also be used for the analysis.

Acknowledgment

This work was supported by Genome Canada, the Natural Sciences and Engineering Research Council of Canada (NSERC), and Canada Research Chairs (CRC) programs.

References

1. Guo K, Li L (2009) Differential C-12/C-13-isotope dansylation labeling and fast liquid chromatography/mass spectrometry for absolute and relative quantification of the metabolome. Anal Chem 81(10):3919–3932. https://doi.org/10.1021/ac900166a

2. Guo K, Li L (2010) High-performance isotope labeling for profiling carboxylic acid-containing metabolites in biofluids by mass spectrometry. Anal Chem 82(21):8789–8793. https://doi.org/10.1021/ac102146g

3. Zhao S, Luo X, Li L (2016) Chemical isotope labeling LC-MS for high coverage and quantitative profiling of the hydroxyl submetabolome in metabolomics. Anal Chem 88 (21):10617–10623. https://doi.org/10.1021/acs.analchem.6b02967

4. Zhao S, Dawe M, Guo K, Li L (2017) Development of high-performance chemical isotope labeling LC-MS for profiling the carbonyl submetabolome. Anal Chem 89:6758–6765. https://doi.org/10.1021/acs.analchem.7b01098

5. Peng J, Chen YT, Chen CL, Li L (2014) Development of a universal metabolome-standard method for long-term LC-MS metabolome profiling and its application for bladder cancer urine-metabolite-biomarker discovery. Anal Chem 86(13):6540–6547. https://doi.org/10.1021/ac5011684

6. Zhou R, Tseng CL, Huan T, Li L (2014) IsoMS: automated processing of LC-MS data generated by a chemical isotope labeling metabolomics platform. Anal Chem 86 (10):4675–4679. https://doi.org/10.1021/ac5009089

7. Huan T, YM W, Tang CQ, Lin GH, Li L (2015) DnsID in MyCompoundID for rapid identification of dansylated amine- and phenol-containing metabolites in LC-MS-based metabolomics. Anal Chem 87(19):9838–9845. https://doi.org/10.1021/acs.analchem.5b02282

8. YM W, Li L (2012) Determination of total concentration of chemically labeled metabolites as a means of metabolome sample normalization and sample loading optimization in mass spectrometry-based metabolomics. Anal Chem 84(24):10723–10731. https://doi.org/10.1021/ac3025625

9. Huan T, Li L (2015) Counting missing values in a metabolite-intensity data set for measuring the analytical performance of a metabolomics platform. Anal Chem 87(2):1306–1313. https://doi.org/10.1021/ac5039994

10. Huan T, Li L (2015) Quantitative metabolome analysis based on chromatographic peak reconstruction in chemical isotope labeling liquid chromatography mass spectrometry. Anal Chem 87(14):7011–7016. https://doi.org/10.1021/acs.analchem.5b01434

11. Li L, Li RH, Zhou JJ, Zuniga A, Stanislaus AE, YM W, Huan T, Zheng JM, Shi Y, Wishart DS, Lin GH (2013) MyCompoundID: using an evidence-based metabolome library for metabolite identification. Anal Chem 85 (6):3401–3408. https://doi.org/10.1021/ac400099b

Direct Infusion-Tandem Mass Spectrometry (DI-MS/MS) Analysis of Complex Lipids in Human Plasma and Serum Using the Lipidyzer™ Platform

Baljit K. Ubhi

Abstract

Lipids play a key role in the signaling pathways of cancer, cardiovascular, diabetic, and inflammatory diseases. A major challenge in the analysis of lipids is the many isobaric interferences present in highly complex samples that confound identification and accurate quantitation. After obtaining the total lipid extract from a sample, differential mobility separation has proven to be a powerful tool for gas-phase fractionation of lipid classes. When combined with mass spectrometry, this allows the unambiguous identification and thus quantification of lipid molecular species. These components, sample extraction, gas-phase separation, and mass spectrometry, form the basis of a novel integrated quantitative lipid analysis platform.

Key words DI-MS/MS, Lipidyzer, Lipidomics, Plasma, Serum

1 Introduction

The metabolome is an important class of molecules to study as it can lend insight into the health and well-being of an organism and the causative elements associated with disease. It integrates the effects of our genes with the impact of our environment by being the product of both internal factors (our genome and proteome) and external factors (our lifestyle and our environment). In addition, the metabolome is extremely dynamic and always changing. By identifying and quantifying changes in specific metabolites, any changes that are observed can be mapped back to specific pathways for detailed interpretation. Thus, quantitative analysis of the metabolome can provide important information that can help predict the onset of a disease or disorder or characterize its progression.

Lipids are a key part of the metabolome and are involved in the formation of important biological elements such as membranes, lipid droplets, and lipoproteins. The study of lipids as a class of

Martin Giera (ed.), *Clinical Metabolomics: Methods and Protocols*, Methods in Molecular Biology, vol. 1730, https://doi.org/10.1007/978-1-4939-7592-1_15, © Springer Science+Business Media, LLC 2018

molecules within the metabolome is important as they act as signaling and inflammatory molecules and are known to be significant players in a number of metabolic disorders, cardiovascular disease, oncology, and other disease areas. However, lipids are an extremely complex and diverse group of compounds. Lipid molecules are polymers that are formed by combining metabolites from different metabolic pathways and result from the combinatorial products of fatty acids and head groups that can often have the same mass. Thus, using phosphatidylcholines (a phospholipid class) as an example, for 40 different fatty acids and five main head groups, these 45 components could account for 8000 unique molecular species.

In order to simplify their analysis and improve the accuracy of identification and quantification, the first step in analyzing lipids by mass spectrometry is efficient extraction of the entire class of lipid compounds from the biological sample. There are several methods which allow the extraction and recovery of total lipid from any kind of organism or matrix. For isolating lipids from a biological sample, there are really only three methods that are readily employed in the field of lipidomics: Folch [1], Bligh and Dyer [2], and the methyl *tert*-butyl ether (MTBE) method [3].

While extraction creates a clean sample of virtually only lipid compounds, the remaining lipid fraction is highly complex. The extracted lipids exhibit tremendous isobaric overlap with many lipids per class, all within a narrow mass range. Separation prior to analysis is essential. Liquid chromatography (LC) has traditionally been used to separate and simplify compounds prior to mass spectrometry. However, because of the extreme complexity of the lipid fraction, one LC method alone doesn't supply the necessary separation power to efficiently resolve all of the lipid compounds for accurate quantification. Traditionally, several methods must be used or a sub-fractionation technique must be added in order to resolve each lipid class and achieve accurate qualitative and quantitative data on the molecular species. This can be done by using solid-phase extraction and separating the polar from the neutral lipid components; however, it is time- and labor-intensive.

Differential mobility spectrometry (DMS) has been used as an effective separation tool to couple to mass spectrometry, used most often to reduce interferences and improve signal-to-noise ratio and therefore sensitivity of detection. For lipid analysis, it has proven very successful as an orthogonal separation technique by allowing lipids to be fractioned in the gas phase by the dipole moment on their head groups. As the device is located at the front of the mass spectrometer, it helps eliminate isobaric overlap often encountered during any lipid analysis, whether applying accurate mass or nominal mass spectrometry. In DMS, an ionized sample enters a mobility cell that is located at atmospheric pressure in front of the mass spectrometer orifice. An RF voltage is applied across the cell that is cycled between high and low fields. As the ions transit the cell, they are separated based on the difference in their mobility between high and low field, which is impacted by numerous molecular properties. A second voltage (a compensation voltage known as COV) is applied to steer analytes through the cell into the mass spectrometer. Now mass isolation, fragmentation, and analysis can be performed on the separated classes yielding simplified, more reliable data for qualitative and quantitative analysis. The DMS is a small planar cell that is added between the source region and the high vacuum region of the instrument.

When DMS is used for lipid analysis, each lipid class will have a different COV that can be used to separate each individual class by head group and allow them to enter into the MS, one class at a time (Fig. 1). This is because each lipid class possesses a different head group with differing dipole moments that are affected by the voltages they encounter within the mobility cell. In a paper by Lintonen et al. [4], a linear relationship between the dipole moment of each lipid class and the COV value was found. By applying a specific COV, each lipid class could be directed through the mobility cell into the MS, and all other classes (and species) will not be transmitted. Alternatively, by ramping the COV, different classes of lipids could, in turn, become stable and transit successfully through the cell, thereby allowing sequential analysis of all lipid classes and respective molecular species. In Fig. 1, the separation of the lipids by class is demonstrated by analyzing the lysophosphatidylcholine/lysophosphatidylethanolamine/phosphatidylcholine/phosphatidylethanolamine (LPC/LPE/PC/PE) classes in the negative ion mode and sphingomyelin (SM) class in the positive ion mode.

Separating lipid classes by DMS allows us to overcome the confounding factor of isobaric overlap of these molecules. When coupled with targeted multiple reaction monitoring (MRM) on a triple quadrupole or QTRAP® mass spectrometer, this simple separation of lipid classes eliminates the need for any up-front liquid chromatography (LC), greatly simplifying the workflow. By selecting specific COV values for lipid class transmission that provide the least amount of lipid class overlap, isobaric interference, inherent in lipid analysis, is significantly reduced.

A new LC-MS/MS solution (Lipidyzer™ Platform, SCIEX) simplifies and automates the high-throughput analysis of lipids by infusion, leveraging the optimized sample preparation procedure and DMS separation power described above. Using flow injection sample introduction, the platform can analyze up to 45 samples per day in a fully automated fashion, including system optimization and data processing. This platform allows for the quantitation of over 1100 lipid molecular species across 13 lipid classes from complex lipid metabolism readily found in human plasma and serum.

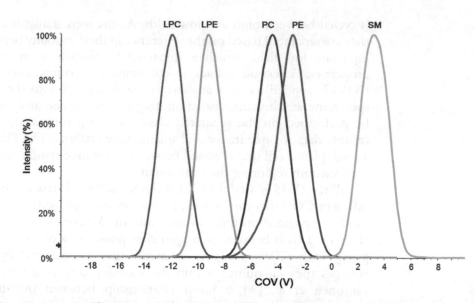

Fig. 1 A differential mobility spectrometry (DMS) ionogram. A compensation voltage (COV) ramp of a mixture of standards infused into the DMS cell. COV ramped from −25 to 10 V separates lipid classes by their head groups in the gas phase. This figure highlights the specificity of the DMS cell to separate lipid classes from complex mixtures. The COV per class is added to the MRM tables in the acquisition method and at any 1 V across the DMS, only that specific lipid class is selected and allowed and passed through to Q1 for subsequent MRM analysis. In this figure, the COV tuning mixture is infused at 7 μL/min, and the Lipidomics Workflow Manager (LWM) software automatically tunes the COV value per class by collecting lysophosphatidylcholine (LPC)/lysophosphatidylethanolamine (LPE)/phosphatidylcholine (PC)/phosphatidylethanolamine (PE) MRMs in the negative ion mode and sphingomyelin (SM) MRMs in the positive ion mode

The platform employs the Bligh and Dyer extraction protocol (modified to use dichloromethane) because of the acceptance in the field and its simplicity. It had readily been shown to extract the larger proportion (~98%) of lipids and is more efficient in its recoveries. The full extraction can be completed in one test tube without any complicated filtration steps. The optimized protocols are included with the platform. Finally, a novel internal standard strategy has been developed for the Lipidyzer Platform that enables more accurate quantification with reduced bias [5]. The chemical kits include over 50 labeled internal standards that cover 13 lipid classes across complex lipid metabolism.

In this chapter, we describe the methods for identification and quantification of complex lipids in plasma and serum using the direct infusion-tandem mass spectrometric (DI-MS/MS) approach employed by the Lipidyzer Platform.

2   Materials

Please be sure to use proper PPE (personal protective equipment). Only HPLC or LC-MS grade solvents should be used.

2.1 Standards

1. SelexION tuning mixture (Part Number 5040141, SCIEX, MA, USA): Aliquot 0.1 mL of the SelexION tuning mixture into a vial. Add 0.9 mL of the 10 mM ammonium acetate in dichloromethane/methanol (50:50) to the vial. Cap the mixture and vortex gently for 5 s.

2. QC Spike Kit (Part Number 5040408, SCIEX, MA, USA).

3. QC control plasma (SCIEX, MA, USA).

4. Lipidyzer internal standards (Part Number 5040156, SCIEX, MA, USA).

2.2 Sample Extraction

1. 10 mM ammonium acetate solution in (50:50) dichloromethane/methanol: Dissolve 770 mg ammonium acetate in 1 L of methanol, and mix 500 mL of this solution with 500 mL dichloromethane.

2. Extraction solvents: 100% methanol, 100% water, and 100% dichloromethane (HPLC grade).

3. For the Bligh and Dyer extraction, borosilicate glass culture test tubes (16 × 125 mm, 10 mL) are used.

2.3 DI-MS/MS

1. All tubing used for the FIA setup is PEEKsil which allows for minimal carry-over between injections (< 0.5%).

2. 10 mM ammonium acetate solution in (50:50) dichloromethane/methanol; *see* Subheading 2.1, **item 4** for preparation.

3. Rinse solvents for lines R0 and R1: (50:50) dichloromethane/methanol; *see* Subheading 2.1, **item 1** for preparation.

4. Rinse solvents for lines R2 and R3: 100% 2-propanol (isopropanol).

5. SelexION chemical modifier: 100% 1-propanol.

6. For sample analysis, 12 × 32 mm 2 mL glass vials are used with 250 µL pulled conical point glass inserts and 9 mm vial caps with pre-slit septa.

3 Methods

3.1 SelexION® Tuning

1. Prepare the SelexION tuning standard by warming one vial from the SelexION Tuning Kit to ambient temperature before opening. If lipids precipitate out of solution during storage, gentle warming followed by vortexing will redissolve lipids.

2. Fill a 1 mL syringe with the SelexION tuning mixture.

3. Prepare a "blank" sample by filling a 2 mL vial with 10 mM ammonium acetate in (50:50) dichloromethane/methanol and cap.

4. Place in the autosampler according to the tray layout (as detailed by the Lipidomics Workflow Manager (LWM) software).

5. Attach the syringe to the PEEK adaptor and set in the MS syringe holder.

6. Attach the tubing from the probe to the PEEK adaptor.

7. In the LWM software, in SelexION tuning **step 2**, click Next.

8. Click Start Run.

9. A window, "Purge Modifier," will automatically pop up. Click on the Purge button. Purging will automatically start and will finish after 4 min. A 30 min equilibration will automatically start after the purging.

10. While the system is purging/equilibrating, press the MS syringe button to start infusing the standard.

11. Complete tuning and save the updated compensation voltages to the MRM tables.

3.2 System Suitability Testing (SST)

For a **Quick Test**, the following steps are necessary:

1. Preparation of the SST LOD sample for the SST Quick Test.

2. Warm one vial from the System Suitability Test Kit to ambient temperature before opening. If lipids precipitate out of solution during storage, gentle warming followed by vortexing will redissolve lipids.

3. Aliquot 0.01 mL of the System Suitability Mixture into a vial.

4. Add 0.99 mL of 10 mM ammonium acetate (50:50) dichloromethane/methanol to the vial.

5. Cap the mixture and vortex gently for 5 s.

6. Transfer 0.25 mL of the mixture into an insert.

7. Place the insert in a vial and cap.

8. Place in the autosampler according to the tray layout (as detailed by the LWM software).

9. The remaining SST LOD mixture can be stored in the $-20\ ^{\circ}C$ freezer for future use.

10. Preparation of a "blank" sample.

11. Fill a 2 mL vial with the 10 mM ammonium acetate (50:50) dichloromethane/methanol and cap.

12. Place in the autosampler according to the tray layout (in LWM software).

For a **Comprehensive Test,** the following steps are necessary:

1. Purge the LC system.

2. Preparation of the SST LOD sample for the SST Comprehensive Test.

3. Warm one vial of System Suitability Test Kit to ambient temperature before opening. If lipids precipitate out of solution during storage, gentle warming followed by vortexing will redissolve lipids.

4. Aliquot 0.01 mL of the system suitability standard into a vial.

5. Add 0.99 mL of the 10 mM ammonium acetate dichloromethane/methanol (50:50) to the vial.

6. Cap the mixture and vortex gently for 5 s.

7. Transfer 0.25 mL of the mixture into an insert.

8. Place the insert in a vial and cap.

9. Place in the autosampler according to the tray layout (in LWM software).

10. The remaining SST LOD mixture can be stored in the −20 °C freezer for future use.

11. Preparation of the two SST RSD samples for the SST Comprehensive Test.

12. Warm a vial of the System Suitability Test Kit to ambient temperature before opening, if not done so already. If lipids precipitate out of solution during storage, gentle warming followed by vortexing will redissolve lipids.

13. Place two inserts into two separate vials.

14. Aliquot 0.05 mL of the System Suitability Mixture into each insert.

15. Add 0.2 mL of the 10 mM ammonium acetate dichloromethane/methanol (50:50) to each insert.

16. Cap both vials and vortex gently for 5 s.

17. Place in the autosampler according to the tray layout.

18. Preparation of a "blank" sample.

19. Fill a 2 mL vial with the 10 mM ammonium acetate dichloromethane/methanol (50:50) and cap.

20. Place in the autosampler according to the tray layout.

3.3 Internal Standard Mixture Preparation

1. Warm each of the internal standard vials to ambient temperature prior to opening. If lipids precipitate out of solution during storage, gentle warming followed by vortexing will redissolve lipids.

2. Before using each syringe, rinse the syringe three times with methanol and then three times with dichloromethane.

3. Vortex the first internal standard lot vial and uncap the vial. Add the appropriate amount of volume according to the Internal Standard Mixture Page in **step 4** in the LWM software.

4. Cap the internal standard lot vial.

5. Repeat **steps 2–4** for each internal standard class you wish to analyze.

6. Concentrate the internal standard mixture solution under nitrogen.

7. Add the calculated final volume (mL) of (50:50) dichloromethane/methanol from the Internal Standard Mixture Page in **step 4** (LWM software) to the vial. Cap the vial.

8. Portion the internal standard mixture into separate vials as needed according to the Internal Standard Mixture Page in **step 4** (in LWM software). Cap the vials.

9. If the mixture will not be used immediately, wrap the capped vial with Teflon tape, and store in a $-20\ ^\circ$C freezer.

10. Store the internal standard lot vial(s) in a $-20\ ^\circ$C freezer.

3.4 Sample Extraction

1. Thaw frozen serum or plasma experimental samples and QC control plasma for 30 min at room temperature.

2. Thaw the internal standard mixture (made in the previous section) and a vial of the QC Spike Kit for 30 min at ambient temperature.

3. Aliquot 0.1 mL of the serum or plasma samples to glass culture tubes.

4. To make the QC samples, aliquot 0.1 mL of the QC control plasma to glass culture tubes.

5. To make the QC Spike samples, aliquot 0.1 mL of the QC control plasma to glass culture tubes. Add 0.05 mL of the QC Spike mixture to each QC Spike sample.

6. To all samples, add 0.9 mL of water, 2 mL of methanol, and 0.9 mL of dichloromethane.

7. The extracts are gently vortexed for 5 s.

8. Add 0.1 mL of the internal standards mixture to the extracts.

9. The extracts are then set on the benchtop at room temperature for 30 min.

10. An additional 1 mL of water and 0.9 mL of dichloromethane are added to the extracts.

11. The extracts are gently vortexed for 5 s and then centrifuged on high speed for 10 min or until the extracts are separated into a bilayer. Note: Over-vortexing samples at this stage will lead to poor-phase partitioning.

12. The bottom organic layer is transferred to a new test tube for each extract.

13. Another 1.8 mL of dichloromethane is added to the original extract test tubes.

14. The original extracts are gently vortexed for 5 s and centrifuged on high speed for 10 min or until the extracts are separated into a bilayer. Note: Over-vortexing samples at this stage will lead to poor-phase partitioning.

15. The bottom layers are taken again and added to the previous aliquots.

16. The combined bottom layers for each sample are concentrated under nitrogen and reconstituted in 0.25 mL of 10 mM ammonium acetate in (50:50) dichloromethane/methanol.

17. The extracts are transferred to inserts and placed in vials for analysis on the Lipidyzer Platform.

3.5 DI-MS/MS Analysis

1. A QTRAP® system with SelexION Technology (SCIEX) is used for targeted profiling (SCIEX, MA, USA).

2. Two methods are used covering 13 lipid classes using a flow injection analysis (FIA): one injection with the SelexION voltages turned ON and another with the SelexION voltages turned OFF.

3. The lipid molecular species are measured using multiple reaction monitoring (MRM) and positive/negative switching. Positive ion mode detected the following lipid classes: SM/DAG/CE/CER/TAG. Negative ion mode detected the following lipid classes: LPE/LPC/PC/PE/FFA.

4. A flow injection analysis (FIA) setup is employed by using the LC to flow at an isocratic rate of 7 μL/min with a ramp up to 30 μL/min for the last 2 min of the experiment to allow for washing.

5. Data acquisition is around 20 min per sample, and 50 μL of the reconstituted sample is infused and the area under the flat infusion line reported and corrected to the appropriate internal standard.

6. Samples are quantified using the LWM software which reports all the detected lipids in nmol/g.

4 Notes

1. Proper storage of lipid standards is crucial to accurate lipid quantitation. It is recommended to store lipid standards at −20 °C.

2. All Lipidyzer kits and standards are stored in (50:50) dichloromethane/methanol which is a highly volatile solvent. Extreme caution should be used to prevent any evaporation during storage. Evaporation will cause the internal standards to degrade. Best practice is to store the standards after opened in amber glass vials wrapped with Teflon tape preferably under argon/nitrogen.

3. Standards should never be stored in plastic or polymer material containers as this will leach impurities from the container.

4. If lipids precipitate out during storage, then gentle warming at ambient temperature followed by mixing on a vortex mixer will redissolve them.

5. Ammonium acetate is hygroscopic, meaning it absorbs water over time. This can lead to insufficient acetate adduct ions being formed and thus impacting accurate quantitation. Store ammonium acetate in a desiccator.

6. The PEEKsil tubing can become blocked; make sure to carefully extract the organic (bottom) layer free from any protein precipitate. Also spinning of the reconstituted sample (in the LC-MS vial) eliminates any particulates in the sample to the bottom of the insert, thus eliminating this problem.

Acknowledgments

The author would like to thank the Lipidyzer development team at Metabolon (Raleigh, NC) and the Lipidyzer development team at SCIEX (Framingham, MA) including Avanti Polar Lipids (Alabaster, AL) for manufacturing the novel internal standards accompanying the Lipidyzer Platform.

References

1. Folch J, Lees M, Stanley GHS (1957) A simple method for the isolation and purification of total lipids from animal tissues. J Biol Chem 226: 497–509

2. Bligh EG, Dyer WJ (1959) A rapid method of total lipid extraction and purification. Can J Biochem Physiol 37(8):911–917

3. Matyash V, Liebisch G, Kurzchalia TV, Shevchenko A, Schwudke D (2008) Lipid extraction by methyl-tert-butyl ether for high-throughput lipidomics. J Lipid Res 49 (5):1137–1146

4. Lintonen TPI, Baker PRS, Suoniemi M, Ubhi BK, Koistinen KM, Duchoslav E, Campbell JL, Ekroos K (2014) Differential mobility spectrometry-driven shotgun lipidomics. Anal Chem 86(19):9662–9669

5. Ubhi BK, Watkins S (2015) High quality, reproducible, lipidomics: making lipid research routinely accessible. Chromatography Today

Part III

GC-MS-Based Metabolomics

Chapter 16

Exploratory GC/MS-Based Metabolomics of Body Fluids

Carole Migné, Stéphanie Durand, and Estelle Pujos-Guillot

Abstract

GC/MS-based metabolomics is a powerful tool for metabolic phenotyping and biomarker discovery from body biofluids. In this chapter, we describe an untargeted metabolomic approach for plasma/serum and fecal water sample profiling. It describes a multistep procedure, from sample preparation, oximation/silylation derivatization, and data acquisition using GC/QToF to data processing consisting in data extraction and identification of metabolites.

Key words Untargeted metabolomics, Gas chromatography, Mass spectrometry, Data processing

1 Introduction

Mass spectrometry (MS) coupled with advanced chromatographic techniques, such as liquid and gas chromatography (GC), has become a powerful metabolomic tool to identify subtle changes in metabolite profiles in biological systems. In particular, GC/MS is a widely used analytical platform in discovery-phase studies for metabolic profiling of various physiological biofluids, such as urine, blood, as well as fecal water samples [1–4]. It has very high separation efficiency, reproducibility, and high sensitivity. Moreover, it provides characteristic, reproducible, and standardized mass spectra, which allow identification when searching against databases. The major disadvantages of GC/MS are a limited analytical coverage (to a set of small volatile biological molecules, thermally stable, or to those that can be derivatized) and a long sample preparation requirement. In fact, in order to provide non-biased data, representative of sample metabolic complexity, GC/MS profiling requires adequate extraction of metabolites from the biological matrix and derivatization for nonvolatile compounds, often necessary to reduce polarity, to increase thermal stability and volatility [5], as well as to improve compound ionization. Because of its ability to derivatize a wide range of metabolites, the most commonly used derivatization method is based on a

Martin Giera (ed.), *Clinical Metabolomics: Methods and Protocols*, Methods in Molecular Biology, vol. 1730,
https://doi.org/10.1007/978-1-4939-7592-1_16, © Springer Science+Business Media, LLC 2018

two-step protocol involving a methoximation reaction followed by a silylation [6], but it requires a nonaqueous environment for the reaction and thereby leads to a more complex sample pretreatment. In most applications, 0.5–2 µL of TMS-derivatized samples are introduced into a heated injector (200–250 °C), and separation of metabolites is typically performed using a fused silica capillary column with a 5% diphenyl cross-linked 95% dimethylpolysiloxane stationary phase (0.25 µm film thickness) using helium as carrier gas [2]. Metabolomics for high-throughput experiments requires fast acquisition rates, which are now available, thanks to recent progress. GC is often coupled to single-quadrupole analyzers, which offer high sensitivity and good dynamic range but operate with slower scan rates and lower resolution compared to time-of-flight (ToF) systems. GC coupled with ToF mass spectrometry is increasingly used for metabolic profiling [7] because of fast acquisition rates, particularly useful for an accurate deconvolution of overlapping peaks obtained from complex mixtures.

The nontargeted approach is based on both annotated and unannotated peak information and therefore requires an important data processing step to handle the large amount of data. In particular, it consists in the alignment of chromatographic peaks along large sample sets, the annotation to keep only one entry for each metabolite [8], and finally metabolite identification. Highly diverse and specialized software solutions for GC/MS data processing have been published. Both automated peak extraction and automated deconvolution of mass spectra are necessary for the comprehensive analysis of GC/MS experiments [9]. The conventional approach in GC/MS is based on mass spectral deconvolution by processing the information present within single chromatograms. One of the most widely used tools is the automated mass spectral deconvolution and identification system (AMDIS; http://chemdata.nist.gov/mass-spc/amdis/overview.html). For GC-ToF data, commercial programs are also commonly used. However, to be able to compare large numbers of samples and to improve data processing throughput, software projects, based on the comprehensive extraction of mass selective peak apex intensities, were initiated by academia, for example, XCMS [10].

Finally, metabolite identification is performed by retention time comparison with pure standard compounds and/or comparison with mass spectral library databases (i.e., NIST). Although these libraries are extensive, they do not yet contain a large number of metabolites that are found in biological metabolic pathways. Public repositories of mass spectra of metabolites (i.e., MassBank [11]) are also used as complementary tools.

In this chapter, we focused on two important human biofluids, plasma/serum and fecal water. The objective is to provide a full description of untargeted metabolomics for biomarker discovery, from sample treatment and analysis to data processing and metabolite identification.

2 Materials

Prepare all solutions using ultrapure water (prepared by purifying deionized water to 18 MΩ-cm at 25 °C) and analytical grade reagents. All solvents should be analytical grade. Prepare and store all reagents and solvents at room temperature (unless indicated otherwise).

2.1 Standards and Preservatives

1. [13C1]-L-valine (99%), a stable isotope-labeled internal standard (IS), is obtained from Cambridge Isotope Laboratories (Andover, MA). Two solutions in ultrapure water are prepared at 100 μg/mL for fecal water extraction and 200 μg/mL for plasma or serum extraction. Dissolve 2 mg of [13C1]-L-valine in 1 mL of ultrapure water. Dilute this solution by 10 or 20 according to the biofluid extracted.

2. Sodium azide (NaN3 ≥ 99.0%) 100 mg/mL in ultrapure water: dissolve 0.5 g of NaN3 in 5 mL of ultrapure water (*see* **Note 1**).

2.2 Extraction

1. Heptane and methanol (HPLC Grade) are used for liquid extraction.

2. An aliquot of 200 mL of methanol is Stored at −20 °C during 24 h.

2.3 Derivatization Reagents for Gas Chromatography Mass Spectrometry

1. *N,O*-bis(trimethylsilyl)trifluoroacetamide (BSTFA) with 1% trimethylchlorosilane (TMCS) is purchased from Thermo Fisher Scientific and stored at 4 °C.

2. Pyridine is stored in a fume hood in amber glass vials in the dark.

3. Methoxylamine solution (15 mg/mL) in pyridine: dissolve 15 mg of methoxylamine in 1 mL of pyridine (*see* **Note 2**).

2.4 Gas Chromatography Mass Spectrometry

1. GC-QToF: use an Agilent 7890B Gas Chromatograph coupled to an Agilent Accurate Mass QTOF 7200, 7693A Injector (SSL) auto-sampler (Agilent Technologies, Inc.).

2. Gas chromatography column: HP-5MS UI 30 m × 0.25 mm i.d. × 0.25 μm film thickness (Agilent J&W Scientific, Folsom, CA, USA) with a deactivated fused silica precolumn 5 m × 0.32 mm i.d. (Agilent J&W Scientific, Folsom, CA, USA).

3. Use N11 crimp caps natural rubber/TEF 1.0 mm (Macherey-Nagel) and micro-vials with crimp neck N 11, amber, conical with a round pedestal glass plate 1.1 mL, 32 × 11.6 mm (Macherey-Nagel).

3   Methods

3.1   Fecal Water, Plasma, and Serum Sample Collection and Storage

1. For fecal water samples, homogenized stool samples (5–8 g) are ultracentrifuged at 4 °C and 171,500 × g using a Beckman Ti 70.1 rotor for 2 h. 2 µL of NaN3 as an antimicrobial agent per gram of fecal water is added, and supernatants are aliquoted as 1 mL samples and kept at −80 °C for a maximum period of 6 months.

2. Plasma is either collected on EDTA, heparin, or citrate. Then, 250 µL aliquots of collected plasma or serum are stored at −80 °C until analysis.

3.2   Processing of Fecal Water Samples

1. Thaw fecal water samples at room temperature.

2. Add 100 µL ice-cold methanol and subsequently 50 µL fecal water to an Eppendorf centrifuge tube (1.5 mL or smaller), vortex, keep at −20 °C for 30 min, and centrifuge (Sigma 3-16PK, Fischer Bioblok Scientific) at 15,493 × g at 4 °C for 10 min.

3. Transfer 90 µL supernatant to high recovery vial with crimp cap.

4. Spike 10 µL of 100 µg/mL IS (13C1-L-valine) to each vial and close. Store the sample vial (in a suitable container) at ≤− 20 °C for more than 30 min.

5. Remove the screw cap, and transfer vial to a freeze drier (precooling) for lyophilization (*see* **Note 3**).

6. Ensure rapid transfer into a drying oven and keep at 50 °C (about 1 min) to avoid absorbing moisture and close sample vial.

7. At the same time, a derivatization control sample (fecal water substituted by Milli-Q water) is prepared in order to determine the background noise produced during sample preparation, derivatization, and GC/MS analysis.

3.3   Processing of Plasma/Serum Samples

Extraction

1. Thaw blood samples at 4 °C overnight.

2. Add 200 µL of ice-cold methanol (−20 °C) to 100 µL **plasma** sample in an Eppendorf tube and vortex, or add 400 µL of ice-cold methanol (−20 °C) to 100 µL **serum** sample in an Eppendorf tube and vortex.

3. After precipitation of the proteins, store at −20 °C for 30 min.

4. Centrifuge the mixture at 15,493 × g for 10 min at 4 °C.

5. Transfer 200 µL supernatant in a brown glass vial of 2 mL, add 10 µL of [13C1]-L-valine (200 µg/mL), and evaporate to dryness using a SpeedVac.

6. At the same time, a derivatization control sample (plasma/serum is substituted by Milli-Q water) is prepared in order to determine the background noise produced during sample preparation, derivatization, and GC/MS analysis.

3.4 Derivatization

Process to the derivatization step straight away after the extraction.

1. Dissolve the dry residue by adding 80 µL of methoxylamine solution to each vial.

2. Vortex vigorously for 1 min and incubate at 37 °C for 24 h (in order to inhibit the cyclization of reducing sugars and the decarboxylation of α-keto acids).

3. Add 80 µL of BSTFA (1%TMCS) to the mixture and derivatize at 70 °C for 60 min.

4. Cool down to room temperature, and transfer 50 µL of derivatized mixture to a glass vial containing 100 µL of heptane, prior to injection.

5. In the same way, a pool sample is prepared from each extracted sample and derivatized in order to monitor the drift of the spectrometer during GC/MS analysis (*see* **Note 4**).

6. Transfer 50 µL of sample pool in a glass vial containing 100 µL of heptane prior to injection; repeat this step eight times (preprocessing analysis) (*see* **Note 5**).

3.5 Gas Chromatography Mass Spectrometry Analysis

1. Analytical sequence: to avoid possible differences between sample batches, randomize the analytical sequence using a Latin square.

2. Four heptane blanks are injected at the beginning of each sequence, followed by four pool samples, and then one pool sample (*see* **Note 4**) and one derivatization control sample after each set of 10 samples.

3. Chromatography method.

Use an Agilent HP-5MS column with helium as the carrier gas at a flow rate of 1 mL/min (ultrahigh purity helium gas (99.9990%) with the following conditions: injector temperature, 250 °C; transfer line heater, 280 °C; and injection volume, 0.5–2 µL in splitless mode (*see* **Note 6**).

The initial oven temperature is 60 °C for 2 min, ramped to 140 °C at a rate of 10 °C/min, to 240 °C at a rate of 4 °C/min, and to 300 °C at a rate of 10 °C/min and finally held at 300 °C for 8 min. The total run time is 49 min. The solvent delay is 5.5 min (*see* **Note 7**).

Agilent "retention time locking" (RTL) is applied to control the reproducibility of retention times (RT). [13C1]-L-valine (IS) is used to lock the GC method.

4. MS method.

Initially tune and calibrate the system using PFTBA. Subsequently, in the analytical sequence, a calibration is done between each sample.

Use the following conditions: electron ionization source temperature, 230 °C; quadrupole temperature, 150 °C; electron energy, 70 eV; ToF settings, acquisition 2GHzEDR with N2 (1.5 mL/min); mass range, m/z 50–800; acquisition rate, 5 spectra/s; acquisition time, 200 ms/spectrum; limits for average PPM error, 3.0, and maximum error, 8.0; and resolution, 8500 (fwhm) at m/z 501.9706.

Raw data files are transformed to mzData files in MassHunter B.07.01 software (Agilent).

3.6 Data Processing: Extraction and Alignment

1. Process the mzData files using XCMS [10, 12] to yield a data matrix containing retention times, accurate masses, and normalized peak intensities. Set XCMS settings (matchfilter algorithm) as follows: fwmh 10–12, step 0.01, Snthreshold 5, bw 10, and mzwid 0.1 (*see* **Note 8**).

2. Import the obtained peak list under the Galaxy web-based platform Worflow4metabolomics [13] (W4M), for quality checks and signal drift correction according to the algorithm described by Van der Kloet et al. [14], to correct for batch effects.

3.7 Metabolite Identification

1. Perform database queries, using the resulting peak list and W4M. If pure compounds (standards) are previously analyzed on the same instrument, use the "bank in-house" tool with a mass error of 0.005 Da and a retention time difference of 0.1 min. If not, tentative identifications can be obtained from comparison of exact masses to those registered in MassBank (https://metlin.scripps.edu).

2. Deconvolute the raw data files using Agilent MassHunter software package (Unknowns Analysis and Qualitative Analysis B.07.01) (or AMDIS, *see* **Note 9**) and compare the mass spectral data against a personal compound database library (Agilent MassHunter PCDL version B.07.00) containing spectra of pure compounds (standards) previously analyzed on the same instrument. If not compare to NIST library.

3. In the last step, integrate the results from both approaches.

4   Notes

1. NaN3 is toxic and should be handled with caution. NaN3 is fatal if swallowed or in contact with the skin: wear protective gloves, protective clothing, and eye protection to prepare this solution.

2. Use BSTFA with 1% TMS and pyridine in a fume hood.

3. Alternatively, it is possible to evaporate the sample to dryness using a SpeedVac.

4. A minimum of 400 μL of pool sample has to be prepared to be able to aliquot it in 8 vials to be analyzed for batch correction. If more pool samples or backup samples are needed, adjust the volume consequently.

5. Heptane should be used in a fume hood to avoid vapors. Care should be taken to avoid contact with the skin.

6. Adjust the injection volume to obtain a good signal-to-noise ratio.

7. Adjust the solvent delay to detect small volatile compounds if necessary.

8. Adjust the XCMS parameters to have a good detection of chromatographic peaks without duplicates.

9. In case of low-resolution data (from GC/MS simple quadrupole), AMDIS can be used alternatively for deconvolution.

Acknowledgments

This work was in part funded by the French National Research Agency within the MetaboHUB infrastructure (ANR-INBS-0010).

References

1. A J, Trygg J, Gullberg J, Johansson AI, Jonsson P, Antti H, Marklund SL, Moritz T (2005) Extraction and GC/MS analysis of the human blood plasma metabolome. Anal Chem 77(24):8086–8094. https://doi.org/10.1021/ac051211v

2. Pasikanti KK, Ho PC, Chan EC (2008) Gas chromatography/mass spectrometry in metabolic profiling of biological fluids. J Chromatogr B Analyt Technol Biomed Life Sci 871(2):202–211. https://doi.org/10.1016/j.jchromb.2008.04.033

3. Gao X, Pujos-Guillot E, Sebedio JL (2010) Development of a quantitative metabolomic approach to study clinical human fecal water metabolome based on trimethylsilylation derivatization and GC/MS analysis. Anal Chem 82(15):6447–6456. https://doi.org/10.1021/ac1006552

4. Tsugawa H, Bamba T, Shinohara M, Nishiumi S, Yoshida M, Fukusaki E (2011) Practical non-targeted gas chromatography/ mass spectrometry-based metabolomics platform for metabolic phenotype analysis. J Biosci Bioeng 112(3):292–298. https://doi.org/10.1016/j.jbiosc.2011.05.001

5. Dettmer K, Aronov PA, Hammock BD (2007) Mass spectrometry-based metabolomics. Mass Spectrom Rev 26(1):51–78. https://doi.org/10.1002/mas.20108

6. Kanani H, Chrysanthopoulos PK, Klapa MI (2008) Standardizing GC-MS metabolomics. J Chromatogr B Analyt Technol Biomed Life Sci 871(2):191–201. https://doi.org/10.1016/j.jchromb.2008.04.049

7. Almstetter MF, Oefner PJ, Dettmer K (2012) Comprehensive two-dimensional gas chromatography in metabolomics. Anal Bioanal Chem 402(6):1993–2013. https://doi.org/10.1007/s00216-011-5630-y

8. Koek MM, Jellema RH, van der Greef J, Tas AC, Hankemeier T (2011) Quantitative metabolomics based on gas chromatography mass spectrometry: status and perspectives.

Metabolomics 7(3):307–328. https://doi.org/10.1007/s11306-010-0254-3

9. Mastrangelo A, Ferrarini A, Rey-Stolle F, Garcia A, Barbas C (2015) From sample treatment to biomarker discovery: a tutorial for untargeted metabolomics based on GC-(EI)-Q-MS. Anal Chim Acta 900:21–35. https://doi.org/10.1016/j.aca.2015.10.001

10. Smith CA, Want EJ, O'Maille G, Abagyan R, Siuzdak G (2006) XCMS: processing mass spectrometry data for metabolite profiling using nonlinear peak alignment, matching, and identification. Anal Chem 78 (3):779–787. https://doi.org/10.1021/ac051437y

11. Horai H, Arita M, Kanaya S, Nihei Y, Ikeda T, Suwa K, Ojima Y, Tanaka K, Tanaka S, Aoshima K, Oda Y, Kakazu Y, Kusano M, Tohge T, Matsuda F, Sawada Y, Hirai MY, Nakanishi H, Ikeda K, Akimoto N, Maoka T, Takahashi H, Ara T, Sakurai N, Suzuki H, Shibata D, Neumann S, Iida T, Tanaka K, Funatsu K, Matsuura F, Soga T, Taguchi R, Saito K, Nishioka T (2010) MassBank: a public repository for sharing mass spectral data for life sciences. J Mass Spectrom: JMS 45 (7):703–714. https://doi.org/10.1002/jms.1777

12. Benton HP, Wong DM, Trauger SA, Siuzdak G (2008) XCMS2: processing tandem mass spectrometry data for metabolite identification and structural characterization. Anal Chem 80 (16):6382–6389. https://doi.org/10.1021/ac800795f

13. Giacomoni F, Le Corguille G, Monsoor M, Landi M, Pericard P, Petera M, Duperier C, Tremblay-Franco M, Martin JF, Jacob D, Goulitquer S, Thevenot EA, Caron C (2015) Workflow4Metabolomics: a collaborative research infrastructure for computational metabolomics. Bioinformatics 31 (9):1493–1495. https://doi.org/10.1093/bioinformatics/btu813

14. van der Kloet FM, Bobeldijk I, Verheij ER, Jellema RH (2009) Analytical error reduction using single point calibration for accurate and precise metabolomic phenotyping. J Proteome Res 8(11):5132–5141. https://doi.org/10.1021/pr900499r

GC-MS Analysis of Short-Chain Fatty Acids in Feces, Cecum Content, and Blood Samples

Lisa R. Hoving, Marieke Heijink, Vanessa van Harmelen, Ko Willems van Dijk, and Martin Giera

Abstract

Short-chain fatty acids, the end products of fermentation of dietary fibers by the gut microbiota, have been shown to exert multiple effects on mammalian metabolism. For the analysis of short-chain fatty acids, gas chromatography–mass spectrometry is a very powerful and reliable method. Here, we describe a fast, reliable, and reproducible method for the separation and quantification of short-chain fatty acids in mouse feces, cecum content, and blood samples (i.e., plasma or serum) using gas chromatography–mass spectrometry. The short-chain fatty acids analyzed include acetic acid, propionic acid, butyric acid, valeric acid, hexanoic acid, and heptanoic acid.

Key words Short-chain fatty acids, Gas chromatography–mass spectrometry, GC-MS, Targeted metabolomics, PFBBr

1 Introduction

Lipids and fatty acids (FAs) are essential molecules in the regulation and control of various biological functions and play a role in the onset and progression of disease [1]. FAs with less than 8 carbon atoms are considered short-chain fatty acids (SCFAs) [2]. SCFAs (predominantly acetic acid, propionic acid, and butyric acid with respectively 2, 3, and 4 carbon atoms) are mainly produced in the colonic lumen after anaerobic fermentation of indigestible carbohydrates by saccharolytic gut bacteria [3]. The link between diet, the gut microbiota, the production of SCFAs and their role in human health and disease is an active area of research [4]. This requires suitable analytical techniques for sensitive and accurate quantification of SCFAs. One technique traditionally used for the analysis of small, volatile molecules is gas chromatography–mass spectrometry (GC-MS). Here, were describe step by step the quantitative analysis of the SCFAs acetic acid, propionic acid, butyric acid, valeric acid, hexanoic acid, and heptanoic acid using GC-MS

Martin Giera (ed.), *Clinical Metabolomics: Methods and Protocols*, Methods in Molecular Biology, vol. 1730, https://doi.org/10.1007/978-1-4939-7592-1_17, © Springer Science+Business Media, LLC 2018

in feces, cecum content, as well as in blood samples (i.e., plasma or serum).

GC-MS is an analytical technique, well suited for the analysis of SCFAs and other (longer) FAs [5]. However, one critical step in the GC-MS analysis of FAs is their conversion into suitable volatile derivatives by derivatization (e.g., by alkylation or silylation) [6]. Traditionally, FAs are being transformed into their methyl ester, or trimethylsilyl ester derivatives in GC-MS analysis [7]. While both approaches work well for longer chain FAs, the intrinsically low boiling point of the SCFA methyl ester or trimethylsilyl ester derivatives results in some issues with their GC-MS based analysis. For example, the trimethylsilyl ester of acetic acid roughly presents the same boiling point as the commonly used derivatization reagents, leading to severe signal overlap.

Alternatively, for SCFA analysis FAs can be derivatized by the alkylation reagent pentafluorobenzyl bromide (PFBBr) [8, 9]. The benzyl bromide group reacts with the carboxylic acid group to form an ester, allowing analysis as pentafluorobenzyl ester. Additionally, this so-formed ester presents ideal properties for electron-capture negative ionization (ECNI), which is a highly selective and sensitive ionization technique. ECNI allows analysis of the negatively charged molecular ions, usually detected in the single ion monitoring mode (SIM) on quadrupole based mass spectrometers.

For targeted analysis of SCFAs, ideally isotopically labeled internal standards (IS) should be used. The use of IS enables quantitative analysis of biological samples and greatly improves specificity [7, 9].

Apart from GC-MS, Nuclear magnetic resonance (NMR) spectroscopy can be used to analyze SCFAs in feces or cecum content. An interplatform comparison performed in our lab, comparing GC-MS and NMR spectroscopy, showed good correlations for the measurements of SCFA concentrations. However, the advantage of GC-MS over NMR is its higher sensitivity, which is essential for the analysis of SCFAs at low concentrations such as SCFAs present in blood.

2   Materials

Use only high purity solvents (preferably LC-MS grade) in order to prevent elevated background signals (*see* **Note 1**). An overview of the amounts of materials and chemicals is provided in Table 1. If vendors of different materials are specifically mentioned in this section, the use of these materials are recommended based on our previous experiences.

Table 1
Chemicals and materials required per sample for the quantification of SCFAs. For every type of biological matrix, take along three blank samples. Blank samples should be processed in exactly the same way as biological samples

Chemical/material	Calibration series sample	Biological sample		
		Feces	Cecum content	Plasma/serum
Biological sample	–	±50 mg	±10 mg	10 µL
Water	250 µL	550 µL	690 µL	250 µL
Acetone	250 µL	250 µL	250 µL	250 µL
1 µg/mL IS solution	10 µL	10 µL	10 µL	10 µL
Standards in EtOH	10 µL[a]	–	–	–
EtOH	–	10 µL	10 µL	10 µL
172 mM PFBBr	100 µL	100 µL	100 µL	100 µL
n-hexane	500 µL	500 µL	500 µL	500 µL
1.5 mL plastic tubes	–	2	3	–
Clean steel beads	–	2	2	–
Glass autosampler vials	2	2	2	2
Glass autosampler inserts	1	1	1	1
Glass autosampler caps	2	2	2	2

[a]For every individual sample of the calibration series, a specific concentration of standards in EtOH is used

2.1 Materials for Sample Preparation

1. Glass autosampler vials, inserts, and caps. It is important to use the highest quality glass ware. We recommend to use the following items: Agilent certified 2 mL vials with screw top; Agilent certified 250 µL inserts with polymer feet; Agilent screw caps with PTFE/red silicone septum.

2. 1 µg/mL IS solution in ethanol (EtOH) (*see* **Note 2**): mix acetic acid-d4, propionic acid-d6 and butyric acid-d8 and dissolve in EtOH to a final concentration of 1 µg/mL. Store at −80 °C. Apart from butyric acid, butyric acid-d8 is also used as the IS for valeric acid, hexanoic acid, and heptanoic acid.

3. Concentration series of SCFA standards in EtOH: use the SCFA standard mixture (Sigma-Aldrich) and dilute with EtOH. Prepare concentrations ranging from the lower limit of quantification (LLOQ) (*see* Table 2) to 1000 µM. Store at −80 °C.

4. 172 mM PFBBr in acetone: add 26.8 µL PFBBr to 1 mL acetone. Prepare fresh daily.

Table 2
Overview of FAs. For each SCFA, an indication of the retention time (RT), the *m/z*-value, an indication of the LLOQ, and the IS to be used are shown

FA	Name	RT (min)	Monitored *m/z* (M⁻)	LLOQ (μM)[a]	IS
FA 2:0-d4	Acetic acid-d4	7.19	62	N/A	N/A
FA 2:0	Acetic acid	7.22	59	20	FA 2:0-d4
FA 3:0-d6	Propionic acid-d6	7.86	78	N/A	N/A
FA 3:0	Propionic acid	7.89	73	5	FA 3:0-d6
FA 4:0-d8	Butyric acid-d8	8.41	94	N/A	N/A
FA 4:0	Butyric acid	8.44	87	2	FA 4:0-d8
FA 5:0	Valeric acid	8.99	101	1	FA 4:0-d8
FA 6:0	Hexanoic acid	9.51	115	5	FA 4:0-d8
FA 7:0	Heptanoic acid	9.99	129	1	FA 4:0-d8

[a]An indication of the lowest concentration to be included in the calibration series. This LLOQ is determined for every individual experiment. The calibration series samples are measured twice. A specific concentration is included if signal/noise >10 and if the accuracy based on the calibration obtained ≥80 and ≤120% for both measurements
N/A not applicable

5. Clean steel beads (only for fibrous biological matrices): rinse 3.2 mm stainless steel beads with methanol and dry at room temperature.

2.2 Materials for GC-MS

1. GC with split/splitless injector, coupled to a quadrupole mass spectrometer with chemical ionization source.

2. Injection: autosampler (recommended).

3. GC column: use an Agilent VF-5 ms column (5% phenyl-methyl; 25 m × 0.25 mm internal diameter; 0.25 μm film thickness).

4. Pure helium (99.9990%) and methane (99.9995%) should be used as carrier and as chemical ionization gas, respectively.

3 Methods

SCFAs are ubiquitous. Hence, extra care has to be taken to prevent sample contamination. Possible sources of contamination include environmental air, pipettes, pipette tips, low quality plastics, and (low purity) solvents (*see* **Note 1**). Interday and intraday repeatability of the method, validated in fetal calf serum (FCS), is provided in Table 3.

Table 3
Interday and intraday repeatability data in FCS. FCS was spiked with 5 µM and 100 µM SCFA. Acetic acid, propionic acid and butyric acid in these samples were quantified in triplicate on three different days using the described method

FA	Day	FCS Mean (µM)	RSD	FCS + 5 µM SCFA Mean (µM)	RSD	FCS + 100 µM SCFA Mean (µM)	RSD	RE[a]
Acetic acid	Intraday 1	108.5	12%	113.4	12%	204.4	3%	−4%
	Intraday 2	126.5	5%	120.8	4%	180.2	3%	−46%
	Intraday 3	117.8	11%	121.4	6%	192.5	6%	−25%
	Interday	117.6	8%	118.5	4%	192.4	6%	−25%
Propionic acid	Intraday 1	<LLOQ	N/A	10.5	18%	101.3	2%	−4%
	Intraday 2	<LLOQ	N/A	9.3	8%	89.4	4%	−16%
	Intraday 3	8.1	24%	13.5	8%	96.5	2%	−13%
	Interday	N/A	N/A	11.1	19%	95.7	6%	−11%
Butyric acid	Intraday 1	<LLOQ	N/A	7.3	24%	96.7	5%	−6%
	Intraday 2	<LLOQ	N/A	5.7	9%	87.9	7%	−14%
	Intraday 3	<LLOQ	N/A	6.7	6%	80.8	10%	−22%
	Interday	<LLOQ	N/A	6.6	12%	88.5	9%	−14%

[a]RE, based on the difference in the determined concentrations. For acetic acid, the RE is determined based on the difference between the FCS and FCS + 100 µM SCFA sample. For propionic acid and butyric acid, the RE is determined based on the difference between the FCS + 5 µM SCFA and FCS + 100 µM SCFA sample
N/A not applicable, *RE* relative error, *RSD* relative standard deviation

3.1 Sample Preparation of Feces, Cecum Content, and Blood

1. Facilitate rapid sampling. Store the samples at −80 °C upon collection if the samples are not prepared immediately (*see* **Note 3**).

2. Matrix dependent preprocessing of feces:

   Prepare an aqueous extract of feces. Weigh feces (approximately 50 mg mouse feces) (*see* **Note 4**) in a 1.5 mL plastic tube with 0.1 mg accuracy and add 300 µL water. Homogenize the sample using a bullet blender: add two clean 3.2 mm steel beads and blend the sample for 5 min. Centrifuge at 1400 × *g* for 10 min. Transfer the supernatant to a fresh 1.5 mL plastic tube.

3. Matrix dependent preprocessing of cecum content:

   Weigh cecum content (approximately 10 mg mouse cecum content) (*see* **Note 4**) in a 1.5 mL plastic tube with 0.1 mg accuracy and add 400 µL water. Homogenize by vortexing. Use a bullet blender if the material is fibrous: add two clean 3.2 mm steel beads and blend the sample for 5 min. Centrifuge at 1400 × *g* for 10 min. Transfer the supernatant to a fresh 1.5 mL plastic tube. Dilute the supernatant 1:5 with water in a total volume of 50 µL using a fresh 1.5 mL plastic tube.

4. Matrix dependent preprocessing of blood:

   Obtain plasma and/or serum. No further preprocessing is needed.

5. Prepare a glass autosampler vial for every sample:

6. For calibration samples, add 250 μL acetone, 10 μL 1 μg/mL IS solution, and 10 μL of the calibration series SCFA standards at the desired concentration. In case of feces or cecum content analysis, add 10 μL water which is preprocessed exactly the same as the biological samples. This includes bullet blending if necessary.

7. For biological samples, add 250 μL acetone (*see* **Note 5**), 10 μL 1 μg/mL IS solution in EtOH (*see* **Note 2**), 10 μL EtOH (*see* **Note 6**), and 10 μL aqueous feces, 10 μL cecum content extract, or 10 μL plasma/serum into a glass autosampler vial.

8. For blank samples, add 250 μL acetone, 10 μL 1 μg/mL IS solution, and 10 μL EtOH into a glass autosampler vial. In case of feces or cecum content analysis, 10 μL water should be added which is preprocessed in exact the same way as the biological samples. This includes bullet blending if necessary. For every type of biological matrix used in an experiment, three blank samples should be included.

9. Vortex all samples.

10. Add 100 μL 172 mM PFBBr in acetone (*see* **Note 7**). Vortex all samples.

11. Heat the samples at 60 °C for 30 min in a laboratory stove. Let the samples cool down to room temperature (approximately 15 min) (*see* **Note 8**).

12. Add 500 μL *n*-hexane and 250 μL water to the samples. Shake the vial in vertical direction for approximately 10 s. Let the samples rest for 1 min at room temperature.

13. Prepare a new empty glass autosampler vial with a glass insert for every sample. Transfer 250 μL of the *n*-hexane (upper layer) into the glass insert.

3.2 GC-MS Analysis

1. Inject 1 μL in the GC-MS, splitless at 280 °C.

2. Use helium as carrier gas at a constant flow rate of 1.20 mL/min.

3. Use the following temperature gradient: 1 min at 40 °C, linear increase at 40 °C/min to 60 °C, held for 3 min at 60 °C, linear increase at 25 °C/min to 210 °C, linear increase at 40 °C/min to 315 °C, and held for 3 min at 315 °C.

4. Set the transfer line temperature at 280 °C.

5. Keep the ionization source temperature at 280 °C.

6. Use methane as chemical ionization gas at approximately 15 psi.

7. Detect ions obtained in the negative mode using SIM (*see* **Notes 9** and **10**). Table 2 provides the *m/z*-values to be monitored and an indication of retention times (RT). As a consequence of small chromatographic differences (e.g., GC column length), the exact RT varies between various GC systems. Hence, calibration using external standards is mandatory.

3.3 Data Analysis

1. Integrate the obtained signal (*see* **Note 11**).

2. Calculate the relative retention time (RRT) and area ratios using the respective IS (*see* Table 2) (*see* **Notes 12** and **13**).

3. Determine the slope and LLOQ for every SCFA by performing linear regression. It is recommended to use a weighing factor of $1/x^2$ [10].

4. Calculate the SCFA concentrations by using the area ratios obtained from the biological samples, average signal of the blank samples as intercept (*see* **Note 14**), and the slopes obtained from the analysis of the calibration series samples. Take into account the sample dilution for feces and cecum content.

4 Notes

1. SCFAs (especially acetic acid and propionic acid) usually show high background signals, resulting in a relatively high LLOQ (*see* Fig. 1). SCFA background signals can be diminished by using high purity solvents (preferably LC-MS grade). Additionally, use glass vials for organic solvents. For plastic tubes, we strongly recommend to use Eppendorf polypropylene tubes.

2. The IS signal should be present in every sample. The IS is used to correct for differences in sample preparation between the samples. Use exactly the same batch of 1 μg/mL IS solution in EtOH for the entire experiment, as minor differences in IS composition potentially translate into systematic under- or overestimation of SCFAs in samples.

3. Collect the biological samples as quick as possible and store the samples at −80 °C. Levels of SCFAs within biological samples are vulnerable for change, especially when the collection is performed slowly or when samples are improperly stored. SCFAs can evaporate from the samples or SCFAs from the air can contaminate the samples.

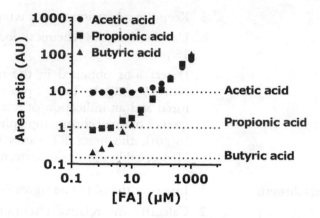

Fig. 1 Background signal of SCFAs. Several SCFAs usually show a high background signal. As a consequence, the LLOQ for these SCFAs is dependent on the background obtained. In our experience, the background signal of acetic acid is higher than that of propionic acid, which in turn is higher than the background signal of butyric acid. The dashed lines in the graph show the extent of the background signals

4. The amount of sample that is required for the analysis of SCFAs might vary between biological samples from different species.

5. Acetone facilitates the precipitation of proteins.

6. The addition of 10 μL EtOH to the samples ensures that the solvents of the biological samples are matched to the solvents in the calibration samples.

7. Within this protocol no base is added to catalyze the derivatization reaction, since the addition of base can severely increase SCFA background signal [8].

8. *n*-Hexane is added after the samples have been cooled down in order to prevent evaporation and spilling.

9. Sensitivity is higher when the MS is operated in SIM mode as compared to the full scan mode. However, the full scan mode can be useful to detect FAs that are not incorporated in the SIM method, or to determine the RT of a specific SCFA. If one decides to operate in full scan mode, an *m/z*-range of 50–150 is recommended to be used.

10. For isotopolog analysis, either *m/z*-values corresponding to isotopologs can be added to the SIM method (e.g., M0, M1, M2, etc. for every SCFA) [8], or the MS can be operated in scheduled scan mode (e.g., scan window including *m/z*-values corresponding to M0, M1, M2, etc., for every SCFA).

11. Pyruvic acid has the same mass and almost the same RT as butyric acid and both acids are derivatized by PFBBr. Consecutively, care has to be taken when both analytes are simultaneously present in the sample (*see* Fig. 2). Particularly in plasma

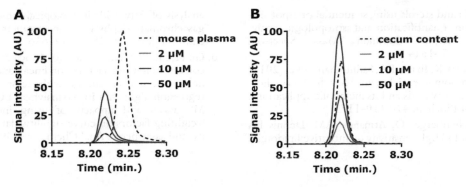

Fig. 2 Signal interference between butyric acid (RT = 8.22 min) and pyruvic acid (RT = 8.25 min). Butyric acid and pyruvic acid are eluting closely while being monitored in the same SIM trace. Particularly in plasma (**a**) and serum samples, pyruvic acid and butyric acid are simultaneously present. In feces and cecum content (**b**), pyruvic acid is usually not detected

and serum samples, pyruvic acid and butyric acid are simultaneously present. In feces and cecum content, pyruvic acid is usually not detected.

12. $$\text{RRT} = \frac{\text{retention } time\ analyte}{\text{retention } time\ IS}$$

13. $$\text{Area ratio} = \frac{\text{area analyte}}{\text{area IS}}$$

14. The blank samples reflect the background signal of the biological samples more accurately than the intercept obtained from the linear regression of the calibration series samples. Therefore, use the average area ratio of the blank samples as background signal/intercept to calculate the concentrations of the biological samples. Use the following formula:

$$\text{Concentration} = \frac{\text{area ratio} - \text{average area ratio blank samples}}{\text{slope}}$$

References

1. Siri-Tarino PW, Chiu S, Bergeron N et al (2015) Saturated fats versus polyunsaturated fats versus carbohydrates for cardiovascular disease prevention and treatment. Annu Rev Nutr 35:517–543

2. Desbois AP, Smith VJ (2010) Antibacterial free fatty acids: activities, mechanisms of action and biotechnological potential. Appl Microbiol Biotechnol 85(6):1629–1642

3. Cummings JH, Pomare EW, Branch WJ et al (1987) Short chain fatty acids in human large intestine, portal, hepatic and venous blood. Gut 28:1221–1227

4. O'Keefe SJD (2016) Diet, microorganisms and their metabolites, and colon cancer. Nat Rev Gastroenterol Hepatol 13:691–706

5. Dołowy M, Pyka A (2015) Chromatographic methods in the separation of long-chain mono- and polyunsaturated fatty acids. J Chem 2015:20

6. Gao P, Xu G (2015) Mass-spectrometry-based microbial metabolomics: recent developments and applications. Anal Bioanal Chem 407:669–680

7. Kloos D-P, Gay E, Lingeman H et al (2014) Comprehensive gas chromatography-electron ionisation mass spectrometric analysis of fatty

acids and sterols using sequential one-pot silylation: quantification and isotopologue analysis. Rapid Commun Mass Spectrom
28:1507–1514

8. Tomcik K, Ibarra RA, Sadhukhan S et al (2011)
Isotopomer enrichment assay for very short
chain fatty acids and its metabolic applications.
Anal Biochem 410:110–117

9. Quehenberger O, Armando AM, Dennis EA
(2011) High sensitivity quantitative lipidomics

analysis of fatty acids in biological samples by
gas chromatography-mass spectrometry. Biochim Biophys Acta 1811:648–656

10. Gu H, Liu G, Wang J et al (2014) Selecting the
correct weighting factors for linear and quadratic calibration curves with least-squares
regression algorithm in bioanalytical LC-MS/
MS assays and impacts of using incorrect
weighting factors on curve stability, data quality, and assay perfo. Anal Chem 86:8959–8966

Chapter 18

GC-MS Analysis of Medium- and Long-Chain Fatty Acids in Blood Samples

Lisa R. Hoving, Marieke Heijink, Vanessa van Harmelen, Ko Willems van Dijk, and Martin Giera

Abstract

Our body contains a wide variety of fatty acids that differ in chain length, the degree of unsaturation, and location of the double bonds. As the various fatty acids play distinct roles in health and disease, methods that can specifically determine the fatty acid profile are needed for fundamental and clinical studies. Here we describe a method for the separation and quantification of fatty acids ranging from 8 to 24 carbon chain lengths in blood samples using gas chromatography-mass spectrometry following derivatization using pentafluorobenzyl bromide. This method quantitatively monitors fatty acid composition in a manner that satisfies the requirements for comprehensiveness, sensitivity, and accuracy.

Key words Medium-chain fatty acids, Long-chain fatty acids, Gas chromatography-mass spectrometry (GC-MS), Targeted metabolomics, PFBBr

1 Introduction

Lipids and fatty acids (FAs) are present in all organisms and constitute essential structural elements of biological membranes, regulate and control cellular function, and are involved in the onset and progression of various diseases [1]. The majority of FAs are present as esters in lipids, such as triacylglycerols, sterol esters, and phospholipids. Only a small fraction is nonesterified, generally termed free FAs (FFAs) [2]. The role of FAs in health and disease has gained extensive interest. FAs differ in chain length, the degree of unsaturation, and location of the double bond(s). As various types of FAs have different associations with disease outcomes, the assessment of the FA composition in biological samples may provide suitable information. Therefore, great effort has been put in the development of comprehensive, sensitive, and reliable methodologies to quantify the FA composition. Here we describe step-by-step the quantitative analysis of the FA profile in plasma using GC-MS, measuring FAs with a chain length ranging from 8 to 24 carbons.

Martin Giera (ed.), *Clinical Metabolomics: Methods and Protocols*, Methods in Molecular Biology, vol. 1730, https://doi.org/10.1007/978-1-4939-7592-1_18, © Springer Science+Business Media, LLC 2018

GC-MS is an analytical technique that is well-suited for the analysis of the total amount of FAs as well as the FA composition within a sample [3]. As blood contains both esterified FAs and FFAs, a separate hydrolysis step is required during sample preparation to determine the total FA composition. For the analysis of FAs by GC-MS, it is of importance to convert FAs into suitable volatile derivatives by derivatization (e.g., alkylation or silylation) [4]. Traditionally in GC-MS analysis, FAs were being transformed into their methyl ester or trimethylsilyl ester derivatives [5]. Alternatively, pentafluorobenzyl bromide (PFBBr) can be used to derivatize FAs. It has been successfully applied for the analysis of FAs with different chain lengths [6–9]. The benzyl bromide group reacts with the carboxylic acid group to form a pentafluorobenzyl ester. In addition, this pentafluorobenzyl ester contains ideal properties for electron capture negative ionization (ECNI), a highly selective and sensitive ionization technique. The combination of PFBBr derivatization and ECNI ionization allows for the analysis of negatively charged molecular ions. These ions are usually detected in the single ion monitoring (SIM) mode on quadrupole-based mass spectrometers. Isotopically labeled internal standards (IS) have to be used for quantitative analysis of FAs by GC-MS. The addition of IS in GC-MS analysis enables quantitative analysis of biological samples and greatly improves detection specificity [10].

2 Materials

Use only high-purity solvents (preferably LC-MS grade) in order to prevent increased background signals (*see* **Note 1**). If vendors from different materials are mentioned in this method section, the use of these chemicals is recommended based on our previous experiences. The use of 10 M NaOH forms an exception. In order for the method to succeed, it is urged to use the specific items mentioned in the Materials section. An overview of the amounts of materials and chemicals is provided in Table 1.

2.1 Materials for Sample Preparation

1. Glass autosampler vials, inserts, and caps. It is recommended to use Agilent certified 2 mL vials with screw top; Agilent certified 250 μL inserts with polymer feet; and Agilent screw caps with PTFE/red silicone septum.

2. 1 μg/mL IS solution in ethanol (EtOH) (*see* **Note 2**): accurately weigh decanoic acid-d19, palmitic acid-d31, and arachidonic acid-d8, and dissolve in EtOH to a final concentration of 1 μg/mL. Store at −80 °C.

3. Concentration series of FA standards in EtOH: use GLC reference standard 85 mix (Nu-Chek Prep), eicosapentaenoic acid (Cayman), docosapentaenoic acid (Cayman), and docosahexaenoic acid (Cayman), and serially dilute using EtOH. Prepare

Table 1
Chemicals and materials needed per sample for the quantification of FAs

Chemical/material	Calibration series sample	Plasma/serum sample
Plasma/serum	–	10 µL
Acetone	250 µL	250 µL
10 M NaOH	–	10 µL
1 µg/mL IS solution	10 µL	10 µL
Standards in EtOH	10 µL[a]	–
EtOH	–	10 µL
172 mM PFBBr	100 µL	100 µL
n-Hexane	500 µL	500 µL
Water	250 µL	250 µL
Glass autosampler vials	2	2
Glass autosampler inserts	1	1
Glass autosampler caps	2	2

For every batch of samples, take along three blank samples. Blank samples should be processed in exactly the same way as biological samples
[a]For every individual sample of the calibration series, a specific concentration of standards in EtOH is used

concentrations ranging from the lower limit of quantification (LLOQ) (*see* Table 2) to 50 µg/mL. Store at −80 °C.

4. 172 mM PFBBr in acetone: add 26.8 µL PFBBr to 1 mL acetone. Prepare fresh daily.

5. 10 M NaOH in water. In order for the method to succeed, it is urged to use a prepared solution from Sigma-Aldrich (Art. No. 72068) (*see* **Note 3**).

2.2 Materials for GC-MS

1. GC with split/splitless injector, coupled to a quadrupole mass spectrometer with chemical ionization source.

2. Injection: autosampler (recommended).

3. GC column: use an Agilent VF-5 ms column (5% phenylmethyl; 25 m × 0.25 mm internal diameter; 0.25 µm film thickness).

4. Pure helium (99.9990%) and methane (99.9995%) are used as carrier and chemical ionization gas, respectively.

3 Methods

Palmitic acid and stearic acid are ubiquitous. Hence, extra care has to be taken to prevent sample contamination. Sources of contamination include low-quality plastics and (low-purity) solvents (*see* **Note 1**).

Table 2
Overview of FAs

FA	Name	RT (min)	Monitored m/z (M⁻)	LLOQ (ng/mL)[a]	IS
FA 08:0	Octanoic acid	10.14	143.1	200	FA 10:0-d19
FA 10:0	Decanoic acid	11.01	171.1	50	FA 10:0-d19
FA 10:0-d19	Decanoic acid-d19	10.92	190.3	N/A	N/A
FA 11:0	Undecanoic acid	11.42	185.2	50	FA 10:0-d19
FA 12:0	Lauric acid	11.87	199.2	100	FA 10:0-d19
FA 13:0	Tridecanoic acid	12.40	213.2	50	FA 10:0-d19
FA 14:0	Myristic acid	13.03	227.2	50	FA 16:0-d31
FA 14:1 (n-5)	Myristoleic acid	12.97	225.2	20	FA 16:0-d31
FA 15:0	Pentadecanoic acid	13.78	241.2	50	FA 16:0-d31
FA 15:1 (n-5)	10-Pentadecenoic acid	13.72	239.2	10	FA 16:0-d31
FA 16:0	Palmitic acid	14.68	255.2	500	FA 16:0-d31
FA 16:0-d31	Palmitic acid-d31	14.43	286.4	N/A	N/A
FA 16:1 (n-7)	Palmitoleic acid	14.50	253.2	50	FA 16:0-d31
FA 17:0	Heptadecanoic acid	15.72	269.3	20	FA 16:0-d31
FA 17:1 (n-7)	10-Heptadecenoic acid	15.53	267.2	10	FA 16:0-d31
FA 18:0	Stearic acid	16.54	283.3	500	FA 16:0-d31
FA 18:1 (n-9) cis	Oleic acid	16.37	281.3	100	FA 16:0-d31
FA 18:1 (n-9) trans	Elaidic acid	16.41	281.3	50	FA 16:0-d31
FA 18:2 (n-6)	Linoleic acid	16.34	279.2	50	FA 16:0-d31
FA 18:3 (n-6)	Gamma-linolenic acid (GLA)	16.15	277.2	50	FA 16:0-d31
FA 18:3 (n-3)	Alpha-linolenic acid (ALA)	16.40	277.2	50	FA 16:0-d31
FA 20:0	Arachidic acid	17.66	311.3	50	FA 20:4-d8
FA 20:1 (n-9)	11-Eicosenoic acid	17.55	309.3	20	FA 20:4-d8
FA 20:2 (n-6)	11,14-Eicosadienoic acid	17.54	307.3	10	FA 20:4-d8
FA 20:3 (n-6)	Homo-Gamma linolenic acid (DGLA)	17.43	305.3	10	FA 20:4-d8
FA 20:3 (n-3)	11,14,17-Eicosatrienoic acid	17.58	305.3	10	FA 20:4-d8
FA 20:4 (n-6)	Arachidonic acid (AA)	17.28	303.2	10	FA 20:4-d8
FA 20:4-d8	Arachidonic acid-d8 (AA-d8)	17.26	311.3	N/A	N/A
FA 20:5 (n-3)	Eicosapentaenoic acid (EPA)	17.33	301.2	10	FA 20:4-d8

(continued)

Table 2
(continued)

FA	Name	RT (min)	Monitored *m/z* (M⁻)	LLOQ (ng/mL)[a]	IS
FA 22:0	Behenic acid	18.48	339.3	50	FA 20:4-d8
FA 22:1 (n-9)	Erucic acid	18.40	337.3	20	FA 20:4-d8
FA 22:2 (n-6)	13,16-Docosadienoic acid	18.39	335.3	10	FA 20:4-d8
FA 22:4 (n-6)	Adrenic acid (AdA)	18.21	331.3	10	FA 20:4-d8
FA 22:5 (n-3)	Docosapentaenoic acid (DPA)	18.25	329.3	10	FA 20:4-d8
FA 22:6 (n-3)	Docosahexaenoic acid (DHA)	18.15	327.2	20	FA 20:4-d8
FA 24:1 (n-9)	Nervonic acid	19.28	365.4	20	FA 20:4-d8

For each FA, an indication of the retention time (RT), the *m/z* value, an indication of the LLOQ, and the IS to be used are shown. *N/A* not applicable
[a]An indication of the lowest concentration to be included in the calibration series. This LLOQ is determined for every individual experiment. The calibration series samples are measured twice. A specific concentration is included if signal/noise >10 and if the accuracy based on the calibration obtained ≥80 and ≤120% for both measurements

3.1 Sample Preparation

1. Facilitate rapid sampling. Store samples at −80 °C upon collection if the samples are not prepared immediately (*see* **Note 4**).

2. Prepare a glass autosampler vial for every sample: for calibration samples, add 250 µL acetone and 10 µL of your calibration series FA standards at the desired concentration. An indication of the expected LLOQ for every FA is provided in Table 2. For biological samples, add 250 µL acetone (*see* **Note 5**), 10 µL EtOH (*see* **Note 6**), and 10 µL plasma or serum into a glass autosampler vial. For blank samples, add 250 µL acetone and 10 µL EtOH into a glass autosampler vial. For every experiment, three blank samples should be included.

3. Hydrolyze the biological and blank samples (*see* **Note 7**): add 10 µL 10 M NaOH to the biological and blank samples. Vortex all samples. Heat the biological and blank samples at 60 °C for 30 min in a laboratory stove. Let the samples cool down to room temperature (approximately 15 min).

4. Add 10 µL 1 µg/mL IS solution (*see* **Notes 2** and **8**) to every sample and vortex all samples.

5. Add 100 µL 172 mM PFBBr in acetone (*see* **Note 9**). Vortex the samples.

6. Heat the samples at 60 °C for 30 min in a laboratory stove. Let the samples cool down to room temperature (approximately 15 min) (*see* **Note 10**).

7. Add 500 µL *n*-hexane and 250 µL water to the samples. Shake vigorously in vertical direction of the vial for 10 s. Let the samples rest for 1 min at room temperature.

8. Prepare a new empty glass autosampler vial with a glass insert for every sample. Transfer 250 μL of the *n*-hexane (upper layer) into the glass insert.

3.2 GC-MS Analysis

1. Inject 1 μL in the GC-MS, splitless at 280 °C.

2. Use helium as carrier gas at a constant flow rate of 1.20 mL/min.

3. Use the following temperature gradient: 1 min at 50 °C, linear increase at 40 °C/min to 60 °C, held for 3 min at 60 °C, linear increase at 25 °C/min to 237 °C, linear increase at 3 °C/min to 250 °C, linear increase at 25 °C/min to 315 °C, held for 1.55 min at 315 °C.

4. Set the transfer line temperature at 280 °C.

5. Keep the ionization source temperature at 280 °C.

6. Use methane as chemical ionization gas at approximately 15 psi.

7. Detect ions obtained in the negative mode using SIM analysis (*see* **Notes 11** and **12**). Table 2 provides the *m/z* values to be monitored and an indication of retention times (RT). As a consequence of small chromatographic differences (e.g., GC column length), the exact RT varies between various GC systems. Hence, calibration using external standards is mandatory.

3.3 Data Analysis

1. Integrate the obtained signal (*see* **Note 13**).

2. Calculate the relative retention times (RRT) and area ratios using the respective IS (*see* Table 2) (*see* **Notes 14** and **15**).

3. Determine the slope and LLOQ for every FA by performing linear regression. It is recommended to use a weighing factor of $1/x^2$ [11].

4. Calculate the FA concentrations by using the area ratios obtained from the biological samples, average signal of the blank samples as intercept (*see* **Note 16**), and the slopes obtained from the analysis of the calibration series samples.

4 Notes

1. Palmitic acid and stearic acid usually give high background signals, resulting in a relatively high LLOQ (Fig. 1). Sources of these FAs include solvents and plastic containers of inferior quality. Background signals of these FAs can be diminished by using high-purity solvents (preferably LC-MS grade). Additionally, use glass vials for organic solvents.

2. The IS signal should be present in every sample. The IS are used to correct for differences in sample preparation between the samples. Use exactly the same batch of 1 μg/mL IS

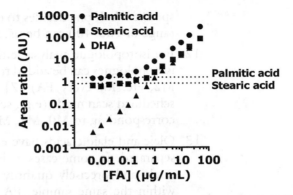

Fig. 1 Background signal of palmitic acid and stearic acid. Palmitic acid and stearic acid usually show a high background signal. As a consequence, the LLOQ for these FAs is higher than for FAs which do not show a high background signal like docosahexaenoic acid (DHA). The *dashed lines* in the graph show the size of the background signals

solution in EtOH for the entire experiment, as minor differences in IS composition might translate into systematic under- or overestimation of FAs in samples.

3. Sodium hydroxide pellets can be heavily contaminated by FAs.

4. Collect the biological samples as quickly as possible, and store the samples at −80 °C. Levels of FAs can change upon sample collection if the collection is performed slowly or when samples are stored improperly by auto-oxidation of polyunsaturated FAs and enzymatic hydroxylation.

5. Acetone facilitates the precipitation of proteins.

6. The addition of 10 μL EtOH to the samples ensures that the solvents of the biological samples are matched to the solvents in the calibration samples.

7. No esterified FAs are present in the calibration series samples. Therefore, no hydrolyzation step is needed.

8. Under highly alkaline conditions, risk of hydrogen–deuterium exchange exists [12]. Therefore, the IS need to be added after hydrolysis.

9. Within this protocol no base is added to catalyze the derivatization reaction, since the addition of base can severely increase FA background [7].

10. *n*-Hexane is added after the samples have been cooled down in order to prevent evaporation and spilling.

11. Sensitivity is higher when the mass spectrometer is operated in SIM mode as compared to the full scan mode. However, the full scan mode can be very useful to detect FAs that are not incorporated in the SIM method or to determine the RT of a

specific FA. If one decides to operate in full scan mode, an m/z range of 100–400 can be used.

12. For isotopologue analysis, either m/z values corresponding to isotopologues can be added to the SIM method (e.g., M0, M1, M2, etc. for every FA) [7] or the MS can be operated in scheduled scan mode (e.g., scan window including m/z values corresponding to M0, M1, M2, etc., for every FA).

13. Oleic and elaidic acids have equal masses and are not baseline separated. In some cases, it is therefore not possible to accurately and precisely quantify one or both of these two FAs within the same sample. FA 20:1 (n-9) and FA 22:1 (n-9), co-elute with, respectively, FA 20:2 (n-6) and FA 22:2 (n-6). As a consequence, the M2 isotopes (containing $2 \times {}^{13}C$ instead of ${}^{12}C$) of FA 20:2 (n-6) and FA 22:2 (n-6) might contaminate the FA 20:1 (n-9) and FA 22:1 (n-9) signal.

14. $RRT = \dfrac{\text{retention time analyte}}{\text{retention time IS}}$.

15. $\text{Area ratio} = \dfrac{\text{area analyte}}{\text{area IS}}$.

16. The blank samples reflect the background signal of the biological samples more accurately than the intercept obtained from the linear regression of the calibration series samples. Therefore, use the average area ratio of the blank samples as background signal/intercept to calculate the concentrations of the biological samples. Use the following formula: $\text{concentration} = \dfrac{\text{area ratio} - \text{average area ratio blank samples}}{\text{slope}}$.

References

1. Siri-Tarino PW, Chiu S, Bergeron N et al (2015) Saturated fats versus polyunsaturated fats versus carbohydrates for cardiovascular disease prevention and treatment. Annu Rev Nutr 35:517–543

2. van Meer G, Voelker DR, Feigenson GW (2008) Membrane lipids: where they are and how they behave. Nat Rev Mol Cell Biol 9:112–124

3. Dołowy M, Pyka A (2015) Chromatographic methods in the separation of long-chain mono- and polyunsaturated fatty acids. J Chem 2015

4. Gao P, Xu G (2015) Mass-spectrometry-based microbial metabolomics: recent developments and applications. Anal Bioanal Chem 407:669–680

5. Kloos D-P, Gay E, Lingeman H et al (2014) Comprehensive gas chromatography-electron ionisation mass spectrometric analysis of fatty acids and sterols using sequential one-pot silylation: quantification and isotopologue analysis. Rapid Commun Mass Spectrom 28:1507–1514

6. Quehenberger O, Armando AM, Dennis EA (2011) High sensitivity quantitative lipidomics analysis of fatty acids in biological samples by gas chromatography-mass spectrometry. Biochim Biophys Acta Mol Cell Biol Lipids 1811:648–656

7. Tomcik K, Ibarra RA, Sadhukhan S et al (2011) Isotopomer enrichment assay for very short chain fatty acids and its metabolic applications. Anal Biochem 410:110–117

8. Quehenberger O, Armando A, Dumlao D et al (2008) Lipidomics analysis of essential fatty acids in macrophages. Prostaglandins Leukot Essent Fat Acids 79:123–129

9. Pawlosky RJ, Sprecher HW, Salem N (1992) High sensitivity negative ion GC-MS method for detection of desaturated and chain-elongated products of deuterated linoleic and linolenic acids. J Lipid Res 33:1711–1717

10. Kloos D, Lingeman H, Mayboroda OA et al (2014) Analysis of biologically-active, endogenous carboxylic acids based on chromatography-mass spectrometry. TrAC Trends Anal Chem 61:17–28

11. Gu H, Liu G, Wang J et al (2014) Selecting the correct weighting factors for linear and quadratic calibration curves with least-squares regression algorithm in bioanalytical LC-MS/MS assays and impacts of using incorrect weighting factors on curve stability, data quality, and assay perfo. Anal Chem 86:8959–8966

12. Lee J, Jang E-S, Kim B (2013) Development of isotope dilution-liquid chromatography/mass spectrometry combined with standard addition techniques for the accurate determination of tocopherols in infant formula. Anal Chim Acta 787:132–139

Chapter 19

Analysis of Oxysterols

Fabien Riols and Justine Bertrand-Michel

Abstract

Oxysterols are oxygenated derivatives of cholesterol formed in the human body or ingested in the diet. By modulating the activity of many proteins (for instance, liver X receptors, oxysterol-binding proteins, some ATP-binding cassette transporters), oxysterols can affect many cellular functions and influence various physiological processes (e.g., cholesterol metabolism, membrane fluidity regulation, intracellular signaling pathways). Due to their crucial role, it is important to be able to quantify them in pathological conditions. The method described here permits to measure the content of oxysterol in plasma, cell, or media using GC-MS.

Key words Oxysterol, Plasma, Gas chromatography mass spectrometry

1 Introduction

Cholesterol has many functions; for instance, it affects biophysical properties of membranes and is a precursor to hormone synthesis. These actions are governed by enzymatic pathways that modify the sterol nucleus or the isooctyl tail [1]. The addition of oxygen to the cholesterol backbone produces its derivatives known as oxysterols. In addition to having an enzymatic origin, oxysterols can be formed in the absence of enzymatic catalysis in a pathway usually termed "autoxidation," [2] which is known for almost a century and observed under various experimental conditions. These metabolic intermediates are involved in cholesterol homeostasis where they play a role as ligands to nuclear and G protein-coupled receptors. These molecular classes are implicated in the etiology of a diverse array of diseases including autoimmune disease, Parkinson's disease, motor neuron disease, breast cancer, the lysosomal storage disease Niemann-Pick type C, and the autosomal recessive disorder Smith-Lemli-Opitz syndrome. For example, the cholesterol metabolites 25-hydroxycholesterol and 7α, 25-dihydroxycholesterol have recently been shown to play important signaling roles in the immune system [3, 4]. 24-hydroxycholesterol has been shown to

Martin Giera (ed.), *Clinical Metabolomics: Methods and Protocols*, Methods in Molecular Biology, vol. 1730, https://doi.org/10.1007/978-1-4939-7592-1_19, © Springer Science+Business Media, LLC 2018

be involved in cholesterol turnover in the brain, and it plays a role in memory [5] and glucose metabolism [6]. In Niemann-Pick type C1 (NPC1) disease which is a rare progressive neurodegenerative disorder characterized by accumulation of cholesterol in the endo-lysosomes, oxysterols are accumulated during the disease progression [7]. A similar case is observed in Smith-Lemli-Opitz syndrome, another inborn error of sterol metabolism [8]. Because of their large implication in physiopathology, they are good potential biomarkers, so they need to be readily measureable in biological fluids, in organs, and in cells. Gas chromatography mass spectrometry (GC-MS) with the use of stable isotope-labeled standards is the gold standard methodology for the analysis of neutral cholesterol metabolites [9] and for acidic oxysterols after appropriate derivatization [10]. As these molecules are often isobaric species and also their steroid nucleus is quite hard to fragment, separation is key. The esterification rate of oxysterols is approximately 70%; in turn a saponification step is needed to measure total oxysterol levels. Due to the possibility of autoxidative oxysterol production, samples should be stored at low temperature (-80 °C), under inert atmosphere, ideally with argon or nitrogen gas. Samples will then be submitted to organic liquid-liquid extraction and to saponification (or not) depending on the pool of oxysterols targeted. Oxysterols are in minority compared to cholesterol; hence, it is important to remove cholesterol before analysis, in order to minimize matrix effects. Extracts are pre-purified by solid-phase extraction (SPE) and finally submitted to derivatization to improve GC-MS separation and detection. Two types of detectors can be used: a single quadrupole where the SIM (single ion monitoring) mode will be chosen to have the best sensitivity or a triple quadrupole where the SRM (single reaction monitoring) mode will be preferred due to a better signal-to-noise ratio.

2   Materials

All solvents have to be prepared with ultrapure water (18 MΩ-cm at 25 °C) and analytical grade reagents. All solvents used are at least of HPLC grade.

2.1 Cell and Media Collection

1. Directly after collection, snap freeze cell or medium samples with liquid nitrogen (see **Note 1**), and store at -80 °C for a maximum period of 1 month.

2. 5 mM ethylene glycol tetraacetic acid (EgTA) solution: dissolve 9.51 mg of EgTA in 10 mL of NaOH 10 M, add 10 mL of water, adjust the pH to 7.4 with concentrated hydrochloric acid (HCl), and complete the volume to 50 mL with water.

Take 10 mL of this solution, and dilute it with 990 mL of water to obtain the 5 mM solution of EgTA.

3. Methanol (MeOH)/EgTA solution (2:1; v/v): 10 mL of EgTA 5 mM is diluted with 20 mL of methanol (*see* **Note 2**).

2.2 Blood Collection

1. Collect blood sample into commercial blood collection tube containing ethylenediaminetetraacetic acid (EDTA). Centrifuge immediately at $1000 \times g$ for 10 min at 4 °C.

2. Collect plasma and snap freeze it with liquid nitrogen (*see* **Note 1**), and store at −80 °C for a maximum of 3 months.

2.3 Standards

1. Primary standards: the following standard materials are needed—7α-hydroxycholesterol, 7β-hydroxycholesterol, 4β-hydroxycholesterol, 5α,6α-epoxycholestanol, 24(S)-hydroxycholesterol, 25-hydroxycholesterol, 7-ketocholesterol, and 27-hydroxycholesterol.

2. Internal standard: 19-hydroxycholesterol serves as the internal standard.

3. Standard stock solutions: dilute each standard in a glass vial to 1000 ng/μL using MeOH, and store at −80 °C after flushing with nitrogen gas.

4. Standard mix solution for calibration: pipet 50 μL of each standard stock solution to a glass vial, and adjust the final volume to 1000 μL with MeOH in order to obtain a final concentration of 50 ng/μL per standard.

5. Internal standard solution for extractions: pipet (or transfer) 50 μL of internal standard stock solution to a glass vial, and adjust the final volume to 1000 μL with MeOH obtaining a final concentration of 50 ng/μL.

2.4 Extraction Material

1. Extraction tube: Pyrex™ Round Bottom Threaded Culture Tubes (13 × 100 mm) with cap PTFE faced rubber insert should be used.

2. Brine (NaCl) solution: prepare a saturated solution of NaCl in water.

2.5 Solid-Phase Extraction

1. Typically use a Waters Positive Pressure 96-SPE Processor.

2. The following solid-phase extraction cartridges should be used (SPE): CHROMABOND® Multi 96 SiOH (96 × 100 mg).

3. For sample collection, use a 96-well 2 mL square collection plate.

4. 0.5% (v/v) 2-propanol in n-heptane: measure 199 mL n-heptane into a 250 mL graduated cylinder, then add 1 mL 2-propanol, and store at ambient temperature in a 250 mL reagent bottle.

5. 30% (v/v) 2-propanol in n-heptane: measure 60 mL n-heptane and 140 mL 2-propanol in a 250 mL graduated cylinder, and store at ambient temperature in a 250 mL reagent bottle.

2.6 Alkaline Hydrolysis and Derivatization

1. 0.5 M potassium hydroxide (KOH) methanolic solution: dissolve 14.03 g of KOH in 500 mL of MeOH.

2. 0.5 M phosphoric acid solution: add 339 μL of phosphoric acid 85% with 9.661 mL of water.

3. 50% (v/v) BSTFA solution: mix 25 μL N,O-Bis(trimethylsilyl) trifluoroacetamide + TMCS (99:1) with 25 μL of acetonitrile (ACN).

2.7 Calibration

1. Serially dilute the calibration stock solution 1:1, starting with 10 μL of standard mix solution (50 ng/μL) mixed with 10 μL of MeOH, and continue seven times.

2. Dry samples under a constant stream of nitrogen, and dissolve in 20 μL of n-heptane; this will result in the concentrations given in Table 1.

2.8 Gas Chromatography Mass Spectrometry

1. GC-MS single quadrupole: exemplary a Thermo Fisher Trace Gas Chromatograph coupled to a Thermo Fisher ISQ mass selective detector; AI 3000 Injector Autosampler (Thermo Fisher Scientific, Waltham, MA, USA) can be used.

2. GC-MS triple quadrupole: exemplary a Thermo Fisher Trace Gas Chromatograph coupled to a Thermo Fisher TSQ 8000 mass selective; AI 1310 Injector Autosampler (Thermo Fisher Scientific, Waltham, MA, USA) can be used.

3. Gas chromatography column: use a HP-5MS capillary column (30 m, 0.25 mm, 0.25 μm phase thickness) (*see* **Note 2**).

Table 1
Calibration standards

Calibration point	[ng/μL] of each oxysterol	Final volume (μL)
8	12.5	20
7	6.25	20
6	3.12	20
5	1.563	20
4	0.781	20
3	0.391	20
2	0.195	20
1	0.098	20

4. Glass screw-top vials: 2 mL, 32 × 12 mm, and 9 mm clear glass screw thread vials purchased from Thermo Fisher Scientific can be used.

5. Glass inserts for screw-top vials: use 0.1 mL micro insert 29 × 5.7 mm clear glass with attached plastic.

6. Caps for screw-top vials: use 9 mm combination seal PP short thread caps.

7. Ultrahigh purity helium gas (99.9990%) and ultrahigh purity ammoniac gas (99.990%) are used as carrier and chemical ionization gas, respectively.

3 Methods

3.1 Sample Processing

1. In case of **cells**, add 1 mL of MeOH/EgTA solution to 5×10^6 cells (*see* **Note 3**), and transfer the suspension to a glass culture tube (13 mm × 100 mm). In case of **media** samples, transfer 200 μL of cell media (*see* **Note 3**) into a glass culture tube. In case of **plasma**, transfer 400 μL of human plasma into a glass culture tube.

2. Add 2.5 mL of dichloromethane, 2.5 mL of MeOH, and 2 mL of water.

3. Add 10 μL of internal standard solution (500 ng).

4. Flush with nitrogen gas (*see* **Note 4**).

5. Vortex for 10 min in capped glass culture tube.

6. Centrifuge the sample for 10 min at $1095 \times g$.

7. Collect the lower organic phase, and transfer to a new glass culture tube (*see* **Note 5**).

8. Dry lipid extract under a constant stream of nitrogen gas. **For free oxysterol extraction, go directly to step 17.**

9. Add 2.5 mL of dichloromethane and 1.75 mL of 0.5 M KOH methanolic solution.

10. Flush with nitrogen gas, and keep for 2 h under stirring at room temperature in capped glass culture tube (*see* **Note 4**).

11. Neutralize with 1.3 mL of 0.5 M phosphoric acid solution (*see* **Note 6**).

12. Add 0.75 mL of MeOH and 0.7 mL of water.

13. Flush with nitrogen gas (*see* **Note 4**).

14. Vortex and then centrifuge for 5 min at $1095 \times g$ in capped glass culture tube.

15. Collect the lower organic phase and transfer to a new tube.

16. Dry lipid extract under a constant stream of nitrogen gas.

17. Condition the 100 mg SiOH SPE plate with 2 mL of n-heptane.

18. Dissolve sample in 1 mL of n-heptane and load onto SPE plate.

19. Wash the plate with 4 mL of 0.5% (v/v) 2-propanol in n-heptane.

20. Elute oxysterols with 2 mL of 30% (v/v) 2-propanol in n-heptane.

21. Dry lipid extract under nitrogen gas.

22. Resolubilize the lipid extract in 2×80 µL of n-heptane, and transfer to a glass vial with insert.

23. Dry lipid extract under nitrogen gas (*see* **Notes 7** and **8**).

24. Add 50% (v/v) BSTFA solution in ACN, and heat for 1 h at 55 °C in a heating block in the capped glass vial (*see* **Note 9**).

25. Cool and dry lipid extract under nitrogen gas.

26. Dissolve the extract in 20 µL of n-heptane and vortex.

27. Immediately analyze sample extract by GC-MS (*see* Fig. 1 for exemplary data).

3.2 Quantification Using SIM or SRM Scan Mode

1. The oven temperature program is as follows: 180 °C for 1 min, 20 °C/min to 250 °C, 5 °C/min to 300 °C where the temperature is kept for 8 min, and then 35 °C/min to 325 °C.

2. The carrier gas is kept at a constant flow rate of 0.8 mL/min.

3. The samples are injected in splitless mode with an injection volume of 1 µL.

4. The injector, transfer line, and source temperature are 270 °C, 280 °C, and 250 °C, respectively.

5. For SIM analysis, the mass spectrometer is operated in electron ionization mode (70 eV).

6. For SRM analysis, the mass spectrometer is operated in positive chemical ionization mode using ammonia as ionization gas at a flow rate of 1.5 mL/min (47.5 cm/s).

7. Ions used are as follows (Table 2):

8. For quantitative analysis construct calibration linear lines (2.7) plotting nominal concentrations against the area ratio obtained for each analyte. Use the weighting equal and ignore the origin. Calculate analyte concentrations using the obtained calibration functions.

4   Notes

1. Liquid nitrogen should be manipulated under natural or mechanical ventilation to prevent oxygen-deficient atmospheres below 19.5%. Use loose fitting thermal insulated or

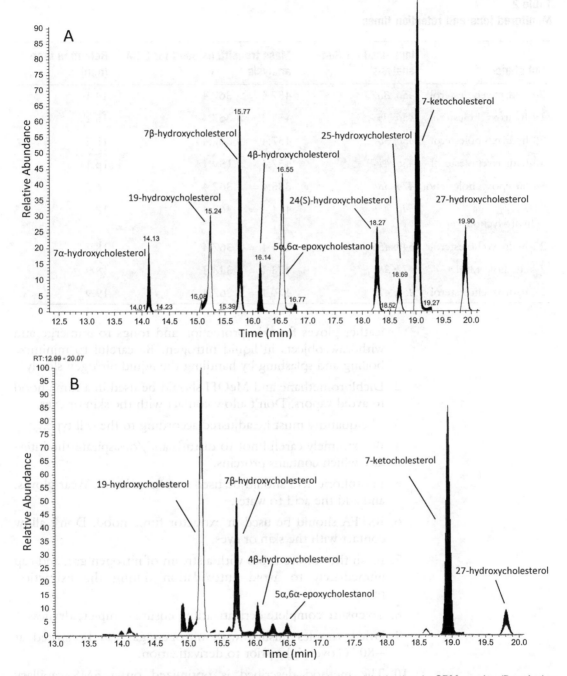

Fig. 1 Oxysterol profile using GC-MS analysis with positive chemical ionization in SRM mode: (Panel **a**) a mixture of nine standards is shown; (Panel **b**) an example analysis of human multipotent adipose-derived stem cells (hMADS) is depicted

Table 2
Monitored ions and retention times

Substance	Ions used for SIM analysis	Mass transitions used for SRM analysis	Retention time (min)
7α-hydroxycholesterol	456; 367	457.4 => 367.4	14.1
19-hydroxycholesterol	456; 353	457.4 => 367.4	15.2
7β-hydroxycholesterol	456; 367	457.4 => 367.4	15.7
4β-hydroxycholesterol	456; 417	457.4 => 159.1	16.1
5α,6α-epoxycholestanol	474; 384	385.4 => 367.4	16.5
24(S)-hydroxycholesterol	546; 456	474.4 => 367.4	18.2
25-hydroxycholesterol	546; 456	474.4 => 367.4	18.7
7-ketocholesterol	472; 382	473.4 => 383.4	18.9
27-hydroxycholesterol	456; 417	474.4 => 367.4	19.9

leather gloves for skin protection and tongs to immerge and withdraw objects in liquid nitrogen. Be careful to minimize boiling and splashing by handling the liquid nitrogen slowly.

2. Dichloromethane and MeOH should be used in a fume hood to avoid vapors. Don't allow contact with the skin or eyes.

3. The quantity must be adjusted according to the cell type.

4. Be extremely careful not to disturb and/or aspirate the interface which contains proteins.

5. Phosphoric acid should be used in a fume hood. Wear gloves, and add the acid to water.

6. BSTFA should be used in extractor fume hood. Don't allow contact with the skin or eyes.

7. Flush the culture tubes with a stream of nitrogen gas, and cap immediately to avoid autoxidation during the extraction process.

8. To ensure complete derivatization, ensure complete dryness.

9. Dried samples flushed with nitrogen gas can be stored at −80 °C overnight prior to derivatization.

10. The method described is optimized on a 5MS capillary column.

Acknowledgments

This work was in part funded by National Health and Medical Research Council of France (Inserm) and the French National Infrastructure MetaboHUB-ANR-11-INBS-0010.

References

1. Russell DW (2003) The enzymes, regulation, and genetics of bile acid synthesis. Annu Rev Biochem 72:137–174. https://doi.org/10.1146/annurev.biochem.72.121801.161712

2. Iuliano L (2011) Pathways of cholesterol oxidation via non-enzymatic mechanisms. Chem Phys Lipids 164(6):457–468. https://doi.org/10.1016/j.chemphyslip.2011.06.006

3. Bauman DR, Bitmansour AD, McDonald JG, Thompson BM, Liang G, Russell DW (2009) 25-hydroxycholesterol secreted by macrophages in response to toll-like receptor activation suppresses immunoglobulin a production. Proc Natl Acad Sci U S A 106 (39):16764–16769. https://doi.org/10.1073/pnas.0909142106

4. Hannedouche S, Zhang J, Yi T, Shen W, Nguyen D, Pereira JP, Guerini D, Baumgarten BU, Roggo S, Wen B, Knochenmuss R, Noel S, Gessier F, Kelly LM, Vanek M, Laurent S, Preuss I, Miault C, Christen I, Karuna R, Li W, Koo DI, Suply T, Schmedt C, Peters EC, Falchetto R, Katopodis A, Spanka C, Roy MO, Detheux M, Chen YA, Schultz PG, Cho CY, Seuwen K, Cyster JG, Sailer AW (2011) Oxysterols direct immune cell migration via EBI2. Nature 475(7357):524–527. https://doi.org/10.1038/nature10280

5. Kotti TJ, Ramirez DM, Pfeiffer BE, Huber KM, Russell DW (2006) Brain cholesterol turnover required for geranylgeraniol production and learning in mice. Proc Natl Acad Sci U S A 103(10):3869–3874. https://doi.org/10.1073/pnas.0600316103

6. Suzuki R, Lee K, Jing E, Biddinger SB, McDonald JG, Montine TJ, Craft S, Kahn CR (2010) Diabetes and insulin in regulation of brain cholesterol metabolism. Cell Metab 12 (6):567–579. https://doi.org/10.1016/j.cmet.2010.11.006

7. Porter FD, Scherrer DE, Lanier MH, Langmade SJ, Molugu V, Gale SE, Olzeski D, Sidhu R, Dietzen DJ, Fu R, Wassif CA, Yanjanin NM, Marso SP, House J, Vite C, Schaffer JE, Ory DS (2010) Cholesterol oxidation products are sensitive and specific blood-based biomarkers for Niemann-pick C1 disease. Sci Transl Med 2(56):56ra81. https://doi.org/10.1126/scitranslmed.3001417

8. Griffiths WJ, Wang Y, Karu K, Samuel E, McDonnell S, Hornshaw M, Shackleton C (2008) Potential of sterol analysis by liquid chromatography-tandem mass spectrometry for the prenatal diagnosis of Smith-Lemli-Opitz syndrome. Clin Chem 54 (8):1317–1324. https://doi.org/10.1373/clinchem.2007.100644

9. Matysik S, Klunemann HH, Schmitz G (2012) Gas chromatography-tandem mass spectrometry method for the simultaneous determination of oxysterols, plant sterols, and cholesterol precursors. Clin Chem 58(11):1557–1564. https://doi.org/10.1373/clinchem.2012.189605

10. Saeed A, Floris F, Andersson U, Pikuleva I, Lovgren-Sandblom A, Bjerke M, Paucar M, Wallin A, Svenningsson P, Bjorkhem I (2014) 7alpha-hydroxy-3-oxo-4-cholestenoic acid in cerebrospinal fluid reflects the integrity of the blood-brain barrier. J Lipid Res 55 (2):313–318. https://doi.org/10.1194/jlr.P044982

Chapter 20

Analysis of Metabolites from the Tricarboxylic Acid Cycle for Yeast and Bacteria Samples Using Gas Chromatography Mass Spectrometry

Reza Maleki Seifar, Angela ten Pierick, and Patricia T.N. van Dam

Abstract

We here explain step by step the implementation of gas chromatography coupled with tandem mass spectrometry for the quantitative analysis of intracellular metabolites from the tricarboxylic acid (TCA) cycle such as citrate, isocitrate, alpha-ketoglutarate, succinate, malate, and fumarate. Isotope dilution is used to correct for potential metabolite losses during sample processing, matrix effects, incomplete derivatization, and liner contamination. All measurements are performed in selected reaction monitoring (SRM) mode. Standards and samples are first diluted with a fixed volume of a mixture of fully ^{13}C-labeled internal standards and then derivatized to give trimethylsilyl-methoxyamine derivatives prior GC-MS/MS analysis.

Key words TCA cycle, Gas chromatography, Mass spectrometry, TMS-MOX derivatives, Metabolomics

1 Introduction

The tricarboxylic acid (TCA) cycle is central to energy production by the oxidation of pyruvic acid to carbon dioxide. The cycle is also important for biosynthesis of amino acids, nucleotide bases, and other metabolites. Alterations in metabolite levels caused by defects in TCA enzymes have been described as underlying hallmarks of cancer [1]. Furthermore, studying the pyruvate metabolism can also help in the construction of important industrial strains and provide tools for controlled cultivation [2]. In this chapter, we describe in detail the application of gas chromatography (GC) coupled with tandem mass spectrometry (MS/MS) for the quantitative analysis of intracellular citrate, isocitrate, alpha-ketoglutarate, succinate, malate, fumarate, and oxaloacetate levels. A global ^{13}C-labeled *Saccharomyces cerevisiae* extract (containing all targeted metabolites as ^{13}C-labeled compounds) is added as internal standard to the sample solutions and standard calibration mixtures,

Martin Giera (ed.), *Clinical Metabolomics: Methods and Protocols*, Methods in Molecular Biology, vol. 1730,
https://doi.org/10.1007/978-1-4939-7592-1_20, © Springer Science+Business Media, LLC 2018

helping to overcome analytical limitations. This extract is produced by cultivation of *S. cerevisiae* on fully labeled ^{13}C-glucose as the exclusive carbon source [3]. Internal standards are added to the samples immediately after quenching or right before extraction of the metabolites.

2 Materials

All chemicals should be of *p.a.* or higher purity.

2.1 Chemicals

1. Succinic acid, alpha-ketoglutaric acid, malic acid, fumaric acid, citric acid, and isocitric acid are needed as calibration standards (*see* **Note 1**).

2. ^{13}C-labeled *S. cerevisiae* cell extract is obtained by growing *S. cerevisiae* on fully labeled ^{13}C-glucose as the exclusive carbon source [3].

3. Silylation reagent mix: trimethylchlorosilane (TMCS) and *N*-methyl-*N*-trimethylsilyltrifuoroacteamide (MSTFA) are used as silylation reagents. Mix 1 mL of MSTFA with 50 µL TMCS (*see* **Note 2**).

4. Methoxyamine hydrochloride (MOX): prepare a solution of (20 mg/mL) MOX in anhydrous pyridine. Vortex and preferably heat for 1 min at the heating block to let MOX completely dissolve (prepare freshly every day).

5. Isooctane and ethyl acetate are used for syringe cleaning.

2.2 Equipment

A freeze dryer, a heating module (e.g., a Reacti-Therm III Heating Module (Thermo Scientific)), a −80 °C freezer, and a centrifuge are needed for sample preparation.

2.3 GC-MS/MS Instrumentation

1. GC system: 7890 A (Agilent Technologies) or equivalent.

2. MS: Triple Quad 7000 (Agilent Technologies) or equivalent.

3. A Zebron ZB-50 capillary column (30 m × 250 µm internal diameter, 0.25 µm film (Phenomenex)) should be used.

4. A straight glass liner, Ultra Inert, single taper with glass wool (Agilent) is recommended (*see* **Note 3**).

5. 10 µL gastight syringe with Teflon-tipped plunger.

6. 2 mL deactivated, silanized vial.

7. Cap screw with PTFE/silicone/PTFE septa (*see* **Note 4**).

8. Short thread screw caps (transparent) with PTFE septa only (Grace).

9. Snap top shell vial, 1 mL, 8 mm, clear (Grace).

10. 250 µL glass inserts.

3   Methods

3.1   Preparation of Intracellular Samples

In the case of yeast and bacteria samples, 1.2 mL broth is rapidly (less than a second) taken from a fermentor into a pre-weighted 50 mL Greiner tube containing 5 mL cold methanol (-40 °C) to freeze whole metabolism of the cells. After weighting the tube once more, the cell pellet is collected by centrifugation for 3 min at $1000 \times g$ at 6 °C. Rotors are kept in a freezer (-20 °C) and installed before the start of centrifugation. Subsequently, the cell pellet is washed with another 5 mL of cold methanol (-40 °C) followed by centrifugation (3 min at $1000 \times g$) at 6 °C. After discarding the supernatant, 120 µL of labeled internal standard is added to the cell pellet (*see* **Note 5**). Extraction of intracellular metabolites is done using hot ethanol [3]. Briefly, 5 mL of preheated (95 °C) ethanol 75% (v/v) is added to the cell pellet and heated for 3.5 min at 95 °C. Then the extract is immediately cooled down using an ice bath followed by another centrifugation step (3 min $1000 \times g$) to remove cell debris. In the next step, the extract is dried using a SpeedVac. The dry matter is reconstituted in 600 µL of deionized pure water and is stored at -80 °C or immediately analyzed. In the case of mammalian cells, the quenching and extraction protocols described by Kostidis et al. [4] can be applied.

3.2   Preparation of Standards for GC-MS

Stock solutions from each targeted compound are prepared. From these stock solutions, a standard mix containing all targeted metabolites is prepared. Aliquots from this standard mix are prepared and stored at -80 °C. For each set of samples, one aliquot is taken for preparation of standard calibration dilution series. Each calibration set includes ten dilution series. A typical standard calibration range for each metabolite is 0.25 µM to 100 µM. The standard calibration range should cover the concentration of the targeted compounds in the sample solutions. A fixed volume of internal standard (^{13}C-labeled extract) is required to be added to the samples and each of the standard calibration mixtures (*see* **Notes 6** and **7**). It should be mentioned that the concentration of ^{13}C-labeled internal standards needs to be in the calibration range of the ^{12}C standards.

1. Pipette the following volumes (Table 1) to GC vials and close them with caps containing Teflon septa.

2. Pipette 100 µL sample (containing 20 µL internal standard) to a GC vial, and close it with a cap containing a Teflon septum.

3. Make two holes in each Teflon septum using a needle.

3.3   Freeze-Drying

1. Freeze samples in a -80°C freezer.

2. Place the frozen samples in a freeze dryer chamber.

3. Start the vacuum pump.

4. Freeze dry overnight.

Table 1
Dilution scheme for preparation of standard calibration mixtures

Dilution series	Standard mix µL	[a]Ultrapure water µL	Internal standard mix µL
CAL 1	1	200	20
CAL 2	2	200	20
CAL 3	5	200	20
CAL 4	10	200	20
CAL 5	20	200	20
CAL 6	40	200	20
CAL 7	60	200	20
CAL 8	100		20
CAL 9	200		20
CAL 10	400		20

[a]200 µL is added to the low standards; otherwise, the required volume for freeze-drying is too little, and samples might be defrosted

3.4 Derivatization

1. Work in a fume hood and wear nitrile gloves.
2. Switch on the heating block (70 °C).
3. Add 50 µL of the MOX solution to each GC vial (*see* **Note 2**).
4. Incubate samples for 50 min at 70 °C using a heating block.
5. Remove from the heating block and let cool down to room temperature.
6. Add 80 µL of the silylation reagent mix to each sample.
7. Incubate samples for 50 min at 70 °C using a heating block.
8. Remove vials from the heating block and let them cool down to room temperature.
9. Transfer samples into the centrifuge glass tubes (1 mL shell vials).
10. Centrifuge for 1 min at 8000 × g.
11. Transfer 70 µL of supernatant of the sample from the centrifuge tube into a glass insert, and place it back into the original GC vial (remove air bubbles if needed).

3.5 GC-MS Settings

1. Set the carrier gas flow (helium 99.9990%) to 1 mL/min.
2. Program the GC oven temperature as follows: keep constant for 1 min at 70 °C followed by a ramp of 10 °C per min up to 220 °C.
3. Set the temperature of the transfer line to 250 °C.

Table 2
GC-MS/MS parameters for TMS-MOX derivatives

Metabolite	Retention time (min)	Precursor ion	^{13}C Precursor ion	[a]Fragment ion	Collision energy
Fumaric acid	9.80	245	249	147	10
Succinic acid	9.87	247	251	147	5
Malic acid	11.46	335	339	147	15
alpha-Ketoglutaric acid	13.65	304	309	147	5
Citric acid	14.97	465	471	147	35
Isocitric acid	15.14	465	471	147	35

[a]Fragment ions for labeled metabolites were the same as unlabeled metabolites

4. Set the injection volume to 1 μL (*see* **Note 8**).

5. If using a multimode inlet (MMI) (Agilent), use the following settings: split-less mode for metabolites with low concentration and in split mode 1:5 for metabolites with high concentrations.

6. Set the MMI temperature: 70 °C (with a split-less time of 1 min if it is needed), after injection temperature is increased to 220 °C with a ramp of 720 °C/min and held for 5 min. Then temperature is increased to 300 °C to clean injector.

7. At the end of the run, backflush the column with five column volumes at 220 °C (*see* **Notes 9** and **10**).

8. Set the MS transfer line to 250 °C.

9. Keep the MS source at 230 °C.

10. Set the helium gas flow of the collision cell to 2.25 mL/min.

11. Set the N_2 collision gas flow to 1.5 mL/min.

12. Use electron ionization operated at 70 eV.

13. The MS is operated in SRM mode.

14. Mass resolution is set to 0.7 mass unit for both quadrupoles.

15. Use the settings shown in Table 2 for GC-MS/MS analysis.

16. Mass Hunter quantitative analysis software (version B.07.00; Agilent) can be used for data processing.

17. Integrate signals and calculate concentrations based on the calibration line as established according to Table 1 (*see* **Note 7**).

4   Notes

1. Use certificate of analysis from chemical providers for preparation of stock solutions.

2. Derivatizing reagents should be stored in a desiccator in a fridge at 4 °C.

3. The liner in GC might need to be replaced based on the sample type after different numbers of injections.

4. For GC vials, silicone rubber should be avoided because of possible sample contamination and reactivity with the derivatization reagents.

5. Internal standards were added to the samples immediately after quenching or right before the extraction step.

6. Fixed volumes of ^{13}C-labeled internal standard (20 μL) were added to samples and each one of the calibration mixtures.

7. Calibration graphs are constructed based on peak height ratios of natural ^{12}C peak height of targeted metabolites to ^{13}C-labeled peak height of internal standards versus concentration of natural isotope ^{12}C standards. $1/x^2$ weighting is used for construction of calibration curves.

8. For multiple injections from the same sample vial, renew the cap after each two injections if it is possible.

9. After about 100 injections, cut 10 cm of the column (injection side) and at the same time replace the liner and inlet septum (gas flow needs to be modified accordingly using retention time locking (RTL), to have reproducible retention times).

10. Run the alkane mix for RTL and system check about every ten samples.

References

1. Cardaci S, Ciriolo MR (2012) TCA cycle defects and cancer: when metabolism tunes redox state. Int J Cell Biol. https://doi.org/10.1155/2012/161837

2. Pronk JT, YDE Steensma H, van Dijken JP (1996) Pyruvate metabolism in Saccharomyces cerevisiae. Yeast 12:1607–1633

3. Wu L, Mashego MR, van Dam JC et al (2005) Quantitative analysis of the microbial metabolome by isotope dilution mass spectrometry using uniformly C-13-labeled cell extracts as internal standards. Anal Biochem 336:164–171

4. Kostidis S, Addie RD, Morreau H et al (2017) Quantitative NMR analysis of intra- and extracellular metabolism of mammalian cells: a tutorial. Anal Chim Acta 980:1–24

Chapter 21

GC-MS Analysis of Lipid Oxidation Products in Blood, Urine, and Tissue Samples

Anne Barden and Trevor A. Mori

Abstract

Oxidant stress has been identified as important in the pathology of many diseases. Oxidation products of polyunsaturated fatty acids collectively termed isoprostanes, neuroprostanes, and isofurans are considered the most reliable measures of in vivo lipid oxidation, and they are widely used to assess oxidant stress in various diseases. Here we describe the measurement of these lipid oxidation products using gas chromatography mass spectrometry with electron capture negative ionization.

Key words Lipid oxidation, F_2-isoprostanes, Neuroprostanes, Isofurans, Gas chromatography mass spectrometry

1 Introduction

Free radicals have been linked to a wide range of human diseases and are known to oxidize lipids, proteins, and DNA [1]. Although a number of measures of lipid peroxidation are available, most involve assays directed against nonspecific or unstable metabolites such as short-chain alkanes, malondialdehyde, lipid hydroperoxides, or ex vivo induced lipoprotein oxidizability [1]. Polyunsaturated fatty acids (PUFAs) react with free radicals to form oxygenated metabolites collectively termed isoprostanes, neuroprostanes, and isofurans [2]. The F_2-isoprostanes (F_2-IsoPs) first reported by Morrow et al. [3] are formed in vivo predominantly by free radical nonenzymatic oxidation of arachidonic acid (AA, C20:4 n-6). The F_2-IsoPs are widely acknowledged as the most reliable in vivo biomarkers of oxidative stress in animals [4] and humans [5, 6]. Importantly, a number of F_2-IsoPs are also biologically active [6]. The presence and quantity of oxygen can alter the lipid peroxidation products formed. For example, under conditions of high oxygen tension, free radical-induced peroxidation of AA leads to formation of the isofurans (IsoFs) [7, 8].

Martin Giera (ed.), *Clinical Metabolomics: Methods and Protocols*, Methods in Molecular Biology, vol. 1730, https://doi.org/10.1007/978-1-4939-7592-1_21, © Springer Science+Business Media, LLC 2018

The n-3 fatty acids eicosapentaenoic acid (EPA, C20:5 n−3) and docosahexaenoic acid (DHA, C22:6 n-3) are oxidized to F_3-isoprostanes (F_3-IsoPs) [9] and F_4-isoprostanes (F_4-IsoPs) [10], respectively. The F_4-IsoPs are commonly referred to as neuroprostanes (NeuroPs). Measurement of lipid oxidation products of EPA and DHA is relevant in the setting of n-3 fatty acid supplementation and in organs that have a high n-3 fatty acid content such as the heart, retina, and brain.

The measurement of in vivo oxidative stress requires samples be collected and stored in a manner that avoids ex vivo autoxidation. This is particularly important for the measurement of lipid oxidation markers in plasma and tissue samples. We have shown that the collection of blood into collection tubes containing dipotassium ethylenediaminetetraacetic acid (EDTA), reduced glutathione (GSH), and butylated hydroxytoluene (BHT) with storage at −80 °C is necessary to minimize ex vivo autoxidation of AA that can lead to artifactual elevation of plasma F_2-IsoP concentrations [11]. Our data show that F_2-IsoP concentrations in samples collected in EDTA alone were significantly elevated compared with those collected into EDTA with the addition of antioxidants [11]. We also showed plasma stored at −20 °C resulted in a marked elevation of plasma F_2-IsoPs irrespective of whether blood collection included EDTA/GSH/BHT.

Measurement of lipid oxidation products in biological samples is complex due to the number of products that can be formed. To date enzyme-linked immunoassays (EIA) are available only for F_2-IsoP, but we have shown these have poor agreement with mass spectrometry [12]. Therefore, mass spectrometry remains the "gold standard" for measurement of lipid oxidation products.

In this chapter, we describe the methods for measurement of IsoPs, NeuroPs, and IsoFs in plasma, urine, and tissue using gas chromatography mass spectrometry (GC-MS) with electron capture negative ionization.

2   Materials

Prepare all solutions using ultrapure water (prepared by purifying deionized water to 18 MΩ-cm at 25 °C) and analytical grade reagents. All solvents should be analytical grade. Prepare and store all reagents and solvents at room temperature (unless indicated otherwise).

2.1 Blood Collection Tubes

1. Dissolve 1 g of dipotassium ethylenediaminetetraacetic acid (EDTA) into 6 mL 0.9% saline. Add 1 g of reduced glutathione, and adjust to pH 7.4 using 5 M NaOH. Make up to a final volume of 10 mL with 0.9% saline. To make 5 M NaOH, dissolve 20 g NaOH in 100 mL water (*see* **Note 1**).

2. Dissolve 200 mg butylated hydroxytoluene (BHT) in 50 mL of ethanol.

3. To each 2.5 mL blood collection tube, add 25 μL of EDTA/reduced glutathione and 10 μL of BHT in ethanol to prevent ex vivo oxidation.

4. Collection tubes can be prepared in advance and kept at 4 °C for a period of months.

2.2 Urine Collection

1. Two 500 μL aliquots of urine from a 24-h collection are stored at −80 °C until analysis. The total volume of the 24-h urine collection and the creatinine measurement are required.

2. If a 24-h collection is not available, a spot urine can be used, and the creatinine concentration should be determined.

2.3 Tissue Collection

1. Liquid nitrogen and porcelain mortar and pestle for processing tissue samples (*see* **Note 2**).

2. Chloroform (CCl_3)/methanol (CH_3OH) (2:1 vol/vol, containing 500 mg/L BHT): Measure 660 mL CCl_3 and 330 mL CH_3OH into a 1 L graduated cylinder. Add 500 mg BHT and store in a 1 L reagent bottle (*see* **Note 3**).

2.4 Standards

1. 8-*iso*-prostaglandin $F_{2\alpha}$-d_4 (8-*iso*-$PGF_{2\alpha}$-d_4) (item № 316,350, Cayman Chemicals, Ann Arbor, MI, USA): Make up as 5 ng/50 μL in CH_3OH, and store at −20 °C.

2. 8,12-*iso*-isoprostane $F_{2\alpha}$-VI-d_{11} (8,12-*iso*-$iPF_{2\alpha}$-VI-d_{11}) (item № 10006878, Cayman Chemicals, Ann Arbor, MI, USA): Make up as 5 ng/50 μL in CH_3OH, and store at −20 °C.

3. F_2-IsoPs: Use 8-*iso*-prostaglandin $F_{2\alpha}$ (8-*iso*-$PGF_{2\alpha}$) (item № 16350, Cayman Chemicals, Ann Arbor, MI, USA) as the standard. Make up as 5 ng/50 μL in CH_3OH, and store at −20 °C.

4. F_3-IsoPs: Use 8-*iso*-prostaglandin $F_{3\alpha}$ (8-*iso*-$PGF_{3\alpha}$) (item № 16,992, Cayman Chemicals, Ann Arbor, MI, USA) as the standard. Make up as 5 ng/50 μL in CH_3OH, and store at −20 °C.

5. IsoF standard prepared as 5 ng/50 mL in CH_3OH and stored at −20 °C was made available as a gift from Professor L.J. Roberts II (Department of Pharmacology, Vanderbilt University, Nashville, TN, USA).

6. NeuroPs: Use 4(RS)-F_{4t}-neuroprostane as the standard. 4(RS)-F_{4t}-neuroprostane was made available as a gift from Dr. Thierry Durand (Faculty of Pharmacy, Institut des Biomolécules Max Mousseron, Montpellier, France) and is prepared as 5 ng/50 mL in CH_3OH and stored at −20 °C.

2.5 Solid-Phase Chromatography

1. Bond Elut™ Certify II cartridges (200 mg, 3 mL) (Agilent Technologies, Santa Clara, CA, USA).

2. Glass culture tubes: Borosilicate disposable culture yubes 13 × 100 mm and caps are purchased from Kimble Chase Life Science and Research Products, Vineland, NJ, USA.

3. 1 M potassium hydroxide (KOH) methanolic: Dissolve 14.03 g of KOH in 250 mL CH_3OH.

4. 100 mM sodium acetate (CH_3COONa), pH 4.6: Dissolve 8.2 g of CH_3COONa (anhydrous) in 1000 mL of water, and adjust to pH 4.6 with concentrated acetic acid (*see* **Note 4**).

5. 100 mM sodium acetate with 5% CH_3OH pH 7.0: Dissolve 8.2 g of CH_3COONa in 950 mL of water. Add 50 mL CH_3OH, mix, and adjust to pH 7.0 with concentrated acetic acid (*see* **Note 4**).

6. 1 M hydrochloric acid (HCl): Fill a 1 L graduated cylinder with 900 mL pure water, and carefully add 100 mL concentrated HCl (10 M) (*see* **Note 4**).

7. CH_3OH/water (1:1): Measure 500 mL CH_3OH and 500 mL water in a 1 L graduated cylinder, and store in a 1 L reagent bottle.

8. Ethyl acetate/*n*-hexane (1:3): Measure 200 mL ethyl acetate and 600 mL hexane in a 1 L graduated cylinder, and store in a 1 L reagent bottle.

9. Ethyl acetate/CH_3OH (9:1): Measure 900 mL ethyl acetate and 100 mL CH_3OH in a 1 L graduated cylinder, and store in a 1 L reagent bottle.

10. Solid-phase extraction (SPE) VacElut 20 Manifold (Agilent Technologies, Santa Clara, CA, USA).

2.6 Derivatization Reagents for Gas Chromatography Mass Spectrometry

1. 10% pentafluorobenzyl bromide (PFBBr) in acetonitrile: PFBBr (0.5 mL) (Sigma, St. Louis, MO, USA; Cat# 101052) is added to 4.5 mL acetonitrile and stored in a dark glass bottle protected from light at 4 °C (*see* **Note 5**).

2. 10% *N,N*-diisopropylethylamine (DIPEA) in acetonitrile: DIPEA (0.5 mL) (Sigma, St. Louis, MO, USA; Cat# D125806) is added to 4.5 mL acetonitrile and stored in a dark glass bottle protected from light in a fume hood (*see* **Note 5**).

3. *N,O*-bis(trimethylsilyl)trifluoroacetamide with 1% trimethylchlorosilane (BSTFA with 1% TMS): Purchased from Sigma, St. Louis, MO, USA (Cat# T6381), and is stored at 4 °C (*see* **Note 5**).

4. Pyridine: Purchased from Sigma, St. Louis, MO, USA, and is stored in a fume hood in a glass vial in the dark (*see* **Note 5**).

5. Isooctane (2,2,4-trimethylpentane): Purchased from Sigma, St. Louis, MO, USA.

6. High purity nitrogen gas.

2.7 Gas Chromatography Mass Spectrometry

1. GC-MS: Uses an Agilent 6890 Gas Chromatograph coupled to an Agilent 5975 Mass Selective Detector and Agilent 6890 7683B Series Injector Autosampler (Agilent Technologies, Santa Clara, CA, USA).

2. Gas chromatography column: DB-5MS 25 m, 0.20 mm, 0.33 μm film thickness (P/N 128–5522) (Agilent Technologies, Santa Clara, CA, USA).

3. Glass screw-top vials: 2 mL, 32 × 12 mm, 9 mm clear glass screw thread vials purchased from Thermo Fisher Scientific (Waltham, MA, USA).

4. Glass inserts for screw-top vials: 0.1 mL micro insert 29 × 5.7 mm clear glass with attached plastic spring purchased from Thermo Fisher Scientific (Waltham, MA, USA).

5. Caps for screw-top vials: 9 mm combination seal PP short thread caps purchased from Thermo Fisher Scientific (Waltham, MA, USA).

6. Ultrahigh purity helium gas (99.999%) and ultrahigh purity methane gas (99.995%).

3 Methods

3.1 Processing of Blood Samples

1. Blood samples are collected into chilled 2.5 mL blood collection tubes and centrifuged immediately at 4 °C and $1000 \times g$ for 10 min.

2. Two 500 μL plasma aliquots are removed and stored at −80 °C until analysis.

3. Plasma (200 μL) + 50 μL (5 ng) 8-*iso*-PGF$_{2\alpha}$-d$_4$ (internal standard) are subjected to base hydrolysis using 1 mL of 1 M KOH in CH$_3$OH in capped glass culture tubes under nitrogen (*see* **Note 6**). Samples are heated for 40 min at 40 °C in a water bath or a heating block.

4. The samples are cooled to room temperature, 2 mL of 100 mM sodium acetate pH 4.6 is added, and samples are adjusted to pH 4.0 with 1 M HCl (*see* **Note 7**).

5. Samples are centrifuged at $1500 \times g$ for 10 min at 4 °C, and the supernatant is removed for solid-phase chromatography.

6. Bond Elut™ Certify II cartridges (200 mg, 3 mL) applied to a VacElut 20 Manifold under vacuum are preconditioned with 2 mL CH$_3$OH, followed by 2 mL 100 mM sodium acetate with 5% CH$_3$OH pH 7 (*see* **Note 8**).

7. The supernatant is applied to the Bond Elut™ Certify II cartridge (*see* **Note 8**).

8. The cartridge is washed sequentially with 2 mL $CH_3OH/$ water (1:1) and 2 mL ethyl acetate/hexane (1:3) (*see* **Note 8**).

9. Oxidation products are eluted with 2 mL ethyl acetate/ CH_3OH (9:1) into a glass culture tube (*see* **Note 8**).

10. Samples are evaporated to dryness under a stream of nitrogen (*see* **Note 9**).

11. 10% PFBBr in acetonitrile (40 µL) and 10% DIPEA in acetonitrile (20 µL) are added to each glass culture tube and left at room temperature for 30 min.

12. Samples are evaporated to dryness under nitrogen (*see* **Note 10**).

13. Add 20 µL BSTFA with 1% TMS and 10 µL pyridine to each sample and then heat for 20 min at 45 °C in a heating block (*see* **Note 10**).

14. Samples are evaporated to dryness under nitrogen (*see* **Note 10**).

15. Reconstitute in 30 µL isooctane in a 0.1 mL glass micro insert in a 2 mL screw-top vial and cap. Samples are ready for GC-MS analysis.

3.2 Processing of Urine Samples

1. To 200 µL urine, add 50 µL (5 ng) 8,12-*iso*-iPF2α-VI-d$_{11}$ (internal standard) (*see* **Note 11**) and 2 mL of 100 mM sodium acetate pH 4.6 buffer, and adjust to pH 4.0 with 1 M HCl.

2. Subject samples to solid-phase chromatography and derivatization as described above for plasma (Subheading 3.1, **steps 6–15**).

3.3 Processing of Tissue Samples

1. A weighed amount of tissue sample (approx. 50 mg) is collected into liquid nitrogen (*see* **Note 2**) and ground to a fine powder with a porcelain mortar and pestle.

2. Tissue samples are placed in glass culture tubes, extracted using the Folch method with 2 mL CCl_3/CH_3OH (2:1) (containing 500 mg/L BHT), and evaporated to dryness in a centrifugal evaporator under vacuum at 40 °C.

3. To each sample, add 50 µL (5 ng) 8-*iso*-PGF$_{2α}$-d$_4$ (internal standard) and 200 µL water.

4. Samples are hydrolyzed using 1 mL of 1 M KOH in CH_3OH in capped glass culture tubes under nitrogen (*see* **Note 6**) with heating for 40 min at 40 °C in a water bath or a heating block.

5. Samples are extracted and subjected to solid-phase chromatography and derivatization as described above for plasma (Subheading 3.1, **steps 4–15**).

3.4 Gas Chromatography Mass Spectrometry Analysis

1. Samples are analyzed by GC-MS. Chromatography uses an Agilent DB-5MS column with helium as the carrier gas at a flow rate of 0.9 mL/min (40 cm/s). The mass spectrometer is operated in electron capture negative ionization mode using methane as the ionizing gas (ion source pressure of 1.8 torr). The injector temperature is 250 °C, the transfer line is 280 °C, and the ion source and quadrupole temperatures are both 150 °C. Injections (2 µL) are made using an Agilent 6890 7683B Series Injector Autosampler in splitless mode for the first 2 min. The initial column temperature is 180 °C, programmed at 20 °C/min from 180 °C to 270 °C, then 5 °C/min from 270 °C to 300 °C, then 2 °C/min from 300 °C to 310 °C, and then 25 °C/min from 310 °C to 320 °C and held at 320 °C for 4 min. The total run time is 20.80 min.

2. Lipid oxidation products are detected by retention time (RT) comparison with authentic standards and selected ion monitoring (SIM) as follows: F_2-IsoPs (8-*iso*-PGF$_{2\alpha}$) at RT 12.66 min and mass-to-charge (*m/z*) 569; 8-*iso*-PGF$_{2\alpha}$-d$_4$ at RT 12.68 min and *m/z* 573; 8,12-*iso*-iPF$_{2\alpha}$-VI-d$_{11}$ at RT 12.85 min and *m/z* 580; F_3-IsoPs (8-*iso*-PGF$_{3\alpha}$) at RT 12.76 min and *m/z* 567; NeuroPs at RT 14.50 min and *m/z* 593; and IsoFs at RT 14.90 min and *m/z* 585.

3. Calculating the concentration of lipid oxidation products: Fig. 1 provides an example for calculating the concentration of 8-*iso*-PGF$_{2\alpha}$ in urine.

4. Identify the peak for 8-*iso*-PGF$_{2\alpha}$ at RT 12.66 min (Peak A) by comparison with the standard for 8-*iso*-PGF$_{2\alpha}$-d$_4$ at RT 12.68 min (Peak B). Integrate the area of Peak A.

5. Identify and integrate the area of the internal standard (Peak C) 8,12-*iso*-iPF$_{2\alpha}$-VI-d$_{11}$ at RT 12.85 min and *m/z* 580.

6. (a) To quantify F_2-IsoP concentration in a 24-h urine collection, use the formula:
   Equation 1: Area Peak A/Area Peak C × 5 ng × 1000 µL/200 µL × 1000 × 1000 × 1/354.5 = pM.
   (b) To calculate the total F_2-IsoP over a 24-h period, use the formula:
   Equation 2: Eq. 1 × 24-h urine volume (L) = pmol/24 h.
   (c) To calculate the total F_2-IsoP in a spot urine, use the formula:
   Equation 3: Eq. 1 × 1/urine creatinine (mmol/L) = pmol/mmol creatinine.

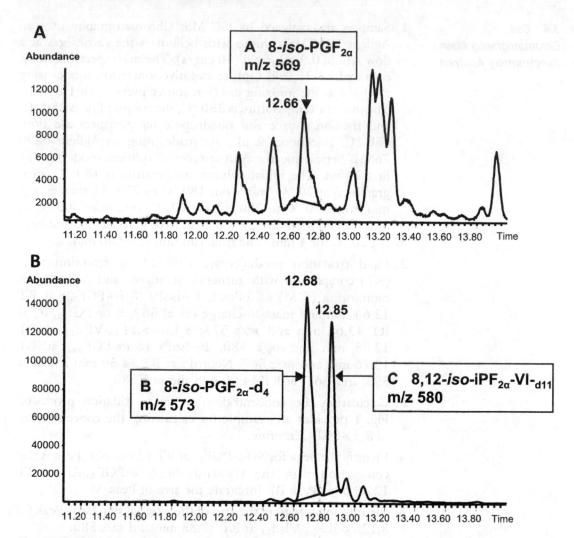

Fig. 1 GC-MS analysis with electron capture negative ionization of (Panel **a**) a urine sample showing the peak at RT = 12.66 min with m/z = 569 corresponding to F_2-IsoPs (8-*iso*-PGF$_{2\alpha}$) and (Panel **b**) the internal standards 8-*iso*-PGF$_{2\alpha}$-d$_4$ at RT = 12.68 and m/z = 573 and 8,12-*iso*-iPF$_{2\alpha}$-VI-d$_{11}$ at RT = 12.85 and m/z = 580

4   Notes

1. NaOH is caustic and should be handled with caution. Special care is required to prepare a solution of NaOH because considerable heat is liberated by the exothermic reaction. Add NaOH to water a little at a time with stirring, and then dilute the solution to make up to a final volume of 100 mL.

2. Liquid nitrogen should be handled in well-ventilated areas. Handle the liquid slowly to minimize boiling and splashing. Use tongs to withdraw objects immersed in liquid nitrogen.

3. Chloroform should be used in a fume hood to avoid vapors. Care should be taken to avoid contact with the skin.

4. Concentrated acetic acid and HCl should be handled with care in a fume hood. Wear gloves when handling, and always add the acid to water not the reverse.

5. Use PFBBr, DIPEA, BSTFA with 1% TMS, and pyridine in a fume hood.

6. Flush the culture tubes with a stream of nitrogen gas and cap immediately.

7. Samples can also be cooled by placing them in ice.

8. Ensure all the solvent has been eluted from the cartridge, and leave the vacuum on between steps.

9. Ensure the glass culture tube is completely dry of solvents. Samples can be stored at -80 °C overnight prior to derivatization.

10. Ensure all solvents are removed prior to commencing the next step.

11. Urine samples use 8,12-iso-iPF2α-VI-d$_{11}$ as internal standard because of an interfering peak in the gas chromatography chromatogram at RT 12.68 min that co-elutes with the 8-iso-PGF$_{2α}$-d$_4$ internal standard.

Acknowledgments

This work was in part funded by the National Health and Medical Research Council of Australia, the National Heart Foundation of Australia, and the Royal Perth Hospital Medical Research Foundation.

References

1. Halliwell B, Gutteridge JMC (2015) Free radicals in biology and medicine. Fifth edition edn. Oxford University Press, Oxford

2. Galano JM, Mas E, Barden A, Mori TA, Signorini C, De Felice C, Barrett A, Opere C, Pinot E, Schwedhelm E, Benndorf R, Roy J, Le Guennec JY, Oger C, Durand T (2013) Isoprostanes and neuroprostanes: total synthesis, biological activity and biomarkers of oxidative stress in humans. Prostaglandins Other Lipid Mediat 107:95–102. https://doi.org/10.1016/j.prostaglandins.2013.04.003

3. Morrow JD, Hill KE, Burk RF, Nammour TM, Badr KF, Roberts LJ (1990) A Series of prostaglandin-F2-like compounds are produced in vivo in humans by a noncyclooxygenase, free radical-catalyzed mechanism. Proc Natl Acad Sci U S A 87(23):9383–9387. https://doi.org/10.1073/pnas.87.23.9383

4. Kadiiska MB, Gladen BC, Baird DD, Germolec D, Graham LB, Parker CE, Nyska A, Wachsman JT, Ames BN, Basu S, Brot N, FitzGerald GA, Floyd RA, George M, Heinecke JW, Hatch GE, Hensley K, Lawson JA, Marnett LJ, Morrow JD, Murray DM, Plastaras J, Roberts LJ, Rokach J, Shigenaga MK, Sohal RS, Sun J, Tice RR, Van Thiel DH, Wellner D, Walter PB, Tomer KB, Mason RP, Barrett JC (2005) Biomarkers of oxidative stress study II. Are oxidation products of lipids, proteins, and DNA markers of CCl4 poisoning? Free Radic Biol Med 38

(6):698–710. https://doi.org/10.1016/j.fre eradbiomed.2004.09.017

5. Galano JM, Lee YY, Durand T, Lee JCY (2015) Special issue on "analytical methods for oxidized biomolecules and antioxidants" the use of isoprostanoids as biomarkers of oxidative damage, and their role in human dietary intervention studies. Free Radic Res 49 (5):583–598. https://doi.org/10.3109/10715762.2015.1007969

6. Milne GL, Dai Q, Roberts LJ (2015) The isoprostanes-25 years later. BBA-Mol Cell Biol L 1851(4):433–445. https://doi.org/10.1016/j.bbalip.2014.10.007

7. Fessel JP, Jackson Roberts L (2005) Isofurans: novel products of lipid peroxidation that define the occurrence of oxidant injury in settings of elevated oxygen tension. Antioxid Redox Signal 7(1–2):202–209. https://doi.org/10.1089/ars.2005.7.202

8. Fessel JP, Porter NA, Moore KP, Sheller JR, Roberts LJ 2nd (2002) Discovery of lipid peroxidation products formed in vivo with a substituted tetrahydrofuran ring (isofurans) that are favored by increased oxygen tension. Proc Natl Acad Sci U S A 99(26):16713–16718. https://doi.org/10.1073/pnas.252649099

9. Barden A, Mas E, Henry P, Durand T, Galano JM, Roberts LJ, Croft KD, Mori TA (2011) The effects of oxidation products of arachidonic acid and n3 fatty acids on vascular and platelet function. Free Radic Res 45 (4):469–476. https://doi.org/10.3109/10715762.2010.544730

10. Nourooz-Zadeh J, Liu EHC, Anggard EE, Halliwell B (1998) F-4-isoprostanes: a novel class of prostanoids formed during peroxidation of docosahexaenoic acid (DHA). Biochem Biophys Res Commun 242(2):338–344. https://doi.org/10.1006/bbrc.1997.7883

11. Barden AE, Mas E, Croft KD, Phillips M, Mori TA (2014) Minimizing artifactual elevation of lipid peroxidation products (F-2-isoprostanes) in plasma during collection and storage. Anal Biochem 449:129–131. https://doi.org/10.1016/J.Ab.2013.12.030

12. Proudfoot J, Barden A, Mori TA, Burke V, Croft KD, Beilin LJ, Puddey IB (1999) Measurement of urinary F(2)-isoprostanes as markers of in vivo lipid peroxidation-a comparison of enzyme immunoassay with gas chromatography/mass spectrometry. Anal Biochem 272 (2):209–215. https://doi.org/10.1006/abio.1999.4187S0003-2697(99)94187-8.

Part IV

CE-MS-Based Metabolomics

Chapter 22

Metabolic Profiling of Urine by Capillary Electrophoresis-Mass Spectrometry Using Non-covalently Coated Capillaries

Rawi Ramautar

Abstract

In the field of metabolomics, capillary electrophoresis-mass spectrometry (CE-MS) can be considered a very useful analytical tool for the profiling of polar and charged metabolites. However, variability of migration time is an important issue in CE. An elegant way to minimize this problem is the use of non-covalently coated capillaries that is dynamic coating of the bare fused-silica capillary with solutions of charged polymers. In this protocol, an improved strategy for the profiling of cationic metabolites in urine by CE-MS using multilayered non-covalent capillary coatings is presented. Capillaries are coated with a bilayer of polybrene (PB) and poly(vinyl sulfonate) (PVS) or with a triple layer of PB, dextran sulfate (DS), and PB. The bilayer- and triple-layer-coated capillaries have a negative and positive outside layer, respectively. It is shown that the use of such capillaries provides very repeatable migration times.

Key words Capillary electrophoresis, Mass spectrometry, Metabolomics, Cationic metabolites, Non-covalently coated capillaries, Urine

1 Introduction

In metabolomics, CE-MS can be considered a useful analytical technique for the global profiling of highly polar and charged metabolites in various biological samples [1–3]. A stable CE-MS method is crucial to obtain reproducible results. For example, the stability of analyte migration times is of utmost importance in metabolomic studies where multiple biological samples have to be profiled and compared [4]. When conventional bare fused-silica capillaries are used, the analysis of biological samples with minimal sample pretreatment may lead to adsorption of matrix components to the capillary wall causing detrimental changes of the electro-osmotic flow (EOF) and, therefore, analyte migration times. Moreover, separation efficiencies may be compromised as a result of adverse analyte-capillary wall interactions.

Martin Giera (ed.), *Clinical Metabolomics: Methods and Protocols*, Methods in Molecular Biology, vol. 1730, https://doi.org/10.1007/978-1-4939-7592-1_22, © Springer Science+Business Media, LLC 2018

Over the past few years, various strategies have been developed to correct for migration time-shifts and for aligning electropherograms in CE-MS-based metabolomic studies [5–7]. However, these procedures are not very effective for aligning metabolites showing strong migration time-shifts among different samples, notably for late-migrating compounds. An attractive way to minimize these issues is the use of non-covalently coated capillaries, i.e., dynamical coating of bare fused-silica capillaries with charged polymers [8]. So far, various CE-MS methods employing non-covalently coated capillaries have been developed for the highly efficient and repeatable analysis of proteins, peptides, and metabolites in various matrices [9–12].

Recently, the utility of CE-MS using non-covalently coated capillaries with layers of charged polymers has been demonstrated for the highly repeatable metabolic profiling of urine [13]. Capillaries were coated with a bilayer of polybrene (PB) and poly(vinyl sulfonate) (PVS) or with a triple layer of PB, dextran sulfate (DS), and PB. The bilayer and triple-layer coatings were evaluated at low and high pH separation settings, thereby providing separation conditions for basic and acidic metabolites. In this chapter, attention is paid to the methodological aspects of the bilayer and triple-layer capillary coatings in CE-MS for the profiling of cationic metabolites in urine from rats using minimal sample pretreatment. It is shown that the use of these easy to produce capillary coatings of charged polymers significantly improves the performance of CE-MS for urinary metabolomic studies.

2 Materials

Prepare all solutions using ultrapure water (prepared by purifying deionized water to obtain a sensitivity of 18 MΩ-cm at 25 °C) and analytical grade reagents.

2.1 Solutions and Samples for Analysis

1. Background electrolyte (BGE) solution: 1 M formic acid, pH 2.0. Add 9.6 mL of water into a 10 mL glass vial and add 0.4 mL of concentrated formic acid to the water in a fume hood. Mix the solution thoroughly using a vortex. Store at 4 °C.

2. Metabolite standard mixture: Prepare stock solutions of 1 mg/mL of creatinine, dopamine, adrenaline, L-phenylalanine, L-tyrosine, glutathione, folic acid, guanosine, and hippuric acid by dissolving appropriate amounts in water. Make aliquots of stock solutions by dilution with BGE to obtain a working solution of 20 μg/mL for each analyte. Store stock and working solutions at −80 °C when not in use.

3. 10% (m/v) polybrene (PB) solution: Add 9.0 mL of water into a 10 mL glass vial and add 1.0 g of PB to the water in a fume hood. Mix the solution thoroughly using a vortex. Store at 4 °C.

4. 3% (m/v) dextran sulfate (DS) solution: Add 9.7 mL of water into a 10 mL glass vial and add 300 mg of DS to the water in a fume hood. Mix the solution thoroughly using a vortex mixer. Store at 4 °C.

5. 5% (v/v) poly(vinyl sulfonate) (PVS) solution: Add 9.5 mL of water into a 10 mL glass vial and add 0.5 mL of PVS to the water in a fume hood. Mix the solution thoroughly using a vortex mixer. Store at 4 °C.

6. Sheath-liquid solution for CE-MS analysis: Mix 50 mL of water with 50 mL methanol. Add 100 μL of concentrated formic acid to this solution. Mix the solution thoroughly using a vortex mixer.

2.2 Analytical Equipment

1. CE-MS: A commercially available CE equipment (Sciex, P/ACE ProteomeLab PA 800) is coupled to MS via a coaxial sheath-liquid interface (Agilent Technologies).

2. CE separation: Commercially available fused-silica capillaries (dimensions: 50 μm ID × 100 cm total length) are coated with charged polymers for electrophoretic separations.

3 Methods

The protocol described here for CE-MS using non-covalently coated capillaries for metabolic profiling studies is for laboratory use only. Prior to using this protocol, consult all relevant material safety data sheets (MSDS). Please use all appropriate laboratory safety procedures, including safety glasses, lab coat, and gloves, when performing the experiments described in this protocol.

3.1 Preparation of Urine Samples

1. Prior to CE-MS analysis, mix the urine sample with BGE (1:1, v/v) and centrifuge for 10 min at 4 °C and $16{,}100 \times g$. Rat urine samples used to obtain the here presented results were kindly provided by AstraZeneca (Department of Drug Metabolism and Pharmacokinetics, Macclesfield, UK) and stored at −80 °C when not in use.

3.2 Preparation of the Bilayer-Coated Capillary

1. Place a new bare fused-silica capillary in the CE instrument and rinse with water at 1380 mbar for 5 min. Assess whether drop formation is observed at the end of the capillary (*see* **Note 1**).

2. Rinse the separation capillary with 1 M NaOH at 1380 mbar for 15 min and then with water at 1380 mbar for 15 min (*see* **Note 2**).

3. Rinse the separation capillary with 10% (m/v) PB solution at 350 mbar for 15 min and then with water at 1380 mbar for 5 min. Subsequently, flush the capillary with 5% (v/v) PVS solution at 350 mbar for 30 min and finally with water at 1380 mbar for 5 min. The PB-PVS-coated capillary is now ready for use (*see* **Note 3**).

3.3 Preparation of the Triple-Layer-Coated Capillary

1. Place a new bare fused-silica capillary in the CE instrument and rinse with water at 1380 mbar for 5 min. Assess whether drop formation is observed at the end of the capillary (*see* **Note 1**).

2. Rinse the separation capillary with 1 M NaOH at 1380 mbar for 15 min and then with water at 1380 mbar for 15 min.

3. Rinse the separation capillary with 10% (m/v) PB solution at 350 mbar for 15 min and then with water at 1380 mbar for 5 min. Subsequently, flush the capillary with 3% (m/v) DS solution at 350 mbar for 15 min, followed by water at 1380 mbar for 5 min.

4. Rinse the separation capillary with 10% (m/v) PB solution at 350 mbar for 15 min and finally with water at 1380 mbar for 5 min. The PB-DS-PB-coated capillary is now ready for use.

3.4 Performance Assessment of Bilayer- and Triple-Layer-Coated Capillaries

1. Prior to CE-MS analysis, assess first in CE-UV mode using absorbance detection at 200 nm the performance of the bilayer- and triple-layer-coated capillaries with the cationic metabolite standards.

2. Add 20 μL of the cationic metabolite standard mixture into an empty 100 μL microvial (PCR vial) which fits into a CE vial and put this vial in the inlet sample tray.

3. Rinse the bilayer- or triple-layer-coated capillary with BGE at 1380 mbar for 5 min followed by sample injection at 35 mbar for 30 s (~15 nL).

4. Apply a voltage of +30 kV with bilayer or −30 kV with triple-coated capillary using a ramp time of 1 min and start acquiring UV absorbance data at 200 nm.

5. For CE analysis with bilayer-coated capillary, assess the recorded data by determining the migration times and the plate numbers of the analyzed cationic metabolite mixture. Check whether the compounds appear in the region between 10 and 18 min (Fig. 1a) and whether the plate numbers are between 100,000 and 300,000 (*see* **Note 4**).

6. For CE analysis with triple-layer-coated capillary, assess the recorded data by determining the migration times and the plate numbers of the analyzed cationic metabolite mixture. Check whether the compounds appear in the region between

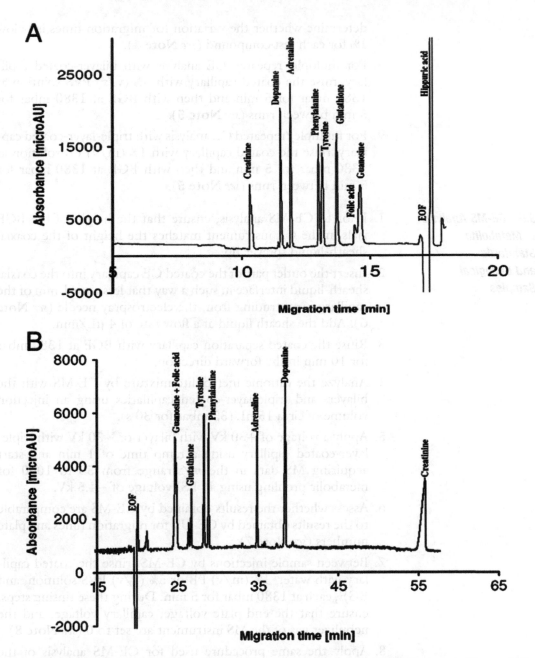

Fig. 1 CE-UV analysis of a test mixture of cationic metabolites using (**a**) a bilayer-coated capillary and (**b**) a triple-layer-coated capillary. Experimental conditions: BGE, 1 M formic acid (pH 2.0); sample injection, 35 mbar for 30 s; detection wavelength, 200 nm (reproduced from ref. 13 with permission)

20 and 60 min (Fig. 1b) and whether the plate numbers are between 100,000 and 300,000 (*see* **Note 4**).

7. Repeat the CE analysis of the cationic metabolite mixture ten times by both the bilayer- and triple-layer-coated capillaries and

determine whether the variation for migration times is below 1% for each test compound (*see* **Note 4**).

8. For multiple/repeated CE analysis with bilayer-coated capillary, rinse the coated capillary with 5% (v/v) PVS solution at 1380 mbar for 5 min and then with BGE at 1380 mbar for 5 min between runs (*see* **Note 5**).

9. For multiple/repeated CE analysis with triple-layer-coated capillary, rinse the coated capillary with 1% (m/v) PB solution at 1380 mbar for 5 min and then with BGE at 1380 mbar for 5 min between runs (*see* **Note 5**).

3.5 CE-MS Analysis of Metabolite Standards and Biological Samples

1. Prior to CE-MS analysis, ensure that the height of the BGE vials in the CE instrument matches the height of the coaxial sheath-liquid sprayer tip.

2. Insert the outlet part of the coated CE capillary into the coaxial sheath-liquid interface in such a way that less than 1 mm of the capillary is protruding from the electrospray needle (*see* **Note 6**). Add the sheath liquid at a flow rate of 4 μL/min.

3. Rinse the coated separation capillary with BGE at 1380 mbar for 10 min in the forward direction.

4. Analyze the cationic metabolite mixture by CE-MS with the bilayer- and triple-layer-coated capillaries using an injection volume of circa 15 nL (35 mbar for 30 s).

5. Apply a voltage of +30 kV with bilayer or −30 kV with triple-layer-coated capillary using a ramp time of 1 min and start acquiring MS data in the m/z range from 50 to 1000 for metabolic profiling using an ESI voltage of −4.5 kV.

6. Assess whether the results obtained by CE-MS are comparable to the results obtained by CE-UV for migration times and plate numbers (*see* **Note 7**).

7. Between sample injections by CE-MS, rinse the coated capillary with water, 1% (m/v) PB or 5% (v/v) PVS solution, and BGE, each at 1380 mbar for 5 min. During these rinsing steps, ensure that the end-plate voltage, capillary voltage, and the nebulizer gas of the MS instrument are set to 0 (*see* **Note 8**).

8. Apply the same procedure used for CE-MS analysis of the metabolite standards to cationic metabolic profiling of (rat) urine samples with both coated capillaries. A typical profile obtained for cationic metabolites in rat urine with the bilayer- and triple-layer-coated capillaries is shown in Fig. 2.

9. Repeat the CE-MS analysis of the (rat) urine sample ten times using both the bilayer- and triple-layer-coated capillaries and determine whether the variation for migration times is below 1% for the following endogenous compounds: creatinine, phenylalanine, and tyrosine (*see* **Note 9**).

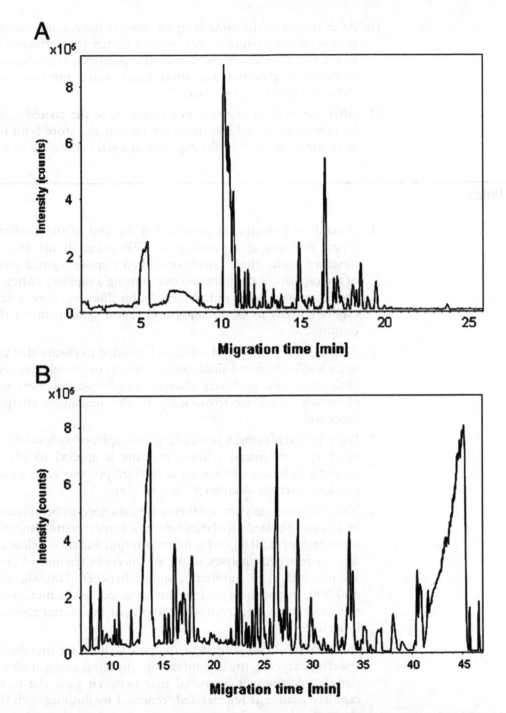

Fig. 2 Metabolic profiles (base peak electropherograms) obtained during CE-MS analysis of rat urine using (**a**) a bilayer-coated capillary and (**b**) a triple-layer-coated capillary. Experimental conditions: BGE, 1 M formic acid (pH 2.0); sample injection, 35 mbar for 30 s; data acquired for mass range from 50 to 1000 *m/z* (reproduced from ref. 13 with permission)

10. After analysis of the urine samples, analyze the cationic metabolite standard mixture to determine whether the performance of the CE-MS method using coated capillaries is still adequate in terms of expected migration times, plate numbers, and detection sensitivity (*see* **Note 7**).

11. After the analyses or when not in use, rinse the coated capillaries with water at 1380 mbar for 15 min and store both the inlet and outlet part of the capillary in a vial containing water.

4   Notes

1. If no drop formation is observed at the end of the capillary, repeat this step at a pressure of 3500 mbar. If no drop is observed under these conditions, then remove a small piece of the capillary at the inlet and outlet using a capillary cutter. If drop formation is still not observed after this procedure, a new capillary needs to be installed and repeat generation of the coatings.

2. Rinsing with 1 M NaOH solution is needed to ensure that the inner wall of the fused-silica capillary is fully negatively charged. Only then, the positively charged polybrene polymers will effectively attach electrostatically to the negatively charged inner wall.

3. For practical reasons, when rinsing the capillary with solutions of charged polymers, a lower pressure is applied to ensure proper attachment of the second or third polymer layer to the previous layer via electrostatic interactions.

4. The following analytical performance data need to be obtained by CE using coated capillaries for cationic metabolite standards (each present at 20 μg/mL): migration time variation below 1% for ten repeated analyses using an injection volume of circa 15 nL, and plate numbers ranging between 100,000 and 300,000. In case these data are not obtained for the metabolite standards, then regeneration/renewing of the coated capillary is needed.

5. In order to obtain consistent migration times for metabolite standards and for urinary metabolic profiling using multiple sample injections, it is crucial that between runs the outer capillary coating is regenerated/renewed by flushing with the charged polymer solution.

6. If the coated capillary is not properly aligned into the coaxial sheath-liquid interface, then an instable MS signal may be observed or electrophoretic current drops.

7. In case no comparable data is obtained, then the MS instrument needs to be tuned and re-calibrated or the CE capillary needs to be renewed.

8. It is important to switch off these MS parameters during the rinsing procedure in order to prevent that the charged polymer solution is entering the vacuum part of the MS instrument. The ion source of the MS instrument needs to be cleaned after 24 h of analysis.

9. In case such data is not obtained, then the MS instrument needs to be tuned and re-calibrated or the CE capillary needs to be renewed.

Acknowledgment

Dr. Rawi Ramautar would like to acknowledge the financial support of the Veni and Vidi grant scheme of the Netherlands Organization for Scientific Research (NWO Veni 722.013.008 and Vidi 723.016.003).

References

1. Ramautar R, Somsen GW, de Jong GJ (2013) The role of CE–MS in metabolomics. In: Metabolomics in practice. Wiley-VCH Verlag GmbH & Co. KGaA, Weinheim, Germany, pp 177–208. https://doi.org/10.1002/9783527655861.ch8

2. Kuehnbaum NL, Britz-McKibbin P (2013) New advances in separation science for metabolomics: resolving chemical diversity in a post-genomic era. Chem Rev 113 (4):2437–2468. https://doi.org/10.1021/cr300484s

3. Hirayama A, Wakayama M, Soga T (2014) Metabolome analysis based on capillary electrophoresis-mass spectrometry. TrAC Trends Anal Chem 61:215–222. https://doi.org/10.1016/j.trac.2014.05.005

4. Ramautar R (2016) Capillary electrophoresis-mass spectrometry for clinical metabolomics. Adv Clin Chem 74:1–34. https://doi.org/10.1016/bs.acc.2015.12.002

5. Nevedomskaya E, Derks R, Deelder AM, Mayboroda OA, Palmblad M (2009) Alignment of capillary electrophoresis-mass spectrometry datasets using accurate mass information. Anal Bioanal Chem 395(8):2527–2533. https://doi.org/10.1007/s00216-009-3166-1

6. Kok MG, Ruijken MM, Swann JR, Wilson ID, Somsen GW, de Jong GJ (2013) Anionic metabolic profiling of urine from antibiotic-treated rats by capillary electrophoresis-mass spectrometry. Anal Bioanal Chem 405 (8):2585–2594. https://doi.org/10.1007/s00216-012-6701-4

7. Garcia-Perez I, Whitfield P, Bartlett A, Angulo S, Legido-Quigley C, Hanna-Brown M, Barbas C (2008) Metabolic fingerprinting of Schistosoma Mansoni infection in mice urine with capillary electrophoresis. Electrophoresis 29(15):3201–3206. https://doi.org/10.1002/elps.200800031

8. Huhn C, Ramautar R, Wuhrer M, Somsen GW (2010) Relevance and use of capillary coatings in capillary electrophoresis-mass spectrometry. Anal Bioanal Chem 396(1):297–314. https://doi.org/10.1007/s00216-009-3193-y

9. Dominguez-Vega E, Haselberg R, Somsen GW (2016) Capillary zone electrophoresis-mass spectrometry of intact proteins. Methods Mol Biol 1466:25–41. https://doi.org/10.1007/978-1-4939-4014-1_3

10. Katayama H, Ishihama Y, Asakawa N (1998) Stable cationic capillary coating with successive multiple ionic polymer layers for capillary electrophoresis. Anal Chem 70(24):5272–5277

11. Catai JR, Torano JS, de Jong GJ, Somsen GW (2006) Efficient and highly reproducible capillary electrophoresis-mass spectrometry of peptides using Polybrene-poly(vinyl sulfonate)-coated capillaries. Electrophoresis 27

(11):2091–2099. https://doi.org/10.1002/elps.200500915

12. Soga T, Ueno Y, Naraoka H, Ohashi Y, Tomita M, Nishioka T (2002) Simultaneous determination of anionic intermediates for Bacillus Subtilis metabolic pathways by capillary electrophoresis electrospray ionization

mass spectrometry. Anal Chem 74 (10):2233–2239

13. Ramautar R, Torano JS, Somsen GW, de Jong GJ (2010) Evaluation of CE methods for global metabolic profiling of urine. Electrophoresis 31(14):2319–2327. https://doi.org/10.1002/elps.200900750

## Chapter 23

CE-MS for the Analysis of Amino Acids

Karina Trevisan Rodrigues, Marina Franco Maggi Tavares, and Ann Van Schepdael

Abstract

Amino acids play an important role in clinical analysis. Capillary electrophoresis-electrospray ionization-mass spectrometry (CE-ESI-MS) has proven to possess several characteristics that make it a powerful and useful tool for the analysis of amino acids in clinical studies. Here we present a method for the separation and quantitative analysis of 27 amino acids in urine based on CE-ESI-MS. The method presents an improved resolution between the isomers Leu, Ile, and *a*Ile, in comparison to other CE-ESI-MS methods in the literature. This method is fast, selective, and simple and has improved sensitivity by applying a pH-mediated stacking strategy, showing that it can be successfully used for amino acid analysis and probably for other small cationic metabolites.

Key words Capillary electrophoresis, Mass spectrometry, Electrospray ionization, Amino acids analysis, Targeted metabolomics, Urine

1 Introduction

It is already well established that amino acids play essential roles in many different fields, including human nutrition, food science, synthesis of drugs, and cosmetics, and play an essential role in metabolomics to name a few. In clinical analysis, the screening of amino acids in biological fluids can help to diagnose and to monitor treatment of diseases. The understanding of the concentration of amino acids is crucial for a precise diagnosis of metabolic disorders [1–3].

Chromatography-based methods (ion chromatography, reversed-phase liquid chromatography, hydrophilic interaction liquid chromatography, or liquid chromatography-mass spectrometry (LC-MS)) have been the most commonly applied techniques for the analysis of amino acids in biological fluids; however, methods reporting quantitative determination of these compounds in urine still have drawbacks. Different column chemistries and derivatization strategies are necessary, and some methods are limited to only a

Martin Giera (ed.), *Clinical Metabolomics: Methods and Protocols*, Methods in Molecular Biology, vol. 1730,
https://doi.org/10.1007/978-1-4939-7592-1_23, © Springer Science+Business Media, LLC 2018

few amino acids. In addition, some chromatographic modes are not able to provide an appropriate retention time for all amino acids. Although LC-MS has shown a great performance for amino acid analysis, ion suppression can often affect quantification accuracy and sensitivity [4–6]. Capillary electrophoresis (CE) has already proven to be a valuable analytical technique for clinical analysis [7]. Capillary electrophoresis coupled to mass spectrometry with electrospray ionization (CE-ESI-MS) has been demonstrated to be a powerful and attractive analytical tool for the determination and quantification of small charged compounds in biological samples [8]. Considering that CE offers a great mass selectivity and sensitivity and MS provides the possibility of structural information, the coupling of CE-MS increases analysis reliability without the need of extensive and complicated sample pretreatment or the use of a specific amino acid analyzer. Several authors have already demonstrated the possibility of amino acid analysis by CE-MS; good separations and solid method development have been reported [1, 9–12]. Recently, our group has also developed a quantitative amino acid analytical method using ion trap mass spectrometry [13]. In CE, the choice of background electrolyte (BGE) strongly affects the separation system. It is well known that the resolving power for charged compounds is the difference in charge to size ratio; thus, at a medium pH range, basic and acidic amino acids migrate in different directions. Thus, in order to achieve simultaneous separation of all amino acids under investigation by CE-MS, either low or high pH BGE can be attempted since the isoelectric points of the amino acids under investigation range from 2.77 (aspartic acid) to 10.76 (arginine). When applying a pH value below 2.77, each amino acid will be positively charged, and consequently they will migrate toward the cathode, which in this work is the terminal pole coupled to the mass spectrometer, and will then be detected with high selectivity and sensitivity. To achieve the simultaneous analysis of amino acids, 0.8 M formic acid was employed. The choice of appropriate sheath-liquid parameters is also very important in a CE-ESI-MS method. The optimized conditions are described in Subheading 3. This chapter explains the technical details of the CE-ESI-MS method for amino acid analysis using urine as sample.

2   Materials

All reagents should be analytical grade. Methanol and formic acid are LC-MS grade. All solutions should be prepared using Milli-Q water by purifying deionized water to attain a resistivity of 18 MΩ cm at 25 °C. Amino acid standard solutions, individual stock solutions of amino acids, and internal standard (methionine sulfone), at a concentration of 10 or 100 mM, are prepared in water, containing

0.1 or 0.2 M hydrochloric acid (HCl). All stock solutions can be stored at 4 °C. However, some of the amino acid solutions should be prepared fresh daily, due to a lack of stability (*see* **Note 1**). The working standard mixture is prepared by diluting the stock solutions with water prior to use.

2.1 CE-MS

1. Running buffer: 0.80 M formic acid containing 15% (v/v) methanol. Initially, prepare 50 mL of 8 M formic acid stock solution by adding 15.094 mL of 99% formic acid reagent to 34.906 mL of water. Mix well and store at 4 °C. Subsequently, prepare the BGE: mix 1 mL of the formic acid stock solution and 1.5 mL of methanol and complete to 10 mL with water (*see* **Note 2**).

2. Sheath liquid (SHL): 0.5% formic acid in 60% (v/v) methanol-water. Mix 60 mL of methanol and 40 mL of a 1.25% formic acid solution prepared in water. Store at 4 °C until use.

3. 12.5% ammonium hydroxide solution (NH_4OH) is prepared by diluting 25% ammonium hydroxide, with an equal volume of water.

2.2 Instrumentation

1. The CE-ESI-MS experiments are performed on a Beckman P/ACE MDQ capillary electrophoresis system coupled to a Bruker HCT Esquire 3000 plus ion trap mass spectrometer system using an Agilent coaxial sheath-liquid CE-MS interface (G1607A Agilent CE-ESI-MS sprayer kit) (*see* **Note 3**).

2. The sheath liquid is delivered through a KD Scientific syringe pump (*see* **Note 4**).

3. The CE system control is performed using the Beckman 32 Karat 8.0 software, while the mass spectrometry control and data acquisition are performed with Bruker Esquire Control™ software and the data evaluation with Bruker ESI Compass Data Analysis 4.1 software.

3 Methods

3.1 Sample Collection and Preparation

1. Following the sample collection, each urine sample is divided in smaller volumes and added to Eppendorf tubes. The tubes are then centrifuged at 14.103 × *g* for 15 min and immediately stored at −80 °C.

2. Before analysis, the selected Eppendorf containing sample is thawed and again centrifuged at 14.103 × *g* for 10 min. Then 750 μL of urine is mixed with 250 μL 0.1% formic acid containing 50 μM methionine sulfone (internal standard) immediately before CE-MS analysis.

3.2 CE-ESI-MS Conditions

1. Before the first use, a new capillary has to be conditioned for at least 20 min with 1 M NaOH, followed by water and BGE (*see* **Note 5**). The capillary should be conditioned daily by applying 20 psi of water for 4 min and BGE for 5 min. Additionally, in between analyses, the capillary is flushed with BGE for 3 min and at the end of the day for 3 min with water and 4 min with air.

2. Uncoated fused silica capillaries (50 μm i.d. × 85 cm total length) are used in all experiments (*see* **Note 6**).

3. For sample injection, a pH-mediated stacking is employed by hydrodynamic injection of a 12.5% NH_4OH solution for 9 s at 0.5 psi, followed by hydrodynamic injection of sample for 20 s at 0.6 psi (*see* **Note 7**). Other CE parameters are 30 kV applied voltage and a capillary temperature of 20 °C.

4. SHL is delivered at a flow rate of 5 μL/min (*see* **Note 8**).

5. The ESI source is operated in positive mode by applying a capillary voltage of 4.5 kV.

6. The drying gas flow is maintained at 5 L/min and the heater temperature at 200 °C. Nitrogen is used as nebulizing gas and the pressure is maintained at 8 psi.

7. Other MS parameters comprise a scan range of *m/z* 50-275 with an accumulation time of 15 ms.

8. Figure 1 shows base peak and extracted ion electropherograms of amino acids analyzed in a pooled urine sample. Table 1 shows linearity results, LOQ, and LOD of each amino acid.

4 Notes

1. All amino acid solutions should be prepared at 100 mM using 0.1 or 0.2 M HCl. Glutamic acid, aspartic acid, and tyrosine are less soluble; thus, 0.2 M HCl has to be used. The concentration of methionine sulfone (internal standard) is 10 mM using 0.1 M HCl as solvent. Methionine and glutamine are very unstable compounds and can suffer from degradation; consequently, the standard solutions of those amino acids should be prepared fresh daily.

2. We recommend preparing the BGE mixture fresh daily, as methanol can evaporate throughout the day possibly compromising method performance.

3. To avoid a siphoning effect, the CE inlet vial has to be positioned at the same height as the sprayer tip of the mass spectrometer. For the same reason, the capillary tip on the inlet should be at a similar height as the capillary tip in the sprayer.

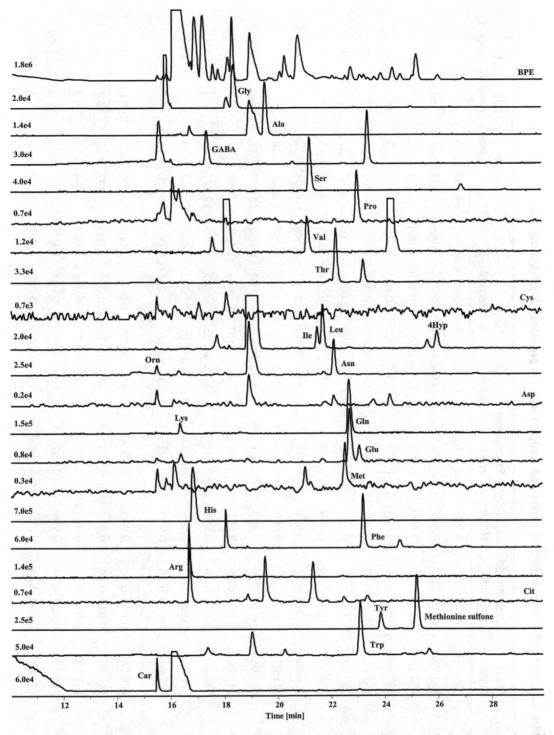

Fig. 1 Selected base peak electropherogram (BPE) of amino acid analysis in a pooled urine sample with extracted ion electropherograms (EIE) of each amino acid detected with the proposed CE-ESI-MS method. Reprinted with permission from 13

Table 1
Linearity results, LOD, and LOQ of each individual amino acid analyzed with the proposed CE-ESI-MS method

Amino acid	Exact mass	Concentration range mM	Linearity (R^2, $n = 5$)	Regression error	Slope	Intercept	F	p-value	LOD (µmol/L)	LOQ (µmol/L)
Alanine	89.048	0.20–1.0	0.9997	0.0004	0.30 ± 0.001	−0.002 ± 0.0005	94,922	0.16	5.7	17
α-isoleucine	131.095	0.020–0.10	0.999	0.0031	5.0 ± 0.071	−0.005 ± 0.004	4941	0.39	2.5	7.5
Arginine	174.112	0.020–0.10	0.999	0.004	8.6 ± 0.081	−0.010 ± 0.004	11,182	0.26	1.7	5.0
Aspartic acid	133.038	0.050–0.25	0.996	0.004	1.2 ± 0.021	−0.003 ± 0.005	3146	0.65	16	47
Asparagine	132.053	0.10–0.50	0.991	0.006	1.8 ± 0.009	0.006 ± 0.007	391	0.55	13	40
b-alanine	89.048	0.20–1.0	0.993	0.0025	0.38 ± 0.006	0.0001 ± 0.003	4467	0.98	27	80
Carnosine	226.107	0.020–0.10	0.999	0.011	7.1 ± 0.11	−0.014 ± 0.014	4549	0.50	6.7	20
Citrulline	175.096	0.050–0.25	0.999	0.007	3.5 ± 0.061	0.007 ± 0.008	3224	0.52	7.7	23
Cysteine	121.020	0.20–1.0	0.999	0.004	0.51 ± 0.008	−0.015 ± 0.004	3756	0.18	29	86
GABA	103.063	0.10–0.50	0.996	0.001	1.3 ± 0.007	−0.004 ± 0.001	28,833	0.25	4.0	12
Glutamic acid	147.053	0.050–0.25	0.998	0.004	2.4 ± 0.040	0.0008 ± 0.005	3483	0.91	7.3	22
Glutamine	146.069	0.050–0.25	0.998	0.002	2.1 ± 0.015	−0.005 ± 0.002	19,813	0.26	3.1	9.4
Glycine	75.032	0.20–1.0	0.998	0.0005	0.13 ± 0.0009	−0.001 ± 0.0006	21,963	0.33	15	45
4-Hydroxy proline	131.058	0.050–0.25	0.999	0.002	2.6 ± 0.026	−0.011 ± 0.003	9910	0.16	3.7	11
Histidine	155.069	0.020–0.10	0.998	0.0008	4.9 ± 0.018	0.005 ± 0.0009	76,690	0.11	0.63	1.9
Isoleucine	131.095	0.020–0.10	0.997	0.004	3.7 ± 0.10	0.004 ± 0.005	1342	0.57	4.7	14
Leucine	131.095	0.020–0.10	0.996	0.007	4.9 ± 0.17	0.005 ± 0.009	790	0.70	6.3	19
Lysine	146.106	0.020–0.10	0.9996	0.002	3.8 ± 0.042	0.001 ± 0.003	8252	0.76	2.4	7.3
Methionine	149.051	0.050–0.25	0.9998	0.005	3.1 ± 0.050	−0.005 ± 0.007	3754	0.60	7.3	22

Ornithine	132.090	0.050–0.25	0.9997	0.002	−0.002 ± 0.003	1.9 ± 0.035	2980	0.64	4.8	15
Phenylalanine	165.079	0.050–0.25	0.996	0.001	−0.015 ± 0.002	7.3 ± 0.013	306,822	0.074	0.6	2.4
Proline	115.063	0.020–0.10	0.999	0.008	0.014 ± 0.010	5.0 ± 0.038	16,965	0.41	6.7	20
Serine	105.043	0.10–0.50	0.9995	0.003	−0.004 ± 0.003	0.90 ± 0.012	5332	0.41	12	36
Threonine	119.058	0.10–0.50	0.996	0.002	−0.001 ± 0.003	0.91 ± 0.010	7859	0.78	10	30
Tryptophan	204.090	0.020–0.10	0.998	0.002	0.001 ± 0.002	5.5 ± 0.045	14,975	0.66	1.4	4.3
Tyrosine	181.074	0.020–0.10	0.998	0.002	−0.007 ± 0.002	4.1 ± 0.034	14,869	0.19	1.8	5.4
Valine	117.079	0.050–0.25	0.997	0.004	−0.005 ± 0.005	3.8 ± 0.037	10,523	0.51	4.3	13

4. If no pulseless HPLC pump is available, we recommend using a syringe pump instead. A non-pulse-free HPLC pump can cause instability in the current, affecting the performance of the CE-MS.

5. We recommend conditioning a new capillary also with sodium hydroxide before first use, but the conditioning should be performed with the capillary outside of the needle source to avoid contamination of the mass spectrometer. This is an important step for a better performance of the method.

6. The electrospray performance depends on the quality of the capillary cut. Jagged edges prevent the formation of a uniform spray and can also act as adsorption sites for sample components. We recommend using a diamond blade cutter.

7. In order to obtain a better focusing of the sample zone, urine samples should be acidified with 0.1% formic acid.

8. Low sheath-liquid flow rates are not recommended when using a syringe pump because current drop can be observed.

References

1. Mayboroda OA, Neusüß C, Pelzing M et al (2007) Amino acid profiling in urine by capillary zone electrophoresis–mass spectrometry. J Chromatogr A 1159:149–153. https://doi.org/10.1016/j.chroma.2007.04.055

2. Bouatra S, Aziat F, Mandal R et al (2013) The human urine metabolome. PLoS One 8: e73076. https://doi.org/10.1371/journal.pone.0073076

3. De Benedetto GE (2008) Biomedical applications of amino acid detection by capillary electrophoresis. In: Capillary Electrophor. Humana Press, Totowa, NJ, pp 457–481

4. Wang C, Zhu H, Pi Z et al (2013) Classification of type 2 diabetes rats based on urine amino acids metabolic profiling by liquid chromatography coupled with tandem mass spectrometry. J Chromatogr B Analyt Technol Biomed Life Sci 935:26–31. https://doi.org/10.1016/j.jchromb.2013.07.016

5. Sakaguchi Y, Kinumi T, Yamazaki T, Takatsu A (2015) A novel amino acid analysis method using derivatization of multiple functional groups followed by liquid chromatography/tandem mass spectrometry. Analyst 140:1965–1973. https://doi.org/10.1039/C4AN01672F

6. Held PK, White L, Pasquali M (2011) Quantitative urine amino acid analysis using liquid chromatography tandem mass spectrometry and aTRAQ reagents. J Chromatogr B Analyt Technol Biomed Life Sci 879:2695–2703. https://doi.org/10.1016/j.jchromb.2011.07.030

7. Jabeen R, Payne D, Wiktorowicz J et al (2006) Capillary electrophoresis and the clinical laboratory. Electrophoresis 27:2413–2438. https://doi.org/10.1002/elps.200500948

8. Zhong X, Zhang Z, Jiang S, Li L (2014) Recent advances in coupling capillary electrophoresis-based separation techniques to ESI and MALDI-MS. Electrophoresis 35:1214–1225. https://doi.org/10.1002/elps.201300451

9. Soga T, Kakazu Y, Robert M et al (2004) Qualitative and quantitative analysis of amino acids by capillary electrophoresis-electrospray ionization-tandem mass spectrometry. Electrophoresis 25:1964–1972. https://doi.org/10.1002/elps.200305791

10. Soga T, Heiger DN (2000) Amino acid analysis by capillary electrophoresis electrospray ionization mass spectrometry. Anal Chem 72:1236–1241. https://doi.org/10.1021/ac990976y

11. Ramautar R, Mayboroda OA, Derks RJE et al (2008) Capillary electrophoresis time of flight mass spectrometry using noncovalently bilayer coated capillaries for the analysis of amino acids in human urine. Electrophoresis 29:2714–2722. https://doi.org/10.1002/elps.200700929

12. Soliman LC, Hui Y, Hewavitharana AK, Chen DDY (2012) Monitoring potential prostate

cancer biomarkers in urine by capillary electrophoresis–tandem mass spectrometry. J Chromatogr A 1267:162–169. https://doi.org/10.1016/j.chroma.2012.07.021

13. Rodrigues KT, Mekahli D, Tavares MFM, Van Schepdael A (2016) Development and validation of a CE-MS method for the targeted assessment of amino acids in urine. Electrophoresis 37:1039–1047. https://doi.org/10.1002/elps.201500534

Part V

NMR-Based Metabolomics

Chapter 24

NMR Analysis of Fecal Samples

Hye Kyong Kim, Sarantos Kostidis, and Young Hae Choi

Abstract

Fecal analysis can generate data that is relevant for the exploration of gut microbiota and their relationship with the host. Nuclear magnetic resonance (NMR) spectroscopy is an excellent tool for the profiling of fecal extracts as it enables the simultaneous detection of various metabolites from a broad range of chemical classes including, among others, short-chain fatty acids, organic acids, amino acids, bile acids, carbohydrates, amines, and alcohols. Compounds present at low μM concentrations can be detected and quantified with a single measurement. Moreover, NMR-based profiling requires a relatively simple sample preparation. Here we describe the three main steps of the general workflow for the NMR-based profiling of feces: sample preparation, NMR data acquisition, and data analysis.

Key words NMR spectroscopy, Feces, Sample preparation, Data analysis, Identification, Gut microbiota

1 Introduction

Metabolomics has been applied to the study of diverse biological samples such as urine, plasma, and cerebrospinal fluid [1]. In recent years, however, the study of feces has gained attention due to the awareness of the importance of the role of gut microbes in human health. For example, gut-microbe interaction is now considered to play a significant role in the regulation of the human immune system [2]. Feces result from a direct contact with the intestine and therefore can be explored to study gut microbial activities and relate these with the gut health status. Gut diseases are often characterized by dysregulation of gut microbiota and their activities. Because gut bacteria are able to break down indigestible food components and produce critical metabolites which cannot be produced by the host, the results of their activity can be reflected in the metabolite profile of the fecal samples. For example, the fecal metabolites of patients with Crohn's disease showed a significant decrease in short-chain fatty acids, methylamine, and trimethylamine [3].

Martin Giera (ed.), *Clinical Metabolomics: Methods and Protocols*, Methods in Molecular Biology, vol. 1730, https://doi.org/10.1007/978-1-4939-7592-1_24, © Springer Science+Business Media, LLC 2018

Generally, nuclear magnetic resonance (NMR) spectroscopy and mass spectrometry (MS) are the most common analytical platforms for metabolomics studies. Each platform possesses unique advantages, and NMR though less sensitive exhibits greater reproducibility and easier sample preparation than MS. Moreover, NMR is capable to provide structural and quantitative information of metabolites across a broad range of chemical classes, with a single measurement and using only one internal standard. Thus, using NMR for the profiling of fecal samples has a great potential and has already been applied to study gastrointestinal diseases such as ulcerative colitis (UC), irritable bowel syndrome (IBS), Crohn's disease, and helminth infection [3–6]. The generated NMR data from fecal extracts is similar in terms of spectral complexity to that of all other body fluids that have been well studied (e.g., urine, blood serum, etc.), and thus, it can be processed and analyzed using the same, univariate, and multivariate statistical methods [7, 8]. It does present some challenges particularly in sampling due to its heterogeneous nature and the need to define meaningful results due to the inherent interindividual variability of the fecal metabolome.

In this chapter, we describe the steps for the NMR-based analysis of fecal samples. Following the sample collection and logistics for human feces, which have both been recently discussed [9], we will focus on the details of the three subsequent steps of fecal analysis: fecal sample preparation, data acquisition, and data analysis. The latter covers data processing for multivariate statistical analysis and metabolites identification. We will not include a comprehensive description of the available statistical methods to analyze the data since it is well beyond the scope of this chapter. Instead, we will focus on the generation of data using our own experience with both human [6] and murine specimen, as well as data available from optimization studies and reviews from other researchers [10–12].

2 Materials

Prepare all solutions with ultrapure deionized water and analytical grade reagents and solvents. Store all reagents and solutions at room temperature unless indicated otherwise.

2.1 Homogenization of Feces Samples

1. Phosphate buffer 0.15 M (K_2HPO_4/KH_2PO_4, pH 7.4): prepare 100 mL solution of 0.15 M K_2HPO_4 and 0.2 mM NaN_3 (solution 1) and 100 mL of 0.15 M KH_2PO_4 and 0.2 mM NaN_3 (solution 2) both in deionized water. Add solution 2 to solution 1, dropwise, until a pH of 7.4 is reached.

2. Mechanical homogenizer: bullet blender with ~ 5 × 1 mm zirconium oxide beads per sample (*see* **Note 1**).

3. Centrifuge with temperature control and capable of speeds >10,000 × g.

4. Microcentrifuge tubes (*see* **Note 2**).

2.2 NMR Sample Preparation

1. Internal standard solution: 4 mM sodium 3-trimethylsilyl (2,2,3,3-d$_4$) propionate (TSP) in D$_2$O (99.9% D).

2. 96 Ritter well plates if a robotic liquid handler is available.

3. 5 or 3 mm NMR tubes (*see* **Note 3**).

2.3 NMR Instrumentation and System Optimization

1. Autosampler integrated to an NMR instrument and software for automation.

2. Methanol-d$_4$ (99.98% D) for temperature calibration: preferably ampules of 750 μL.

3. Aqueous solution with reference compounds for shimming optimization: generally provided by the manufacturer. For Bruker instruments, the standard sucrose sample in 9:1 H$_2$O/D$_2$O, including sodium 3-(trimethylsilyl) propanesulfonic-d_6 acid (DSS-d$_6$) as internal standard, can be used.

3 Methods

3.1 Processing of Fecal Samples

1. Defrost samples at room temperature (about 1 h).

2. Weigh approximately 300 mg into a microcentrifuge tube and add ~ 5 × 1 mm zirconium oxide beads (*see* **Note 4**).

3. Add sufficient phosphate buffer 0.15 M, pH 7.4, to obtain a ratio of fecal sample (mg) to volume of buffer (μL) 1:2 (*see* **Note 5**).

4. Homogenize the sample by bead beating using a bullet blender for 30 s. Observe the samples and repeat this step if fecal slurry is not homogeneous.

5. Centrifuge at 16,000 × g for 15 min, at 4 °C (*see* **Note 6**).

6. Transfer 600 μL of the supernatant (*see* **Note 7**) of each sample to a new microcentrifuge tube and store at −80 °C until NMR analysis. If NMR measurements can be performed immediately, then proceed to the next section.

3.2 NMR Sample Preparation

1. Defrost samples at room temperature.

2. Add 60 μL of the internal standard solution (4 mM TSP in D$_2$O) to each aqueous fecal sample to reach the final concentration of TSP in the NMR sample of 0.4 mM. D$_2$O will be used for the field lock and TSP as the chemical shift reference and quantification standard (*see* **Note 8**).

3. Vortex for 10 s.

4. Centrifuge at 16,000 × g for 10 min, at 4 °C.

5. Transfer 560 μL to a 5 mm NMR tubes using a pipette or a robotic liquid handler. If 3 mm tubes are used, transfer 165 μL with either a syringe or a robotic liquid handler (*see* **Note 9**).

6. Place tubes in the autosampler for NMR measurements. If temperature regulation of the autosampler system is not available, then use batches of a small number of samples (e.g., up to 6–8), and store the remaining samples temporarily at 4 °C for no longer than 24 h (*see* **Note 10**).

7. Mix all leftover solutions from **step 5** to prepare pool samples. Ideally, one pool sample per group of samples of the study should be made. Vortex for 10 s and store all of them at −80 °C, keeping one sample that will be used for the NMR spectrometer optimization.

3.3 NMR Instrument Setup

1. Optimize all axes (3D) shims using the aqueous reference solution, generally provided by the instrument manufacturer. The internal standard (TSP or DSS) peak half-height linewidth without line broadening should be <0.7 Hz for acceptable resolution.

2. Calibrate probe temperature with a fresh sample of pure methanol-d_4 to obtain an actual sample temperature of 300 K (i.e., 27 °C). To achieve this, acquire a short single-pulse ^{1}H NMR experiment (two scans). Apply a line broadening of 3 Hz and calculate the actual temperature from the distance between the two methanol peaks (CH_3 and OH) (*see* **Note 11**). Repeat the experiment with appropriate adjustments of the temperature setting until an exact temperature of 300 K (27 °C) is achieved. Save the temperature setting.

3. Optimize shims for the fecal aqueous samples using a sample of similar matrix. It is preferable to use a pooled sample made from the samples of the study. Store the optimized shims file and use it as the starting point prior to the shimming of each sample during the automatic acquisition process.

4. Use the 1D ^{1}H NOESY experiment with presaturation to optimize the frequency in which the presaturation pulse for the water signal suppression will be applied. For aqueous fecal samples, a bandwidth of 50 Hz is required to sufficiently suppress the water signal. Store the optimum frequency value (parameter O1 for Bruker instruments) at which the residual water peak does not affect the spectral baseline outside of 4.65–5.1 ppm when a cryoprobe is used or 4.7–4.9 for the room-temperature operating inverse probes.

5. Optimize profiling parameters using one of the pooled samples. Two experiments per sample should be used for fecal extract profiling, a 1D ^{1}H NOESY (preferably with z-gradients; *see* **Note 12**) and a 2D ^{1}H *J*-resolved (JRES). Except for the

water irradiation frequency, described in the previous step, the experimental time and the receiver gain (RG) should be adjusted in this step. The optimum RG value is automatically calculated. The number of scans (NS) should be manually adjusted based on a number of factors. The most important are the signal-to-noise ratio (SNR) of the weakest signals observed (e.g., the singlet of fumarate or aromatic compounds can be used) and the aim of the study (*see* **Note 13**). Typically, for a 600 MHz NMR instrument equipped with a cryoprobe, 64–128 scans (7–15 min) can be used for the 1D NOESY experiment. Other standard parameters are 20 ppm spectral width, 10 ms mixing time, 4 s relaxation delay, and 64 K FID data points (*see* **Note 14**). The 2D JRES experiment should be acquired using the same pulses as the 1D NOESY (90-degree pulse and presaturation) and the same RG. The relaxation delay should be 2 s, 12,288 data points in the direct dimension (F2), 2 transients over 40 increments in the indirect dimension (F1), 10,000 Hz spectral width in F2, and 78 Hz in F1. Store the parameters of the two experiments so that they can be used for the automatic profiling.

3.4 NMR Profiling of Fecal Extracts and Data Processing

1. Prior to data acquisition, each sample should be allowed to rest in the probe for 5 min to reach the required temperature with less than 0.1 K fluctuation.

2. The following steps should be performed automatically using the software routines provided by the manufacturer of the instrument: load the stored shim file settings, tune and match the probe, field lock, 1D (z-axis) shimming, autophase of the lock signal, and calculation of the 90-degree pulse length [13]. The calculated pulse is automatically implemented in the two saved profiling experiments (1D NOESY and JRES) and the acquisition starts.

3. Use the automation routines provided by the software to process the spectra: Fourier transformation, phase and baseline correction, calibration of chemical shift scale to TSP signal at 0.00 ppm, and apodization (exponential multiplication of FID) with line broadening of 0.3 Hz.

4. Acquire 2D NMR spectra, using the pooled samples. Typically, the homonuclear ^{1}H-^{1}H total correlation spectroscopy (TOCSY) and the heteronuclear ^{1}H-^{13}C heteronuclear single quantum correlation (HSQC) experiments are collected in addition to the profiling experiments used for all samples (*see* **Note 15**).

3.5 NMR Data Analysis

1. Observe all spectra and if necessary apply manual baseline correction. A typical ^{1}H NMR spectrum of human feces is shown in Fig. 1.

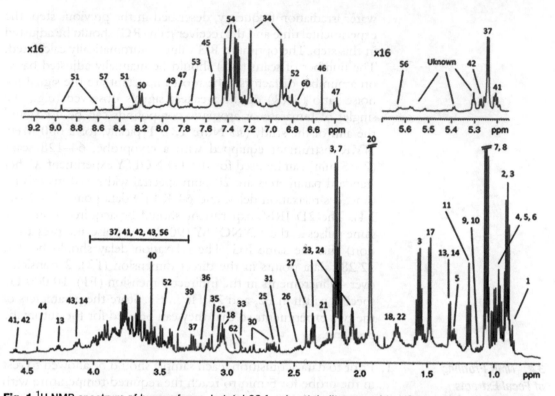

Fig. 1 ^{1}H NMR spectrum of human feces. In total 60 fecal metabolites were identified as 1, 2-methylbutyrate; 2, valerate; 3, n-butyrate; 4, leucine; 5, isoleucine; 6, valine; 7, propionate; 8, isobutyrate; 9, 3-methyl-2-oxoisovalerate; 10, 2-oxoisovalerate; 11, ethanol; 12, 3-hydroxybutyrate; 13, threonine; 14, lactate; 15, -2-hydroxyisobutyrate; 16, 3-hydroxy-2-butanone; 17, alanine; 18, lysine; 19, thymine; 20, acetate; 21, -5-aminopentanoate; 22, ornithine; 23, proline; 24, glutamate; 25, methionine; 26, glutamine; 27, succinate; 28, 2-oxoglutarate; 29, 3-phenylpropionate; 30, aspartate; 31, methylamine; 32, malate; 33, trimethylamine; 34, tyrosine; 35, malonate; 36, choline; 37, D-glucose; 38, taurine; 39, methanol; 40, glycine; 41, D-xylose; 42, D-galactose; 43, fructose; 44, dihydroxyacetone; 45, uracil; 46, fumarate; 47, urocanate; 48, ethanolamine; 49, xanthine; 50, hypoxanthine; 51, nicotinate; 52, 3-hydroxyphenylacetate; 53, tryptophan; 54, phenylalanine; 55, orotate; 56, UDP-glucuronate; 57, formate; 58, benzoate; 59, 4-aminohippurate; 60, homovanillate; 61, putrescine; 62, asparagine. Adapted from [7] with great acknowledgment

2. Identify the regions that do not contain relevant information, e.g., the water region.

3. Convert NMR data to a format suitable for multivariate data analysis using either proprietary (e.g., AMIX from Bruker BioSpin) or freely available NMR processing software within MATLAB or the R statistical environment [14]. In this step, the dimensions of the data will be reduced by dividing it into small integrated regions (bins). The integral of each bin is calculated and used as a new variable. The same region, which might contain one or more NMR peaks, is calculated for all samples, resulting in a new data set of a number of rows equal to the number of samples and a number of columns equal to the number of bins. The bin size ranges from 0.0025 to

0.04 ppm and can either be constant or of a variable size (*see* **Note 16**).

4. Normalize the data. Since the weight of each fecal sample has been recorded, the total spectral area, i.e., the sum of bins, can be normalized to that weight. This can be performed in Excel or by routines that are implemented with the software used for statistical analysis (*see* **Note 17**).

5. Scale the data so that all variables have an equal importance for the development of the statistical model. Usually the autoscale method or Pareto scale is used, i.e., each variable is divided by its standard deviation or the root of the standard deviation, respectively. This can be performed in Excel or by routines that can be implemented with the software used for statistical analysis, e.g., SIMCA (Umetrics, Umea, Sweden) (*see* **Note 17**).

6. Perform principal component analysis (PCA), using available software such as AMIX, SIMCA, MATLAB, or freely available packages within the R statistical environment. PCA is an excellent tool to reveal underlying trends of such complex data sets such as grouping of samples. Possible outliers, e.g., measurement errors or sample outliers, can also be detected with PCA (*see* **Note 18**).

7. Depending on the study design, supervised statistical methods such as partial least square discriminant analysis (PLS-DA) or orthogonal PLS-DA (OPLS-DA) can also be performed to get an overview of the data set (*see* **Note 18**).

8. Perform the validation of the statistical model using either a test set, consisting of un-modeled samples (cross-validation), or permutation tests (*see* **Note 19**).

9. Analyze the score plots and loading plots of the first two to three components (e.g., those that explain most of the total explained variance) to identify the bins that contribute to group separation.

3.6 Identification of Fecal Metabolites

1. Select the significant bins emerging from the data analysis.

2. Compare the NMR signals included in these bins to available web databases and the literature (*see* **Notes 20** and **21**).

3. Verify the assigned metabolites with 2D–NMR method spectra of the pooled samples. Use the TOCSY data to connect the ^{1}H-^{1}H correlations and the HSQC to annotate singlets and severely overlapped multiplets using the ^{1}H and ^{13}C chemical shifts of compounds.

3.7 Quantification of Metabolites

1. Quantify the selected metabolites from the statistical analysis using deconvolution. Some proprietary software packages for this task are Chenomx NMR suite (Chenomx, Edmonton Inc.)

or MestreNova (MNova software, Mestrelab Research SL). Follow the instructions provided by the software companies. If such software is not available, look for other free options within the R environment, such as Batman and BQuant [15–17].

2. Calculate the concentration in each fecal sample by correcting the measured concentration with the weight of each fecal sample.

3. If needed, annotate the full spectra using the methods described in Subheading 3.6 and quantify as many compounds as possible by repeating steps 1 and 2.

4 Notes

1. We have successfully used the bead-beating method for crude fecal homogenization. However, other efficient methods are also available which can be used for this step such as tissue lysers. Ultrasonication or vortexing can replace the homogenization step.

2. When the bead-beating homogenization method is used, carefully select the microcentrifuge tubes compatible with the bullet blender machine. Check this with the manufacturer instructions to avoid leakage while bead beating.

3. The choice of 5 or 3 mm NMR tubes depends on the probe equipment in the first place and secondly on the amounts of available samples. A 5 mm cryoprobe can support the analysis with 3 mm tubes or even the smaller 1.7 mm tube due to the reduction in thermal noise. However, the conventional 5 mm ambient temperature operating probes typically exhibit best signal-to-noise ratio (SNR) with 5 mm tubes.

   The smaller tubes are ideal if a limited amount of sample is available as they require much less volume, 165 μL and 30 μL for 3 mm and 1.7 mm tubes, respectively, compared to the 560 μL needed for a 5 mm tube.

4. The weight can be changed depending on the availability of the sample. If murine stool specimens are to be analyzed, the weight of available material may be lower. Also, the amount needed per sample depends on the available instrumentation; see Note 3.

   In general the ratio of weight of fresh fecal sample to buffer (W_f/V_b) 1:2 is recommended [10].

5. Human fecal samples can be very heterogeneous and their structure and composition also depends on the diet. Sometimes, the proposed W_f/V_b ratio of 1:2 might not be sufficient to achieve the homogenization of the fecal slurry and the

separation of a fecal aqueous extract after centrifugation. In those cases, the volume of buffers should be increased to W_f/V_b ratios of 1:3 to 1:10. The same is necessary if there is a limited quantity of available material (*see* also **Notes 3** and **4**).

6. Repeat centrifugation if it is not possible to separate the fecal water from insoluble particles. If the problem persists, homogenization should be repeated with additional buffer.

7. For 3 mm tubes, a volume of 200–250 μL is sufficient for the NMR sample, with a residual solution that can be used to prepare the pooled samples.

8. It is important to ensure the accuracy of the volume of internal standard that is added, since the integral of its peak will be used for the quantitation of the metabolites. If a robotic liquid sampler is available, it should be preferred for this step.

9. If a syringe is used, wash it twice with distilled water of analytical grade between each sample preparation.

10. Fecal aqueous samples are not stable at room temperature for longer than 5 h [9]. Therefore, depending on the time of analysis required for each sample, small batches of samples should be placed in the autosampler. If refrigerated rack positions are available, like with the Bruker SampleJet system (Bruker BioSpin Ltd.), then sample stability is extended to 24 h.

11. The actual sample temperature can be calculated by measuring the distance ($\Delta\delta$, in ppm) between the methanol peaks in the NMR spectrum of a 100% methanol solution. The measured $\Delta\delta$ is then used in the equation, $T\ (°C) = 130.00 - 29.53\Delta\delta - 23.87\Delta\delta^2$, to extract the actual temperature [18]. In Bruker's TopSpin, this is automatically done with the "calctemp" command.

12. Gradients result in better water suppression, and compared to the non-gradient version of 1D NOESY (pulse program: noesy1dphpr for Bruker systems), the benefit in SNR can be significant with fecal water samples.

13. In some cases, the aim of studies might be the quantification of specific metabolites rather than the full profiling of each sample. In these cases, the total acquisition time is adjusted to the molecular targets, so that the resulted SNR is sufficient for accurate quantification.

14. These profiling parameters are based on the pulse sequence -RD − $g_{z,1}$–90° - t - 90° - t_m − $g_{z,2}$–90°—ACQ. RD is the relaxation delay of 4 s, $g_{z,1}$ and $g_{z,2}$ are gradients with 1 ms duration, t is a short delay of 3 μs between pulses, 90° is the 90° radiofrequency pulse (RF), t_m is the mixing time of 10 ms, and ACQ is the acquisition of the free induction decays (FIDs), set up to 2.72 s. The water resonance is suppressed with a

continuous RF of 50 Hz, applied during the relaxation delay and the mixing time.

15. 2D NMR spectroscopy can be applied for further identification of metabolites. Among various 2D NMR methods, HSQC (heteronuclear single quantum coherence spectroscopy) is particularly useful for this purpose. HSQC detects any carbon atom directly attached to a proton. Consequently, the chemical shift of a ^{13}C atom connected to a specific proton can be obtained. Standard Bruker pulse sequences for HSQC measurement (with gradients and water presaturation) are as follows: HSQC hsqcetgpprsisp2.2 2048 × 256, 12 ppm (1H) × 165.7 ppm (13C); spectra are processed with zero filling in both dimensions and standard windows (900 phase-shifted QSINE for phase-sensitive HSQC). Another useful method is TOCSY (^{1}H-^{1}H total correlation spectroscopy), which creates correlations between all protons within a given spin system. It provides similar information to a COSY (correlation spectroscopy) experiment with regard to directly coupled hydrogens, but provides further structural information by identifying interconnected groups of indirectly spin-coupled hydrogens. Useful pulse sequences for TOCSY measurement are as follows: TPPI phase-sensitive mode, with water presaturation during relaxation delay, a spectral width 6 kHz in both dimensions, a 2 s relaxation delay, an 80 ms mixing time, 1 K data points in f2, and 512 increments in f1. Zero filling in f1 to 1 K real data points.

16. Instead of bins with constant width, it is often very helpful to apply adaptive binning, i.e., the width of the bin is adjusted to the position of peaks and this position varies from sample to sample [19]. Further details about binning of NMR spectra can be found in the review by Forshed [20].

17. The normalization of the data is a row operation (i.e., sample-wise), while scaling is a column operation (variable-wise). Both can be performed with Excel. However, it is more convenient and faster to perform the processing steps prior to statistical analysis using dedicated software that can be found in packages within the R statistical environment (Metabolomic package, MUMA, etc.) or MATLAB (The MathWorks Inc., Natick, MA). AMIX (Bruker) also offers the option of normalization, while SIMCA (Umea, Sweden) provides the option for several scaling methods (autoscaling, Pareto, logarithmic transform, etc.) to an already normalized data set.

18. Different types of multivariate data analysis can be applied, as well as reviewed in two papers [21, 22]. In general, unsupervised methods such as PCA and hierarchical cluster analysis (HCA) are the first choice to obtain an overview of samples

including their distribution and identification of outliers. The next step is to apply supervised methods to exploit prior information and assess the distributing metabolites for class or group separation. Multivariate data analysis can be performed using commercially available software such as SIMCA (Umetrics, Umea) and the PLS toolbox from MATLAB (The MathWorks Inc., Natick, MA). A review on the different packages is available [14].

19. When applying a PLS-DA model or PLS modeling, validation must be performed. Without proper validation, there is a high risk of over-fitting with these methods. The two most widely used validation methods are cross-validation and test-set validation. Test-set validation involves having a separate data set to check if the results of the statistical analysis are valid on new data and how well they fit [21]. Permutation is another way of validating the classification model [23].

20. Several free online databases can be used to assist in the identification of metabolites including the HMDB spectral database (the human metabolome database, http://www.hmdb.ca), BMRB peak server (Biological Magnetic Resonance Data Bank, http://www.bmrb.wisc.edu/metabolomics/), and the PRIMe server (the Platform for RIKEN Metabolomics, http://prime.psc.riken.jp). AMIX (Bruker) and the Chenomx NMR suite software (Canada) package also offer metabolite databases or tools to identify metabolites. The pros and cons of these databases are discussed in another review [24].

21. Some databases include spectra recorded at a different pH so that it is essential to ensure that the queries match the experimental conditions, in this case that they are recorded at a neutral pH (7.0–7.4).

Acknowledgments

The authors thank Dr. E.G. Wilson for her comments and review of the manuscript.

References

1. Beckonert O, Keun HC, Ebbels TMD, Bundy J, Holmes E, Lindon JC, Nicholson JK (2007) Metabolic profiling, metabolomic and metabonomic procedures for NMR spectroscopy of urine, plasma, serum and tissue extracts. Nat Protoc 2:2692–2703

2. Nicholson JK, Holmes E, Kinross J, Burcelin R, Gibson G, Jia W, Pettersson S (2012) Host-gut microbiota metabolic interactions. Science 336:1262–1267

3. Bjerrum JT, Wang Y, Hao F, Coskun M, Ludwig C, Günther U, Nielsen OH (2015) Metabonomics of human fecal extracts characterize ulcerative colitis, Crohn's disease and healthy individuals. Metabolomics 11:122–133

4. Le Gall G, Noor SO, Ridgway K, Scovell L, Janieson C, Johnson IT, Colquhoun IJ, Kemsley EK, Narbad A (2011) Metabolomics of fecal extracts detects altered metabolic activity of gut microbiota in ulcerative colitis and irritable bowel syndrome. J Proteome Res 10:4208–4218

5. Marchesi JR, Holmes E, Khan F, Kochhar S, Scanlan P, Shanahan F, Wilson ID, Wang Y (2007) Rapid and noninvasive metabonomic characterization of inflammatory bowel disease. J Proteome Res 6:546–551

6. Kostidis S, Kokova D, Dementeva N, Saltykova KHK, Choi YH, Mayboroda OA (2017) ^{1}H-NMR analysis of feces: new possibilities in the helminthes infections research. BMC Infect Dis 17:275

7. Saccenti E, Hoefsloot HCJ, Smilde AK, Westerhuis JA, Hendriks MMWB (2014) Reflections on univariate and multivariate analysis of metabolomics data. Metabolomics 10:361–374

8. McKenzie JS, Donarski JA, Wilson JC, Charlton AJ (2011) Analysis of complex mixtures using high-resolution nuclear magnetic resonance spectroscopy and chemometrics. Prog Nucl Magn Reson Spectrosc 59:336–359

9. Gratton J, Phetcharaburanin J, Mullish BH, Williams HRT, Mark Thursz M, Nicholson JK, Holmes E, Marchesi JR, Li JV (2016) Optimized sample handling strategy for metabolic profiling of human feces. Anal Chem 88:4661–4668

10. Lamichhane S, Yde CC, Schmedes MS, Jensen HM, Meier S, Bertram HC (2015) Strategy for nuclear magnetic resonance-based metabolomics of human feces. Anal Chem 87:5930–5937

11. Deda O, Gika HG, Wilson ID, Theodoridis GA (2015) An overview of fecal sample preparation for global metabolic profiling. J Pharm Biomed Anal 113:137–150

12. Wu J, An Y, Yao J, Wang Y, Tang H (2010) An optimized sample preparation method for NMR-based faecal metabonomic analysis. Analyst 135:1023–1030

13. Wu P, Gottfried O (2005) Rapid pulse length determination in high-resolution NMR. J Magn Reson 176:115–119

14. Izquierdo-García JL, Villa P, Kyriazis A, del Puerto-Nevado L, Pérez-Rial S, Rodriguez I, Hernandez N, Ruiz-Cabello J (2011) Descriptive review of current NMR-based metabolomic data analysis packages. Prog Nucl Magn Reson Spectrosc 59:263–270

15. Astle W, De Iorio M, Richardson S, Stephens D, Ebbels T (2012) A Bayesian model of NMR spectra for the deconvolution and quantification of metabolites in complex biological mixtures. J Am Stat Assoc 107:1259–1271

16. Hao J, Liebeke M, Astle W, Maria De Iorio M, Bundy JG, Ebbels T (2014) Bayesian deconvolution and quantification of metabolites in complex [1]D NMR spectra using BATMAN. Nat Protoc 9:1416–1427

17. Zheng C, Zhang S, Ragg S, Raftery D, Vitek O (2011) Identification and quantification of metabolites in ^{1}H NMR spectra by Bayesian model selection. Bioinformatics 27:1637–1644

18. Findeisen M, Brand T, Berger S (2007) A ^{1}H-NMR thermometer suitable for cryoprobes. Magn Reson Chem 45:175–178

19. De Meyer T, Sinnaeve D, Van Gasse B, Tsiporkova E, Rietzschel ER, De Buyzere ML, Gillebert TC, Bekaert S, Martins JC, Van Criekinge W (2008) NMR-based characterization of metabolic alterations in hypertension using an adaptive intelligent binning algorithm. Anal Chem 80:3783–3790

20. Forshed J, Torgrip RJO, Åberg KM, Karlberg B, Lindberg J, Jacobsson SP (2005) A comparison of methods for alignment of NMR peaks in the context of cluster analysis. J Pharma Biomed Anal 38:824–832

21. Liland KH (2011) Multivariate methods in metabolomics - from pre-processing to dimension reduction and statistical analysis. Trends Analyt Chem 30:827–841

22. Cevallos-Cevallos JM, Reyes-De-Corcuera JI, Etxeberria E, Danyluk MD, Rodrick GE (2009) Metabolomic analysis in food science: a review. Trends Food Sci Technol 20:557–566

23. Smolinska A, Blanchet L, Buydens LMC, Wijmenga SS (2012) NMR and pattern recognition methods in metabolomics: from data acquisition to biomarker discovery: a review. Anal Chim Acta 750:82–97

24. Lutz NW, Sweedler JV, Wevers RA (2013) Ch. 1. Exploring the human metabolome by NMR spectroscopy and mass spectrometry, Wishart DS. In: Methodologies for metabolomics: experimental strategies and techniques. Cambridge University Press, London, pp 3–29

## Chapter 25

Quantitative Analysis of Central Energy Metabolism in Cell Culture Samples

Sarantos Kostidis

Abstract

Nuclear magnetic resonance (NMR) is one of the key analytical platforms used in the analysis of intracellular and extracellular metabolites. Despite the technological advances that allow for the production of high-quality data, the sampling procedures of cultured cells are less well standardized. Different cell lines and culture media composition require adjustments of the protocols to result meaningful quantitative information. Here we provide the workflow for obtaining quantitative metabolic data from adherent mammalian cells using NMR spectroscopy. The robustness of NMR allows for the implementation of the here described protocol to other cell types with only minor adjustments.

Key words NMR spectroscopy, Mammalian cells, Metabolomics, Sample preparation, Fingerprint, Footprint

1 Introduction

In vitro cell-based metabolomics studies have found widespread use in many areas of research, like toxicology [1], cancer metabolism [2], regenerative medicine [3, 4], immune metabolism [5], and many more. The main goal in such studies is to understand the influence of metabolism on biological effects and mechanisms and, ultimately, integrate this information onto metabolic maps [6]. Such information is obtained by extracting quantitative metabolic data from both the intracellular and extracellular compartments. Several targeted metabolomics approaches have been developed to accomplish this goal, including methods based on mass spectrometry (MS) [7, 8] and nuclear magnetic resonance (NMR) [9, 10] spectroscopy. MS-based techniques provide a broader molecular window due to the superior sensitivity compared to NMR, but they are often jeopardized by ionization suppression, matrix effects, and linearity issues. On the other hand, NMR-based analysis is capable to provide quantitative information of the core metabolism (i.e., amino acids, glycolysis, and TCA cycle

Martin Giera (ed.), *Clinical Metabolomics: Methods and Protocols*, Methods in Molecular Biology, vol. 1730,
https://doi.org/10.1007/978-1-4939-7592-1_25, © Springer Science+Business Media, LLC 2018

intermediates), spanning as much as six orders of magnitude despite the method's inherent lack of sensitivity. Moreover, NMR is a nondestructive and robust platform, while quantification can be carried out without the need for specific internal standards.

In order to acquire quantitative cellular data, either by MS or NMR, the metabolites have to be extracted from the cells and the culture medium and reconstituted in a proper solvent [11]. For NMR-based analysis, cultured cells should grow to populations in the order of a million or more before being harvested. The main steps of the sample preparation process are the separation of the cells from the medium and the quenching of metabolism, followed by the extraction and recovery of intracellular and extracellular components. Although several studies have been conducted, aiming to optimize the conditions of these main steps [12–17], there is not any standard operating procedure that can be applied to all cell types. Prior to every study, several evaluation steps have to be performed in order to establish the optimum conditions for cells sampling.

In the present chapter, we provide a workflow for sampling and analyzing adherent mammalian cells using NMR spectroscopy. Emphasis is given to the key steps of effective quenching, metabolic recovery by extraction, and quantification of metabolites in the NMR data. The potential pitfalls that often accompany each of these steps are highlighted and discussed in Subheading 4. Using the described method, a total of about 60–70 extracellular and intracellular metabolites can be quantified from adherent mammalian cells.

2 Materials

2.1 Sampling of Culture Medium

1. Centrifuge equipment capable to operate at temperatures of 4 °C or below and 2 mL microcentrifuge tubes.

2. Cold LC-grade 100% methanol solution, stored overnight at −80 °C.

3. Dry-ice.

2.2 Quenching Intracellular Metabolism of Adherent Cells

1. Solution of warm PBS at 37 °C: 1.9 mM Na_2HPO_4, 8.1 mM NaH_2PO_4, 150 mM NaCl. For every 10 cm culture dish, about 5 mL of PBS is needed.

2. Small cryostorage container with liquid nitrogen (LN_2): for every 10 cm culture dish, about 10 mL of LN_2 are needed (see **Note 1**).

3. Aspiration pump.

4. Disposable sterile plastic serological pipettes (10 mL) and pipettors.

5. Dry-ice in a large box to accommodate the frozen culture petri dishes post quenching.

6. 2 mL microcentrifuge tubes.

2.3 Extraction of Intracellular Metabolites

1. Microcentrifuge tubes (2 mL) and centrifuge operating at temperatures of 4 °C or below.

2. Cell scrapers.

3. Cold extraction solvent: methanol/chloroform/water, 8.1:0.9:1 (v/v/v) (*see* **Note 2**), previously stored at −80 °C. About 1.5 mL is required per culture dish of 10 cm (*see* **Note 3**).

4. Nitrogen gas or other drying system.

5. Spectrophotometer for total protein quantitation.

6. Pierce BCA protein assay kit for total protein estimation.

7. Cell counter device: automatic or manual (if BCA kit is not an option).

2.4 NMR Spectroscopy

1. pH meter.

2. Phosphate buffer in D_2O (K_2HPO_4/KH_2PO_4, pH 7.4): make two solutions of 50 mL in deuterated water, the first consisting of 1.5 M K_2HPO_4 and 2 mM NaN_3 (1) and the second 1.5 M KH_2PO_4 and 2 mM NaN_3 (2). Add 2–1, dropwise, until pH is 7.4. Note the final volume and add 9 volumes of D_2O (dilute 10× to 0.15 M concentration). Divide the final solution in two parts and to the first part add sodium 3-trimethylsilyl (2,2,3,3-d_4) propionate (TSP) to reach a concentration of 0.4 mM (buffer for culture medium), while in the second part add TSP to 0.02–0.04 mM (cell extracts buffer) (*see* **Note 4**).

3. 96 Ritter well plates (if a robotic liquid handler is available).

4. 3 mm NMR tubes for Bruker SampleJet (Bruker Biospin, Ltd.) in 96-tube racks (*see* **Note 5**).

3 Methods

3.1 Cell Culture Medium Sample

1. Aspirate 100–300 μL of culture medium (*see* **Note 6**).

2. Mix immediately with two volumes of cold 100% methanol (*see* **Note 7**) in a 2 mL microcentrifuge tube.

3. Place the tube on dry-ice and repeat **steps 1–3** until all samples are collected.

4. Vortex samples for 10 s.

5. Transfer samples to freezer (−20 °C or lower) and allow proteins to precipitate for at least 20 min (*see* **Note 8**).

6. Centrifuge at $16,000 \times g$ for 20 min at 4 °C.

7. Collect supernatants and dry under gentle nitrogen gas stream (*see* **Note 9**).

8. Repeat **steps 1–7** for cell-free culture medium (3× samples).

3.2 Quenching

1. Take out the culture dishes from the incubator and in the fume hood; immediately aspirate the culture medium using the aspiration pump.

2. Add 5 mL of warm PBS and shake gently for 1 s.

3. Aspirate PBS and immediately add 10 mL LN2 (*see* **Note 10**).

4. Allow LN2 to evaporate (about 1 min) and transfer the dish immediately to a box of dry-ice. It is critical that the frozen cells are not allowed to thaw after quenching.

5. Repeat **steps 1–4** until all samples are quenched. Keep all samples frozen until extraction.

3.3 Extraction of Intracellular Metabolites

1. Transfer all samples in a cold room (4 °C) if available.

2. Add 1.5 mL of the cold extraction solvent (*see* **Note 11**).

3. Detach cells from the culture dish surface using a cell scraper (*see* **Note 12**).

4. Transfer the mixture of cell extract and debris to a 2 mL microcentrifuge tube and place it on dry-ice until all samples are extracted.

5. Centrifuge at $16,000 \times g$ for 10 min at 4 °C.

6. Collect supernatants and dry under a gentle stream of nitrogen gas.

7. Use the precipitated cell biomass that is remained after the centrifugation to calculate each sample's protein content. The Pierce BCA protein assay kit is used for total protein quantification [18]. Use the materials provided by the kit and follow the accompanied instructions to calculate the protein content using a spectrophotometer and the absorbance at 562 nm. Alternatively, estimate the cell numbers of each cell line with a cell counter (*see* **Note 13**).

3.4 NMR Sample Preparation

1. Reconstitute the dried extracts and culture medium samples with 220–250 μL of the phosphate buffer in D_2O (*see* **Notes 14–16**).

2. Vortex for 10 s.

3. Transfer the samples to 96 Ritter well plate with a pipette (*see* **Note 17**).

4. Transfer 190 μL of each sample to 3 mm NMR tubes using a liquid robotic sampler if available or a syringe.

5. From the leftovers of each sample, prepare pool samples, if possible, one pool sample per biological group of samples (*see* **Note 18**).

6. Place all samples on a cooling position in the SampleJet (*see* **Note 19**).

3.5 NMR Spectroscopy

1. Load a pure methanol sample and calibrate the temperature setting to get an actual sample temperature in the probe of 300 K (27 °C), using a short single-pulse NMR experiment [19]. The temperature is estimated by the distance $\Delta\delta$ (ppm) between the two methanol peaks and using the formula T (°C) = $130.00 - 29.53\Delta\delta - 23.87\Delta\delta^2$. In Topspin (Bruker Biospin Ltd.), this step can be performed automatically with the "*calctemp*" command (*see* **Note 20**).

2. Store the calibrated temperature setting and use it throughout the remaining study.

3. Load one of the pool samples. Optimize shims and store shim file (*see* **Note 21**).

4. Run the ^{1}H 1D NOESY experiment (*see* **Note 22**), and set up the number of scans based on the signal to noise ratio (SNR) of less abundant peaks (e.g., aromatic protons signals, fumarate, etc.) (*see* **Note 23**).

5. Calculate the receiver gain (RG) to be used for all experiments in the study.

6. Store the 1D NOESY profiling experiment in the software's library (for Topspin: command: *wpar NAME_OF_EXPERIMENT*).

7. Run a 2D J-resolved experiment (JRES) using the same RG as the one used for 1D NOESY (*see* **Note 24**). Store the JRES profiling parameters as before.

8. Start automatic measurements of all other samples using the two saved NMR experiments. For each sample, the following steps should occur automatically: loading of the sample into the probe, wait for 5 min to adopt the 300 K temperature, tuning and matching of the probe, field lock to D_2O, load stored shim file (from the pool sample), on axis (z-axis) shimming, automatic 90° pulse calibration [20], and acquisition of 1D NOESY and JRES (*see* **Note 25**).

9. All collected NMR data in the form of free induction decays (FIDs) are automatically Fourier transformed, baseline and phase corrected, and referenced to TSP at 0.00 ppm. For the 1D NOESY, the FIDs are also zero filled to 128 k points (factor of 2) and weighed using a line broadening factor of 0.3 Hz. The baseline of all 1D spectra is evaluated and further corrected

manually, using the Topspin (or other available software routines) (*see* **Note 26**).

10. Once all samples have been measured, use the aliquots of the stored pool samples to acquire 2D NMR experiments. For each sample, the profiling experiments should be repeated, and in addition, the 2D ^{1}H-^{1}H correlation spectroscopy (COSY) and total correlation spectroscopy (TOCSY) and the ^{1}H-^{13}C heteronuclear single quantum correlation (HSQC) and heteronuclear multiple quantum correlation (HMBC) experiments are collected (*see* **Note 27**).

3.6 Metabolites Identification and Quantification

1. First, try to identify as many metabolites as possible using online databases and data from the literature. The most commonly used and freely available metabolite databases are the Human Metabolome Database (HMDB; http://hmdb.ca) [21], the Biological Magnetic Resonance Data Bank (BMRB; http://www.bmrb.wisc.edu), [22] and the Birmingham Metabolite Library (BML; http://www.bml-nmr.org/) [23]. Other well-known commercial databases are the Bbiorefcode (Bruker Biospin Ltd.) and Chenomx library (Chenomx Inc.).

2. Use the 2D NMR data of pool samples to verify all assignments (*see* **Note 28**).

3. Import data to a NMR deconvolution software to quantify the detected metabolites. The proprietary Chenomx NMR suite is considered as the gold standard for this task. It uses an internal database of more than 300 metabolites and an interphase allowing for interactive fitting (*see* **Note 29**). Other options are the also proprietary MNOVA (MestreLab, SL, Spain) and the freely available BATMAN [24], which runs within the R statistical environment (http://www.R-project.org/) (*see* **Note 30**). Detailed instructions and application notes are provided by the developers of these software packages (*see* **Note 31**).

4. Following the here described protocol, a panel of more than 60 metabolites can be identified in mammalian cultured cell lines. Use the information of selected peaks and their chemical shifts provided in Table 1, as a basis for the assignment of the intracellular and extracellular metabolites and as targets for quantification.

5. Correct the relative or absolute concentrations of quantified metabolites to the protein content of each sample (or alternatively to the cell numbers).

6. The process is repeated for the extracellular metabolites. Compare the results with the concentrations of the same metabolites in cell-free medium samples to export uptake and release of metabolites from and by the cells, respectively.

Table 1
Characteristic proton chemical shifts (at pH 7.4) of intracellular and extracellular metabolites of cultured mammalian cells. The listed resonances can be used for quantification by deconvolution

Metabolite	δ ^{1}H (ppm)	Metabolite	δ ^{1}H (ppm)
1-Methylnicotinamide	4.48, 8.18, 8.9, 8.97, 9.28	Malate	2.34, 2.65, 4.29
α-Ketoglutarate	2.43, 2.99	Methionine	2.11, 2.12, 2.18, 2.63, 3.84
2-Oxoisocaproate	0.92, 2.60	Methylmalonate	1.23, 3.16
3-Methyl-2-oxovalerate	0.88, 1.08, 1.44, 1.68, 2.91	Methylsuccinate	1.07, 2.11, 2.51, 2.61
ATP	4.25, 4.39, 4.59, 6.13, 8.26, 8.52	N,N-Dimethylglycine	2.90
Acetate	1.90	N-Acetylaspartate	2.00, 2.48, 2.67, 4.38
Alanine	1.48, 3.78	N-Acetylglutamine	1.91, 2.02, 2.31, 4.15
Arginine	1.64, 1.72, 1.91, 3.23, 3.76	NAD+	4.22, 4.39, 4.53, 6.02, 6.08, 8.16, 8.18, 8.41, 8.82, 9.13, 9.32
Asparagine	2.84, 2.94, 3.98	Nicotinate adenine dinucleotide	4.40, 4.51, 6.03, 8.04, 8.14, 8.42, 8.73, 8.99, 9.12
Aspartate	2.67, 2.80, 3.88	O-Phosphocholine	3.20, 3.58, 4.15
Betaine	3.25, 3.89	Ornithine	1.78, 1.92, 3.04, 3.77
Choline	3.19, 3.50, 4.05	Pantothenate	0.88, 0.92, 2.40, 3.97
Citrate	2.53, 2.64	Phenylalanine	3.12, 3.27, 3.98, 7.32, 7.36, 7.41
Creatine	3.02, 3.91	Proline	1.98, 2.05, 2.34, 3.33, 3.40, 4.12
Phosphocreatine	3.03, 3.93	Pyroglutamate	2.02, 2.39, 2.49, 4.16
Formate	8.44	Pyruvate	2.35
Fructose	3.55, 3.66, 3.69, 3.79, 3.88, 3.98, 4.01, 4.10	Serine	3.83, 3.93, 3.98
Fumarate	6.50	Succinate	2.38
GTP	5.93, 8.13	Taurine	3.25, 3.41
Galactitol	3.68, 3.97	Threonine	1.31, 3.57, 4.24
Galactose	3.48, 3.63, 3.69, 3.72, 3.79, 3.84, 3.92, 3.98, 4.07, 4.58, 5.26	Tryptophan	4.05, 7.19, 7.27, 7.31, 7.53, 7.72

(continued)

Table 1
(continued)

Metabolite	δ ^{1}H (ppm)	Metabolite	δ ^{1}H (ppm)
Glucose	3.23, 3.39, 3.45, 3.50, 3.71, 3.81, 3.88, 4.63, 5.22	Tyrosine	3.05, 3.18, 3.92, 6.89, 7.18
Glutamate	2.04, 2.11, 2.34, 3.74	UDP-N-acetylglucosamine	2.06, 3.54, 4.27, 4.36, 5.50, 5.96, 7.93
Glutamine	2.13, 2.44, 3.76	UDP-glucuronate	4.38, 5.62, 5.97, 7.93
Glutathione (reduced)	2.15, 2.54, 2.94, 3.76, 4.55	UDP-glucose	4.19, 4.23, 4.27, 4.36, 5.59, 5.97, 7.94
Glycine	3.55	UDP-galactose	4.27, 4.35, 5.63, 5.97, 7.95
Histidine	3.12, 3.22, 3.98, 7.08, 7.87	UMP	4.34, 4.41, 5.98, 8.09
Hypotaurine	2.63, 3.34	Valine	0.98, 1.03, 2.26, 3.59
Isoleucine	0.93, 1.00, 1.25, 1.46, 1.97, 3.66	Myoinositol	3.26, 3.52, 3.61, 4.05
Lactate	1.31, 4.09	sn-Glycero-3-phosphocholine	3.22, 3.60, 3.64, 4.31
Leucine	0.94, 0.95, 1.70, 3.73	β-alanine	2.54, 3.16
Lysine	1.46, 1.72, 1.89, 3.01, 3.74	CoA	0.72, 0.84, 2.60
Cystine	3.18, 3.38, 4.08	2-Hydroxyisobutyrate	1.35
Hypoxanthine	8.18, 8.20	3-Hydroxybutyrate	1.19, 2.30, 2.40, 4.14
Pyridoxine	2.45, 7.64	Niacinamide	7.58, 8.24, 8.70, 8.93
Trans-4-Hydroxy-L-proline	2.14, 2.42, 3.35, 3.47, 4.33	Galactose-1-phosphate	3.72, 3.76, 3.90, 3.99, 5.48

4   Notes

1. LN$_2$ temperature is below its boiling point of -195.56 °C and will cause cold burns in case of direct contact with the skin. Always use protective gloves and glasses when handling LN2. Pour LN$_2$ into the culture dish using a Falcon tube attached to tweezers to avoid direct contact with the tube.

2. Based on the fact that some residual water will remain in the culture dish after quenching, it was estimated that the 90% (v/v) methanol/chloroform 9:1 solution is actually close to

75% (v/v) methanol/chloroform 9:1 when added to the dish [14]. Adjust volumes appropriately if different sizes of culture dishes are used.

3. The selection of the extraction solvent is based on the resulting metabolic recovery. The more polar the solvent system is, the better the recovery of the polar metabolites will be. Unfortunately, this comes at the expense of effective protein removal, which is desired for the highest possible quality of the NMR spectra. Proteins induce several intense and broad resonances in the proton spectrum, which hamper the accuracy of metabolites quantification. Therefore, a compromise between metabolic recovery and protein removal has to be made by adjusting the polarity of the extraction solvent. The here, proposed mixture was also proposed by Lorentz et al. [14] and in general has worked well in our hands with several cell lines (endothelial cells, cancer cells, and macrophages). We recommend an evaluation of metabolic recovery using several proposed extraction solvents (50% MeOH, 70% MeOH, 75% MeOH/CHCl$_3$, and 100% MeOH) before each study.

4. The intensity of the TSP signal in the NMR spectra should be comparable to the other components of the sample. Since the cell extracts are much less concentrated than the culture medium samples, it is preferable to use different quantities of internal standard among these two types of samples.

5. The 3 mm NMR tubes are selected because they require less volume (190 μL) than the commonly used 5 mm tubes (560 μL). This way, the dried extract is dissolved in less volume, leading to more concentrated solution and increase NMR sensitivity. However, this configuration works well for 5 mm cryoprobes, but may suffer from increased thermal noise if a 5 mm room temperature probe is used. In that case, we recommend a trial experiment with both 3 mm and 5 mm tubes with the corresponding volumes of sample. Then, a direct comparison of the SNR between the two measurements will provide the optimum setup for the study.

6. The volume of medium depends on the total culture medium, i.e., the size of the culture dish. In general, the withdrawn medium quantity should not induce any stress to the cells.

7. The extraction solvent methanol/chloroform/water, 8.1:0.9:1 (v/v/v) can also be used for this step. We encourage testing the best system using blank culture medium with 10% fetal bovine serum (FBS) prior to the actual study.

8. The samples can be stored for longer times if a −80 °C freezer is available. This can be practical since, typically, the quenching and extraction steps of the intracellular metabolites should be performed right after the medium collection. This way the medium samples can be processed at later times (even several days later).

9. Best is to perform the drying step at the day of analysis to avoid unnecessary moisture in the samples.

10. The speed of the washing and quenching process is critical to avoid the turnover of labile metabolites, for example, ATP, NADH, and others. For optimum results, these steps should be performed by two persons. In that case, our experience shows that the cells can be quenched in about 5 s after the medium is aspirated.

11. Keep the extraction solvent cold by placing it in a box with dry-ice throughout the whole process.

12. An indication of the complete detachment of the cells from the dish surface is when the scraping becomes smooth with no resistance from anomalies of the surface.

13. In case that the Pierce BCA protein assay kit is not available and cell counting is preferred, this cannot be performed on the same samples used for extraction. Therefore, it is recommended to culture two additional samples under identical conditions and time and estimate the cell numbers in these samples using either an automatic or a manual cell counter method.

14. Although the 3 mm NMR tubes require just 190 µL of sample volume, we recommend dissolving the dried extracts in more volume so there is a sufficient quantity of leftovers to prepare pool samples. The latter are very useful in two ways. First, to prepare the NMR profiling parameters prior to automation and, second, to be used for 2D NMR for assignment of metabolites.

15. In case that 5 mm tubes are used, the dried extract should be dissolved in 650 µL.

16. Use the buffer with 0.02–0.04 mM TSP for cell extracts and the one with 0.4 mM TSP for culture medium samples.

17. If a robotic liquid handler is not available, then the samples are transferred manually to the NMR tubes with a syringe.

18. Mix the collected leftovers so aliquots per biological group are obtained (e.g., 3× samples of 60 µL), vortex 10 s, and store at −80 °C until analysis. Directly use one of the aliquots for setting up the NMR measurements.

19. If a SampleJet or other similar cooling setup is not available, do not place all samples in an autosampler, as the sample integrity might be affected after prolonged time at room temperature. Instead, use only small batches of samples (e.g., 4–6) and store the remaining ones in the fringe.

20. For accurate temperature calibration, the methanol sample should be shimmed well. Process the FID using a line broadening of 3 Hz before applying *calctemp*.

21. Use the peak of the internal standard to assess shimming quality. When no apodization is applied (i.e., no line broadening), the width of TSP at half height should be less than 0.7 Hz. Use *hwcal* command (Topspin, Bruker) to automatically calculate it.

22. The 1D NOESY experiment's pulse sequence has the form -RD—$g_{z,1}$—$90°$—t—$90°$—t_m—$g_{z,2}$—$90°$—ACQ. RD is the relaxation delay of 4 s, $g_{z,1}$ and $g_{z,2}$ are gradients with 1 ms duration, t is a short delay of 3 μs between pulses, $90°$ is the $90°$ radiofrequency pulse (RF), t_m is the mixing time of 10 ms, and ACQ is the acquisition of the free induction decays ($FIDs$), set up to 2.72 s. The residual water resonance is suppressed with a continuous RF of 50 Hz (presaturation), applied during the relaxation delay and the mixing time.

23. While a SNR of 3 is considered as the limit of detection (LOD), the limit of quantification (LOQ) is dependent on how precisely a resonance can be quantified. This can be estimated by calculating the coefficient of variation (CV %). We recommend to use triplicates of blank culture media, for which the available volume is not a restriction, and perform some 1D experiments to determine the optimum SNR in order to get an acceptable CV (often CV < 15–20% is sufficient).

24. The 2D JRES pulse sequence has the form -RD—$90°$—t_1—$180°$—t_1—ACQ. RD is set to 2 s, and t_1 increment between pulses and ACQ is set to 12.8 ms. 12,288 data points are collected in direct dimension ($F2$) with 2 transients over 40 increments in the indirect dimension ($F1$). The spectral width is set to 16.66 ppm in $F2$ and 78 Hz in $F1$. The same presaturation pulse as in the 1D NOESY is applied during the relaxation delay.

25. The total experimental time can vary from 20 min to even 1 h per sample for less concentrated samples.

26. At this point, it is critical to evaluate the quality of the spectra. First, the TSP singlet is used to assess the shimming quality. Poor shimming will reduce resolution and subsequently affect the quality of quantification. Second, since the quantity of the internal standard TSP is used as a reference for the quantification of all other metabolites, it should be the same across all samples (recall that the same quantity was added during sample preparation). If inconsistent TSP intensities are observed, try to correct the quantification results of those samples if possible, or remove them from further analysis. With modern NMR instruments, it is possible to generate artificial reference signals

(e.g., ERETIC2 [25]) with which any TSP addition inconsistencies can be corrected.

27. For the COSY experiment, use the parameters: the *cosygpprqf* pulse program (Topspin, Bruker), 2048 points in the *F2* and 256 increments in *F1* with 16 scans per increment, 13.35 ppm spectral width for both dimensions, and 2 s relaxation delay. For the TOCSY experiment, use the dipsi2gpphpr pulse program, which uses the DIPSI2 sequence for the homonuclear Hartmann-Hahn transfer during the spinlock period of 60 ms. Acquire 2048 points in the F2, with 32 scans and 256 increments in *F1*, 13 ppm spectral width for both dimensions, and 2 s relaxation delay. For the HSQC experiment, use the hsqce-detgpprsisp2.2 pulse program. This pulse sequence includes multiplicity selection for primary, secondary, and tertiary carbons. Collect 1024 data points in *F2*, with 128 scans and 180 increments. Set the spectra width to 13.01 ppm for ^{1}H (*F2*) and 165.65 ppm for ^{13}C (*F1*) and the relaxation delay to 1.5 s. For the HMBC experiment, use the hmbcgplpndprqf pulse program. It includes a low-pass J-coupling filter to exclude one-bond correlations (use ^{1}J$_{CH}$ = 145 Hz) and set the average for the long-range couplings at 7 Hz. Collect 2048 data points in *F2*, with 164 scans and 256 increments. Set the spectral width to 13 ppm for ^{1}H (*F2*) and 240 ppm for ^{13}C (*F1*), and use a relaxation delay to 1.5 s. Presaturation of 25–50 Hz can be applied in all 2D experiments.

28. We strongly recommend to verify all assignments made by the available databases with 2D NMR data. Use either the ^{1}H-^{1}H or the ^{1}H-^{13}C correlations to make sure the annotated metabolites are correctly identified, and note their chemical shifts. Use this information for quantification. In case of ambiguities, perform spiking experiments with pure compounds rather than reporting an ambiguous annotation. In order to better resolve overlapped peaks, use also the 2D JRES spectra which are collected for all samples.

29. When quantifying a compound with Chenomx, all of its protons are fitted. This method combined with the knowledge of peak positions (from the 2D data), and tools like the sum spectral line can have a beneficial influence in quantification accuracy. However, some sources of variation have been reported from user to user or for specific compounds [26]. In order to avoid erroneous quantification results, we recommend the following: study carefully and work with the application notes provided on the website of the developer and practice with samples of known composition until a good fitting methodology is obtained. Ideal candidates for practice are the blank culture medium, for which the composition is usually known

and their content is similar regarding the metabolites found in the intracellular matrix.

30. In contrast to Chenomx, BATMAN only fits the resonances which are defined by the user. As input files, it uses the information of about 2500 peaks, freely available by HMDB. The necessary inputs for each peak are its exact chemical shift, the J-coupling, the number of protons that contribute to this peak (e.g., 3 protons from a methyl group), and the compound ID. The latter can also be unknown or a new compound (not in the HMDB database) specified by the user. J-couplings can be extracted from the JRES spectra. Based on our experience with BATMAN, the fitting requires more computation time and is less precise compared to Chenomx, especially in the cases of overlapped peaks or complex multiplets. However, the fact that it is freely available makes it an attractive alternative if Chenomx is not available.

31. The extracted quantitative data is generally considered as relative concentrations of metabolites. In order to transform it to absolute values (if needed), some corrections are necessary (e.g., adjust for the relaxation times of each fitted peak) [25]. On the other hand, Chenomx provides the option to acquire data based on 1D NOESY experiments with specific parameters and without gradients (see software's available documentation and application notes). In that case, absolute concentrations are extracted based on the modeled compounds in Chenomx library without any further corrections. It should be noted, however, that the here proposed gradient-based NOESY experiment provides better SNR and water suppression. Moreover, we obtain comparable results when we compare the data from the two experiments. Based on this, the selection of the profiling experiment is based on what software will be used for quantification and how concentrated are the samples of the study. The only critical point is that this selection should be consistent throughout the study.

References

1. Liu W, Deng Y, Liu Y et al (2013) Stem cell models for drug discovery and toxicology studies. J Biochem Mol Toxicol 27:17–27

2. Benjamin DI, Cravatt BF, Nomura DK (2012) Global profiling strategies for mapping Dysregulated metabolic pathways in cancer. Cell Metab 16:565–577

3. Schlegel M, Köhler D, Körner A et al (2016) The neuroimmune guidance cue netrin-1 controls resolution programs and promotes liver regeneration. Hepatology 63:1689–1705

4. McNamara LE, Sjostrom T, Meek RMD et al (2012) Metabolomics: a valuable tool for stem cell monitoring in regenerative medicine. J R Soc Interface 9:1713–1724

5. Kelly B, O'Neill LA (2015) Metabolic reprogramming in macrophages and dendritic cells in innate immunity. Cell Res 25:771–784

6. Johnson CH, Ivanisevic J, Siuzdak G (2016) Metabolomics: beyond biomarkers and towards mechanisms. Nat Rev Mol Cell Biol 17:451–459

7. Bennett BD, Yuan J, Kimball EH et al (2008) Absolute quantitation of intracellular metabolite concentrations by an isotope ratio-based approach. Nat Protoc 3:1299–1311

8. Paglia G, Hrafnsdóttir S, Magnúsdóttir M et al (2012) Monitoring metabolites consumption and secretion in cultured cells using ultra-performance liquid chromatography quadrupole–time of flight mass spectrometry (UPLC–Q–ToF-MS). Anal Bioanal Chem 402:1183–1198

9. Nagana Gowda GA, Abell L, Lee CF et al (2016) Simultaneous analysis of major coenzymes of cellular redox reactions and energy using ex vivo 1 H NMR spectroscopy. Anal Chem 88:4817–4824

10. Goldoni L, Beringhelli T, Rocchia W et al (2016) A simple and accurate protocol for absolute polar metabolite quantification in cell cultures using quantitative nuclear magnetic resonance. Anal Biochem 501:26–34

11. León Z, García-Cañaveras JC, Donato MT et al (2013) Mammalian cell metabolomics: experimental design and sample preparation. Electrophoresis 34:2762–2775

12. Dietmair S, Timmins NE, Gray PP et al (2010) Towards quantitative metabolomics of mammalian cells: development of a metabolite extraction protocol. Anal Biochem 404:155–164

13. Sellick CA, Hansen R, Stephens GM et al (2011) Metabolite extraction from suspension-cultured mammalian cells for global metabolite profiling. Nat Protoc 6:1241–1249

14. Lorenz MA, Burant CF, Kennedy RT (2011) Reducing time and increasing sensitivity in sample preparation for adherent mammalian cell metabolomics. Anal Chem 83:3406–3414

15. Dettmer K, Nürnberger N, Kaspar H et al (2011) Metabolite extraction from adherently growing mammalian cells for metabolomics studies: optimization of harvesting and extraction protocols. Anal Bioanal Chem 399:1127–1139

16. Bi H, Krausz KW, Manna SK et al (2013) Optimization of harvesting, extraction, and analytical protocols for UPLC-ESI-MS-based metabolomic analysis of adherent mammalian cancer cells. Anal Bioanal Chem 405:5279–5289

17. Ser Z, Liu X, Tang NN et al (2015) Extraction parameters for metabolomics from cultured cells. Anal Biochem 475:22–28

18. Smith PK, Krohn RI, Hermanson GT et al (1985) Measurement of protein using bicinchoninic acid. Anal Biochem 150:76–85

19. Findeisen M, Brand T, Berger S (2007) A 1H-NMR thermometer suitable for cryoprobes. Magn Reson Chem 45:175–178

20. Wu PSC, Otting G (2005) Rapid pulse length determination in high-resolution NMR. J Magn Reson 176:115–119

21. Wishart DS, Jewison T, Guo AC et al (2013) HMDB 3.0–the human Metabolome database in 2013. Nucleic Acids Res 41:D801–D807

22. Ulrich EL, Akutsu H, Doreleijers JF et al (2008) BioMagResBank. Nucleic Acids Res 36:D402–D408

23. Ludwig C, Easton JM, Lodi A et al (2012) Birmingham metabolite library: a publicly accessible database of 1-D 1H and 2-D 1H J-resolved NMR spectra of authentic metabolite standards (BML-NMR). Metabolomics 8:8–18

24. Hao J, Liebeke M, Astle W et al (2014) Bayesian deconvolution and quantification of metabolites in complex 1D NMR spectra using BATMAN. Nat Protoc 9:1416–1427

25. Bharti SK, Roy R (2012) Quantitative 1H NMR spectroscopy. TrAC Trends Anal Chem 35:5–26

26. Sokolenko S, Blondeel EJMM, Azlah N et al (2014) Profiling convoluted single-dimension proton NMR spectra: a Plackett–Burman approach for assessing quantification error of metabolites in complex mixtures with application to cell culture. Anal Chem 86:3330–3337

Part VI

MALDI-Based Techniques and Mass Spectrometry Imaging of Clinical Samples

Chapter 26

Mass Spectrometry Imaging of Metabolites

Benjamin Balluff and Liam A. McDonnell

Abstract

Mass spectrometry imaging (MSI) is a technique which is gaining increasing interest in biomedical research due to its capacity to visualize molecules in tissues. First applied to the field of clinical proteomics, its potential for metabolite imaging in biomedical studies is now being recognized. Here we describe how to set up experiments for mass spectrometry imaging of metabolites in clinical tissues and how to tackle most of the obstacles in the subsequent analysis of the data.

Key words Mass spectrometry imaging, Tissue metabolomics, Biomarker discovery, Data analysis

1   Introduction

Mass spectrometry and NMR spectroscopy are the workhorse technologies in metabolomic studies [1]. Among mass spectrometry techniques, mass spectrometry imaging (MSI) is unique in that it allows visualizing metabolite distributions in biological samples such as tissues. MSI involves the local extraction, desorption, and ionization of the metabolites in a spatially correlated manner, for example using matrix-assisted laser desorption/ionization (MALDI) or desorption electrospray ionization (DESI) [2].

MALDI MSI can be used for imaging many different kinds of compounds, ranging from proteins to small molecules including metabolites or exogenous compounds. The latter application has led to the widespread application of MALDI MSI in the pharmaceutical industry, for the determination of the abundances and locations of drugs, their metabolized products, and interacting molecules in the target and nontarget tissues [3].

Due to its histological specificity (the maximum resolution is nowadays at a single cell level), MSI has also had a big impact for biomarker discovery in complex solid tissues such as cancer [2]. Although most biomarker studies have investigated proteins, clinical MSI is slowly being expanded to the investigation of metabolites [4].

Martin Giera (ed.), *Clinical Metabolomics: Methods and Protocols*, Methods in Molecular Biology, vol. 1730,
https://doi.org/10.1007/978-1-4939-7592-1_26, © Springer Science+Business Media, LLC 2018

MSI of metabolites benefited from the introduction of new sample preparation protocols, including new MALDI matrices [5, 6], chemical stabilization of tissue [7], chemical derivatization of target compounds [8], methods for the retrieval of metabolic information from archived tissues [9], and the availability of high mass resolution MSI instruments. The latter has proven important to resolve specific metabolite ions from the sometimes complex chemical background obtained from a tissue section [10].

A successful biomedical metabolomic study using MSI depends on many factors: the sample selection, sample handling, sample preparation, type of instrumentation—defining mass accuracy, mass resolution power, and sensitivity—and last but not least also the appropriate data processing and analysis procedures [11].

In this chapter we describe all the mentioned steps of how MSI can be used for metabolite imaging and data analysis in clinical tissues.

2 Materials

Prepare all solutions using ultrapure water and, where possible, use analytical grade reagents. Prepare and store all reagents at room temperature unless indicated otherwise. For clinical or preclinical research, it is imperative that ethical approval is sought and obtained prior to undertaking any of the steps described below.

2.1 Tissue Collection

It is well established that the metabolome changes rapidly after tissue sampling/animal sacrifice. Several methods have been reported to maintain metabolic endogenous metabolite levels, ranging from cold perfusion [12]/in situ freezing [7] to "freeze" metabolic activity prior to organ excision, to heat–/microwave-based denaturing of metabolic enzymes (see Note 1 for more information, including published comparisons of the different methods). Here for reasons of its broad applicability, we refer to already excised tissue samples, specifically fresh frozen tissues isolated via rapid tissue excision and freezing (preferably <1 min postmortem time).

2.2 Tissue Sectioning

1. Clean a stock of indium-tin-oxide (ITO)-coated glass slides (Bruker Daltonics, Bremen, Germany) by washing with deionized water (3×) and methanol (3×); use a multimeter to determine which side is conductive. Coat the ITO slides with 0.05% poly-L-lysine (poly-L-lysine coating used for greater adherence of tissue sections, protocol used as reported in Aichler et al. [13]). Once the poly-L-lysine coating is dry, mark the conductive side with a Tippex pen for easy identification.

2. Prepare a stock of microscope slides for histological assessment of the tissue sections.

3. Precool the ITO slides and slidebox by placing them within a precooled (−20C) cryostat chamber.

4. Prepare stock solutions of hematoxylin and Eosin. For example, using preprepared hematoxylin solution from Klinipath BV, article code 4085.9002; Eosin Y 1% alcohol solution from Klinipath BV, article code 640375.

2.3 Tissue Preparation

1. Prepare a stock solution of 2 mg/mL 9-aminoacridine in 70% methanol. Measure 2 mg of 9-aminoacridine using a mass balance that is accurate to 0.1 mg. Measure 0.7 mL methanol and 0.3 mL deionized water. Combine the 0.7 mL methanol with the 0.3 mL deionized water in a 1.5 mL Eppendorf tube, and then dissolve the 2 mg 9-aminoacridine in the 7:3 methanol/water solvent. Use a vortex mixer to ensure the matrix solution is free of particulates.

2. Prepare a stock solution of 5 mg/mL 2,3-diphenyl-pyranylium tetrafluoroborate (DPP-TFB) in 100% methanol. Measure 5 mg of (DPP-TFB) using a mass balance that is accurate to 0.1 mg. Measure 1 mL methanol and place in a 1.5 mL Eppendorf tube. Then dissolve the 5 mg DPP-TFB in the methanol. Use a vortex mixer to ensure the DPP-TFB solution is free of particulates.

3. Prepare a stock solution of 30 mg/mL 2,5-dihydroxybenzoic acid (DHB) in 70% MeOH and 0.1% TFA. Measure 30 mg of DHB using a mass balance that is accurate to 0.1 mg. Measure 0.7 mL methanol and 0.3 mL 0.1% TFA and place in a 1.5 mL Eppendorf tube. Dissolve the 30 mg DHB in the methanol/0.1% TFA solvent. Use a vortex mixer to ensure the DPP-TFB solution is free of particulates.

4. Prepare a solution of 70% methanol.

5. Prepare a solution of 20 mg/mL 2,4,6-trihydroxyacetophenone [THAP] in 50/50 ethanol/water.

6. Prepare 20 μM solutions of YGGFL, angiotensin II, [glu1]-fibrinopeptide B (all available from Sigma-Aldrich) in 50/50 ethanol/water.

2.4 Data Acquisition

1. Prepare a MALDI-TOF/TOF instrument, such as RapiFlex (Bruker Daltonics, Bremen, Germany) for positive ion and negative ion MALDI, reflectron mode detection, m/z range 50–1000, using a laser spot size of 50 μm. **Note 2** describes how to determine laser spot size using a thin matrix layer preparation. Ensure the mass spectrometer's m/z scale is calibrated using a mixture of YGGFL, angiotensin II, and [glu1]-fibrinopeptide B for positive mode calibration, or matrix clusters of THAP for negative mode calibration.

3   Methods

3.1   Sample Preparation

1. Precool a cryostat microtome (e.g., Leica CM1950) to the desired cutting temperature, e.g., −21 °C. Mount the excised tissue onto a precooled cryostat specimen holder. This may be achieved using a small amount of water, gelatin solution, or Tissue-Tek™ solution. On contact with the precooled specimen holder and frozen tissue block, the solution rapidly freezes, to effectively fix the tissue specimen to the specimen holder. Note: if using Tissue-Tek™, great care should be taken to ensure only a small, localized droplet is used to "glue" the tissue to the specimen holder. Tissue-Tek™ is known to cause strong ionization bias effects, and so its presence should be avoided on or near the tissue surfaces of interest.

2. Depending on the tissue specimen of interest, the tissue specimen may need to be "trimmed" to reach the desired location. For example, if analyzing coronal tissue sections of a mouse brain, then the desired region of the brain, e.g., between −0.10 and +0.40 mm from bregma, needs to be reached before the analytical tissue sections can be cut. Trimming may be performed manually or automatically depending on cryostat model. Brain location can be determined by expert comparison of histological images (cut at 5 μm thickness, mounted onto standard microscope slides and histologically stained) with a tissue atlas, e.g., those contained in the Allen Brain Atlas. For tumor specimens histopathological assessment during sectioning is essential to ensure the analytical tissue sections contain sufficient tumor material of the desired grade and cellularity and free of excessive necrosis.

3. Cut tissue sections of 12 μm thick using the cryostat microtome. Rapidly thaw-mount the tissue sections onto the marked (conductive) side of the precooled ITO slide within the cryostat chamber. Rapid thaw-mounting is achieved by localized warming of the reverse side of the MALDI target using a finger for maximum 3 s. Then place the slide + tissue into the precooled slide holder. Store the slide-mounted tissues at −80 °C until use.

4. For the analysis of adenylate and TCA cycle metabolites, the matrix 9-AA is used. First rapidly freeze-dry the slide-mounted tissue sections and then place into an automated matrix deposition system. Here we refer to the SunChrom SunCollect system, but other spray-based systems may be used with minor modification to the method. The 2 mg/mL 9-AA solution is sprayed onto the tissue section using two layers at 5 μL/min followed by six layers at 10 μL/min, with complete drying between layers (Y-offset = 1.0 mm, Z (needle height) = 25 mm,

speed X = speed Y = medium 1). After the matrix coating is dry, use a Tippex pen to add fiducial markers at four corners surrounding the tissue section (but not in contact with it).

5. For the analysis of amino metabolites and neurotransmitters, the metabolites are first derivatized using DPP-TFB. First rapidly freeze-dry the slide-mounted tissue sections and then place into an automated matrix deposition system. The 5 mg/mL solution of DPP-TFB in 100% MeOH is then deposited onto the tissue in 5 layers at a flow rate of 10 μL/min (Y-offset = 1.0 mm, Z (needle height) = 25 mm, speed X = speed Y = medium 1). For complete derivatization, the brain sections are kept overnight at room temperature and pressure. Following complete derivatization the MALDI matrix DHB is deposited onto the tissue section. The 30 mg/mL DHB solution in 70% MeOH and 0.1% TFA is sprayed onto the derivatized tissue section using two layers at 5 μL/min followed by six layers at 10 μL/min, with complete drying between layers. After the matrix coating is dry, use a Tippex pen to add fiducial markers at four corners surrounding the tissue section (but not in contact with it).

3.2 Data Acquisition

MALDI MSI experiments can be performed using many different types of mass spectrometer, including MALDI-TOF/TOF, MALDI-Q-TOF, MALDI-IMS-TOF, MALDI-FT-ICR, and MALDI-Orbitrap systems. Here we refer to the MALDI-TOF/TOF instruments from Bruker Daltonics, to date the most common systems used for MALDI MSI.

1. First acquire high-resolution optical images of the matrix-coated tissue sections using a flatbed scanner, e.g., an optical image recorded at 254 dpi corresponds to a pixel size of 100 μm. Mount the matrix-coated tissue section into the microscope slide holder, being careful to ensure good electrical contact between the stainless steel slide holder and the ITO coating of the conductive surface (note: it may be necessary to wipe the matrix coating from the edge of the slide to ensure a good contact).

2. Start a new MSI experiment in FlexImaging. Import the optical image of the matrix-coated tissue section into FlexImaging, and align the instrument stage holder to the optical image by, in-turn, selecting the fiducial markers in the imported optical image and the system's sample visualization camera. Once aligned the MSI experiment may be defined. Select the tissue section to be analyzed as the first measurement region, as well as small region close to the tissue section to serve as a control (useful for assessing potential analyte delocalization and the ready identification of background ions). If more than one

tissue section is present on the slide, they may be selected as separate measurement regions. For each measurement region, the spatial resolution then needs to be defined (e.g., 100 μm), the number of laser shots (e.g., 300), and the FlexControl method selected. If adenylate and TCA cycle intermediate are analyzed, the FlexControl method corresponds to that for the analysis of negative ions; if amino metabolites and neurotransmitters are analyzed, the FlexControl method corresponds to that for the analysis of positive ions.

3. Test the performance of the selected FlexControl method on the tissue section and vary the laser power until high-quality metabolite signals are obtained, past MALDI threshold but not too high to avoid fragmentation of the metabolites and excessive background noise. A data processing method may be selected to preprocess each mass spectrum of the MALDI MSI dataset; if such preprocessing is performed during data acquisition, select a validated FlexAnalysis method for the metabolite peaks of interest. Once the FlexImaging settings have been finalized, data acquisition can be initiated.

4. After MALDI MSI data acquisition, any residual MALDI matrix is removed by washing the slide-mounted tissue sections in 70% ethanol (2 × 30 s dips). The tissue section is then histologically stained, using hematoxylin and Eosin (tissue tissue) or Nissl (mouse brain). Lou et al. recently reported a method by which the histological staining can be optimized for the thicker tissue sections analyzed by MALDI MSI [11].

5. A high-resolution histological image of the histologically stained tissue section is then recorded using a digital slide scanner (e.g., 3D Histech Pannoramic desk). This histological image is then imported into FlexImaging for co-registration with the MALDI MSI results, through registering the histological image (high-resolution optical image of the histology-stained tissue section) to the optical image of the matrix-coated tissue section, from which the experiment was defined. The registration is performed by selecting the same fiducial markers in both images.

3.3 Data Analysis

There are many data analysis strategies, but the analysis of metabolomic MSI datasets, which contain multiple samples (from different patients and experiments), usually relies on a recurrent data processing routine, which is summarized in Fig. 1. To compare different MSI datasets, it is necessary to compile them into one MSI dataset. During the compilation, a minimum of data preprocessing steps is considered important to enable and increase comparability between metabolic MSI datasets. As we focus on Bruker instrumentation and data, it is worth mentioning that most of the functionality to merge and compare multiple MSI datasets is available through the

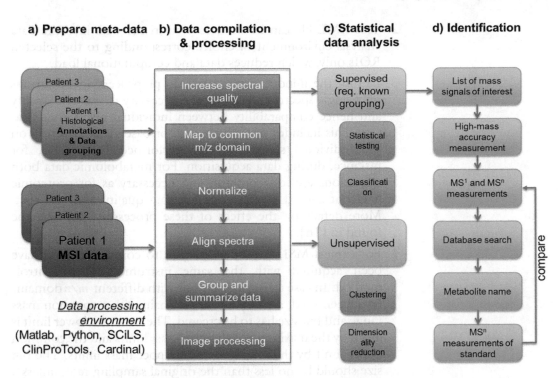

a) Prepare meta-data   **b) Data compilation & processing**   **c) Statistical data analysis**   **d) Identification**

Fig. 1 Data analysis workflow for mass spectrometry imaging (MSI) data from multiple experiments (patients). The pure data analysis workflow can be divided into three major steps, and which is normally followed by the identification step. The three pure data analysis steps consist of (**a**) preparing the metadata, including the annotation of the co-registered histological images, which is then passed to the data processing environment together with the acquired MSI data. (**b**) data processing and compilation, in which a series of operations are applied to increase the comparability between the samples (**c**) statistical analysis. Once the data has been processed, summarized, and grouped, statistical analyses can be performed in two ways: supervised, which requires a previous knowledge on the grouping of the data, and unsupervised, which requires no prior information about the grouping of the data. The latter can be used to find new groups using clustering techniques or to reduce the dimensionality of MSI datasets. Supervised analyses can be done to create classifiers based on discriminatory metabolite patterns. This list of molecules is usually of interest for a biochemical interpretation. However, MSI requires a separate step to assign names to that list of molecules (**d**)

MSI data analysis software SCiLS Lab. However, these functions will be described in order to enable researchers to repeat those steps in noncommercial data analysis environments such as Python, Matlab (bioinformatics toolbox), or R (Cardinal MSI package).

1. As MSI provides spatially resolved molecular profiles, a careful histological annotation using the co-registered microscopic image for restriction to the relevant areas of the tissue can increase specificity and sensitivity of the subsequent data analysis procedure [14, 15]. The preciseness of annotation is naturally limited by the pixel size of the MSI data. In FlexImaging, for example, the annotations can be saved and exported as XML files, which contain the names of spectra belonging to each defined region of interest (ROI).

2. This XML file can be used to read in the data into the data analysis environment of choice corresponding to the selected ROIs only, which reduces data and computational load.

3. During the import, some basic MS preprocessing operations on a single spectrum level may be necessary to increase quality and hence comparability between individual pixels and datasets. This includes smoothing and/or baseline subtraction on the individual spectra, if not performed beforehand, for instance, during data acquisition. For metabolomic data both operations are less common and necessary as for proteomic data but should be performed if data quality indicates that. More details on the effect of these processing steps can be found in [16].

4. For clinical MSI, the set of samples to compare should have been acquired with the same instrument (FlexControl) method. In case different methods with different m/z domains have been used, a common m/z domain with a common mass range and bin size has to be created. The common lower limit is given by the maximum of all lower mass limits and the common upper limit by the minimum of all upper mass limits. The bin size should be no less than the original sampling rate, unless a certain smoothing effect is desired by a down-sampling. However, in TOF/TOF-based metabolomic MSI, a linear bin size of $\Delta m/z$ 0.01 should be sufficient. Nonlinear binning is supported by SciLS Lab, which allows a more efficient way of distributing the data points to describe a mass spectrum.

5. An important pillar for ensuring comparability between MSI datasets is normalization of single pixels. Deininger et al. have discussed several normalization techniques for MALDI MSI, among them total-ion-count (TIC), root-mean-square (RMS), and median normalization [17]. TIC is used when the baseline is strongly influenced by chemical noise as in linear TOF. RMS, in contrast, is determined by real peak intensities and therefore popular for FT-ICR data. As metabolomic TOF/TOF data exhibits less chemical baseline than linear TOF MSI, but more than in FT-ICR, the application of the normalization methods depends on the quality of the single spectra.

6. When comparing multiple MSI datasets from different experiments, mass shifts might be observed (Fig. 2a, c) that can have a strong impact on the final result outcome. This can be overcome by spectral alignment which is less well described in MSI literature. Alignment (or recalibration) of spectra is usually done on the representative spectra of the different datasets, since mass shifts within datasets are usually considered negligible. For this, a set of reference peaks has to be defined that fulfill two criteria: they should be present in almost all of the

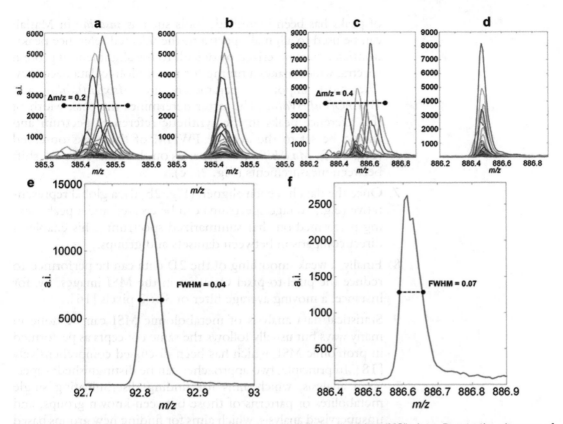

Fig. 2 Mass shifts and spectral alignment of mass spectrometry imaging (MSI) data. Due to the absence of internal standards, MSI data can exhibit mass shifts within and between experiments. Within-experiment mass shifts refer to the mass shifts occurring from pixel to pixel, which are usually very small and therefore negligible. However, between-experiment mass shifts can be significant and therefore have a strong negative effect on comparability between datasets. Panels (**a, c**) show examples of observed mass shifts (0.2–0.4 Da) from 59 MSI experiments performed with the same method [4]. Alignment of the spectra can be performed by a recalibration of the data using reference peaks. Matlab functions, such as *msalign*, require the definition of two parameters. One that describes the peak width, which can be estimated by looking at the average FWHM values of the detected peaks in the MSI experiment. Panels (**e, f**) show examples from the study of Lou et al. where the peak width was estimated to be 0.05 Da. Panels (**b, d**) show the peaks of (**a, c**), respectively, after successful alignment with *msalign*

spectra that are to be aligned, and they should cover the mass range of interest. Useful reference peaks are matrix peaks or molecules that are likely to be omnipresent such as AMP. However, in case no a priori reference peaks are available, a peak picking can be performed on the average spectrum of each sample. The found peaks are clustered with a certain tolerance which can be mass range-dependent, for instance, 1500, 800, and 600 ppm for the m/z blocks 0–200, 201–500, and 501–1000, respectively [4]. For alignment, we highly recommend peak detection rates >90% and a minimum of five peaks which are distributed across the whole mass range. Once a set of peaks has been identified, tools such as *msalign* in Matlab can be used to align all spectra to the selected reference peaks. *Msalign* is parameterized by default to the alignment of protein spectra, which makes a tuning for metabolomic data necessary. The most important parameters are *MaxShiftValue* and *WidthOfPulsesValue*. The latter determines the peak width of the reference peaks in the synthetic reference spectrum and should be set to the average FWHM of the peaks observed (Fig. 2e, f). *MaxShiftValue* depends on the observed peak shift between measurements (Fig. 2a, c).

7. Once the data has been aligned (Fig. 2b, d), a global representative (e.g., average spectrum) can be created and a peak picking performed on that summarized spectrum. This enables a direct comparison between datasets and groups.

8. Finally, a weak smoothing of the 2D data can be performed to reduce the pixel-to-pixel variation in the MSI images, by, for instance, a moving average filter of 3×3 pixels [16].

9. Statistical data analysis of metabolomic MSI can be done in many ways but usually follows the same concepts as performed in proteomic MSI, which has been discussed comprehensively [18]. In principle, two approaches can be distinguished: supervised analysis, which aims for finding discriminating single metabolites or patterns of those between known groups, and unsupervised analysis, which aims for finding new groups based on the detected metabolic profiles.

3.4 Metabolite Identification

Determination of the identity of differentiating metabolomic mass signals is usually of high interest to get an understanding of the biological meaning of the metabolite(s) in the context of the project or to enable the detection of the molecule(s) using other more targeted techniques. The most common approach for MSI data is depicted in Fig. 1.

1. Due to lack of comprehensive MALDI MS/MS databases and search algorithms metabolites, a highly accurate mass is an important indication of identity. FT-based mass spectrometers provide the highest mass accuracy (<1 ppm) but offer lower throughput. A common approach is therefore to perform screening on high-throughput instruments, such as MALDI-TOF/TOF systems, and determine the accurate mass of the metabolites to interest in single subsequent experiments [4, 10]. The accurate molecular weight can then be matched with a small mass tolerance (1–3 ppm) to a public database, such as HMDB (http://www.hmdb.ca/), METLIN (https://metlin.scripps.edu), or lipid maps (http://www.lipidmaps.org/). Furthermore, it can be checked if the observed fragmentation spectra are in line with the structure of the proposed

match [9, 19]. Ultimate confirmation can be obtained by comparing the fragmentation pattern with one from a corresponding standard compound [20].

2. Further biological interpretation can be offered by pathway analysis software, such as Reactome (http://www.reactome.org/) or MetaboAnalyst (http://www.metaboanalyst.ca) [1]. The latter also integrates a comprehensive set of statistical data analysis tools.

4   Notes

1. Several strategies have been reported for limiting postmortem changes of the tissue metabolome:

   Heat stabilization of ex vivo tissues (HS)—enzymes are inactivated by rapidly heating the tissue to 95 °C using high-power heating blocks. It has been demonstrated to halt postmortem degradation of adenine nucleotides that otherwise occurs in ex vivo snap-frozen tissue [21].

   In situ freezing (ISF) under anesthesia has been demonstrated to be superior to ex vivo snap-frozen tissue [22] and heat-stabilized ex vivo tissues for maintaining metabolite integrity [7], including adenine nucleotides.

   In situ focused microwave irradiation (FMW) uses focused microwaves to rapidly heat (<2 s) the tissue and denature/deactivate enzymes. It is the most rapid deactivation method and has been reported that, while ISF and FMW provide similar results for most metabolites, there are several metabolites that are best analyzed using FMW because of their very rapid postmortem changes [23].

2. On-target laser spot size may be estimated by preparing a thin layer of the matrix onto an ITO slide, by using a dry matrix deposition method and a matrix solution containing a high concentration of volatile solvent. A raster with a pixel-to-pitch much larger than the laser spot diameter, and a large number of laser shots (sufficient to ablate all matrix), results in a laser burn pattern in the matrix coating. Through using laser irradiation conditions identical to that used for MALDI MSI, the on-target burn pattern provides an estimate of the on-target laser spot size. Through the analysis of a raster of burn spots, an average spot size can be determined.

Acknowledgments

This work has been made possible with the support of the Dutch Province of Limburg. BB thanks the European Union (ERA-NET: TRANSCAN 2), ITEA, and RVO (ITEA 151003/ITEA 14001) for their financial support.

References

1. Gika HG, Wilson ID, Theodoridis GA (2014) The role of mass spectrometry in nontargeted metabolomics. In: Simó C, Cifuentes A, García-Cañas V (eds) Fundamentals of advanced Omics technologies: from genes to metabolites. Elsevier, Amsterdam

2. Addie RD, Balluff B, Bovee JV et al (2015) Current state and future challenges of mass spectrometry imaging for clinical research. Anal Chem 87(13):6426–6433. https://doi.org/10.1021/acs.analchem.5b00416

3. Nilsson A, Goodwin RJ, Shariatgorji M et al (2015) Mass spectrometry imaging in drug development. Anal Chem 87(3):1437–1455. https://doi.org/10.1021/ac504734s

4. Lou S, Balluff B, Cleven AH et al (2017) Prognostic metabolite biomarkers for soft tissue sarcomas discovered by mass spectrometry imaging. J Am Soc Mass Spectrom 28 (2):376–383. https://doi.org/10.1007/s13361-016-1544-4

5. Zhou D, Guo S, Zhang M et al (2017) Mass spectrometry imaging of small molecules in biological tissues using graphene oxide as a matrix. Anal Chim Acta 962:52–59. https://doi.org/10.1016/j.aca.2017.01.043

6. Wu Q, Chu JL, Rubakhin SS et al (2017) Dopamine-modified TiO2 monolith-assisted LDI MS imaging for simultaneous localization of small metabolites and lipids in mouse brain tissue with enhanced detection selectivity and sensitivity. Chem Sci 8(5):3926–3938. https://doi.org/10.1039/c7sc00937b

7. Mulder IA, Esteve C, Wermer MJ et al (2016) Funnel-freezing versus heat-stabilization for the visualization of metabolites by mass spectrometry imaging in a mouse stroke model. Proteomics 16(11–12):1652–1659. https://doi.org/10.1002/pmic.201500402

8. Shariatgorji M, Nilsson A, Goodwin RJ et al (2014) Direct targeted quantitative molecular imaging of neurotransmitters in brain tissue sections. Neuron 84(4):697–707. https://doi.org/10.1016/j.neuron.2014.10.011

9. Ly A, Buck A, Balluff B et al (2016) High-mass-resolution MALDI mass spectrometry imaging of metabolites from formalin-fixed paraffin-embedded tissue. Nat Protoc 11 (8):1428–1443. https://doi.org/10.1038/nprot.2016.081

10. Buck A, Balluff B, Voss A et al (2016) How suitable is matrix-assisted laser desorption/ionization-time-of-flight for metabolite imaging from clinical formalin-fixed and paraffin-embedded tissue samples in comparison to matrix-assisted laser desorption/ionization-Fourier transform ion cyclotron resonance mass spectrometry? Anal Chem 88 (10):5281–5289. https://doi.org/10.1021/acs.analchem.6b00460

11. Lou S, Balluff B, Cleven AH et al (2016) An experimental guideline for the analysis of histologically heterogeneous tumors by MALDI-TOF mass spectrometry imaging. Biochim Biophys Acta 1865:957–966. https://doi.org/10.1016/j.bbapap.2016.09.020

12. Shariatgorji M, Nilsson A, Bonta M et al (2016) Direct imaging of elemental distributions in tissue sections by laser ablation mass spectrometry. Methods 104:86–92. https://doi.org/10.1016/j.ymeth.2016.05.021

13. Aichler M, Elsner M, Ludyga N et al (2013) Clinical response to chemotherapy in oesophageal adenocarcinoma patients is linked to defects in mitochondria. J Pathol 230 (4):410–419. https://doi.org/10.1002/path.4199

14. Rauser S, Deininger SO, Suckau D et al (2010) Approaching MALDI molecular imaging for clinical proteomic research: current state and fields of application. Expert Rev Proteomics 7 (6):927–941. https://doi.org/10.1586/epr.10.83

15. Schwamborn K (2017) The importance of histology and pathology in mass spectrometry imaging. Adv Cancer Res 134:1–26. https://doi.org/10.1016/bs.acr.2016.11.001

16. McDonnell LA, van Remoortere A, van Zeijl RJ et al (2008) Mass spectrometry image correlation: quantifying colocalization. J Proteome Res 7(8):3619–3627. https://doi.org/10.1021/pr800214d

17. Deininger SO, Cornett DS, Paape R et al (2011) Normalization in MALDI-TOF imaging datasets of proteins: practical considerations. Anal Bioanal Chem 401(1):167–181. https://doi.org/10.1007/s00216-011-4929-z

18. Jones EA, Deininger SO, Hogendoorn PC et al (2012) Imaging mass spectrometry statistical analysis. J Proteome 75(16):4962–4989. https://doi.org/10.1016/j.jprot.2012.06.014

19. Dekker TJ, Jones EA, Corver WE et al (2015) Towards imaging metabolic pathways in tissues. Anal Bioanal Chem 407(8):2167–2176. https://doi.org/10.1007/s00216-014-8305-7

20. Buck A, Ly A, Balluff B et al (2015) High-resolution MALDI-FT-ICR MS imaging for

the analysis of metabolites from formalin-fixed, paraffin-embedded clinical tissue samples. J Pathol 237(1):123–132. https://doi.org/10.1002/path.4560

21. Blatherwick EQ, Svensson CI, Frenguelli BG et al (2013) Localisation of adenine nucleotides in heat-stabilised mouse brains using ion mobility enabled MALDI imaging. Int J Mass Spectrom 345:19–27. https://doi.org/10.1016/j.ijms.2013.02.004

22. Hattori K, Kajimura M, Hishiki T et al (2010) Paradoxical ATP elevation in ischemic penumbra revealed by quantitative imaging mass spectrometry. Antioxid Redox Signal 13 (8):1157–1167. https://doi.org/10.1089/ars.2010.3290

23. Sugiura Y, Honda K, Kajimura M et al (2014) Visualization and quantification of cerebral metabolic fluxes of glucose in awake mice. Proteomics 14(7–8):829–838. https://doi.org/10.1002/pmic.201300047

Part VII

Study Design, Data Analysis, and Bioinformatics

Chapter 27

Quality-Assured Biobanking: The Leiden University Medical Center Model

Rianne Haumann and Hein W. Verspaget

Abstract

Prospective or "de novo" biobanking is becoming increasingly popular. Biobanks are installed to provide large collections of biological materials for future medical research. Quality assurance of biobank samples is an important aspect of biobanking. Therefore, it is vital that all samples are collected and processed in a similar manner according to standardized procedures to ensure high-quality samples and reduce variability in the analytical process. We describe the processes of the centralized biobanking facility at the Leiden University Medical Center (LUMC).

   Key words Biobanking, Storage, Biological materials, Blood, Quality assurance, Quality control, Pre-analytical variations, Standard operating procedures

1   Introduction

A biobank is a collection of biological materials coupled to associated data consisting of donor information and sample processing information [1]. The biobank ideally provides an infrastructure for the collection, processing, and storage of biological materials together with a supportive information technology (IT) system for the registration and documentation of associated data [2, 3]. Traditionally, biobanks consisted of tissue collections from pathology departments that remained after the diagnostic process and were accessible for researchers for secondary scientific use under restricted conditions. Nowadays, many biobanks are more specialized in "de novo" collection of various biological samples and associated data obtained with an informed consent of the participant. The establishment of a biobank is a relatively complex process in which many aspects from sample collection to IT systems have to be organized and coordinated [2].

   Biobanking is of great importance for medical research in order to guarantee the availability of patient material since these biological materials are solely collected for research purposes [4]. "De novo"

Martin Giera (ed.), *Clinical Metabolomics: Methods and Protocols*, Methods in Molecular Biology, vol. 1730,
https://doi.org/10.1007/978-1-4939-7592-1_27, © Springer Science+Business Media, LLC 2018

biobanks collect biological materials prospectively for yet undefined research purposes. Biological material that remains after diagnostic process is not allowed to be directly used for research purposes unless it is used anonymously and without associated clinical data or when obtained with an informed consent. Centralizing the collection and storage of biological material in biobanks for research purposes is more efficient than setting up multiple small collections, and it also reduces the patient burden. Biobanks have the sole purpose of collecting biological materials for multiple research purposes, which are undefined at the time of collection, using biological materials more efficiently. Quality of samples is usually better controlled because biological materials are processed in a similar manner. Therefore, biobanking is becoming increasingly popular and more important for biomedical research. In addition, it is quite difficult for individual research groups or laboratories to establish large collections of biomaterials. Consequently, centralized biobank facilities are advocated because they are better equipped for sample collection of large cohorts.

The importance of large numbers of high-quality samples is emphasized by reported difficulties with sample collection and biomarker validation. Due to heterogeneity of diseases and in order to achieve statistical power, large numbers of samples are needed. The development of high-throughput analysis enables simultaneous analysis of hundreds of samples, but the availability of sufficient samples is often difficult without biobanks [3, 5, 6]. Poste [5] reported that over 150,000 papers documented potentially interesting biomarkers; however, only less than 100 biomarkers have eventually been validated for clinical use. In addition, documentation of the standardization of sample collection is often lacking, and researchers cannot obtain samples with uniform quality hampering high-quality research. Analysis of samples which in retrospect were found to be of suboptimal or even bad quality is costs ineffective and should be avoided. Freedman et al. [7] calculated that almost 50% of costs spent for biomedical research were wasted due to irreproducible results in preclinical research related to poor sample or analytical quality. These observations show an enormous amount of research associated with high costs and little high-quality output for improvement of diagnosis and treatment of patients. Therefore, sample quality should be better monitored and registered resulting in a better research output. In addition, journals nowadays also require documented sample handling and storage information for the publication of research on biomarkers in biological samples [8].

The process of sample analysis can be divided into three phases: the pre-analytical, the analytical, and the post-analytical phase. The pre-analytical phase comprises the sample collection, processing, and storage of biological samples. The analytical phase consists of the biochemical analysis of the samples followed by the post-analytical phase in which the data is analyzed. The pre-analytical phase is particularly prone to errors with the highest reported error rates ranging from 31% to 75%. During the analytical phase, the prevalence of errors is reported to be around 13–32%, and during the post-analytical phase, an error rate of 9–31% is thought to occur, both of which are considerably lower than the pre-analytical phase [9, 10]. Sample errors in the pre-analytical phase are, for example, hemolysis, precentrifugation delay, wrong centrifuge settings, postcentrifugation processing delay, and wrong temperature settings [9]. It is very important to reduce the pre-analytical errors because it may lead to unnecessary patient burden, higher costs, low-quality samples, and faulty interpretation of analytical results. Green [11] estimated that a single pre-analytical error costs approximately € 175. Knowledge of the effect of a pre-analytical error on sample quality is therefore essential. Samples that contain pre-analytical errors might still be suitable for a range of analytical techniques, depending on the analyte to be assessed, and thus still valuable for research and biobanks.

Sample processing procedures should be described in standard operating procedures (SOPs) to enable uniform processing of samples. Unfortunately, there is not a worldwide consensus on the best method for sample processing that could be used by every laboratory or biobank. The National Cancer Institute (NCI) and the International Society for Biological and Environmental Repositories (ISBER) are examples of organizations that propose best practices for biobanking [12, 13]. One of the reasons of the publication of multiple best practices is the limited fundamental research of the influence of pre-analytical variables on sample quality. Research has shown that variations in pre-analytical processing can have a pronounced effect on the results of the analysis of biological samples [9, 10]. Thus, although challenging, it is imperative to establish large collections of uniformly processed and stored biological samples, and therefore it is important to document the sample processing and storage conditions in the biobanking process. Consequently, documentation of sample handling should be coupled to the individual biological sample. Sample track-and-trace data are important and can be registered in, for example, a Biobank Information Management System (BIMS), which will be explained in the following paragraph.

The Parelsnoer Institute (www.parelsnoer.org) is one of the largest clinical biobank organizations in the Netherlands. The Parelsnoer Institute is a federative biobank of the eight university medical centers in the Netherlands, which collect biological materials and associated data from patients of specific disease cohorts. The Parelsnoer Institute provides an infrastructure for the collection of samples and data capture. At the moment, the Parelsnoer collection encompasses 15 disease cohorts: abdominal aortic aneurysm, congenital heart disease, cerebrovascular accident, diabetes, hereditary colorectal cancer, inflammatory bowel disease, ischemic heart disease, leukemia-myeloma-lymphoma, multiple endocrine neoplasia, neurodegenerative diseases, esophageal and gastric cancer, pancreatic cancer and pancreatitis, renal failure, rheumatoid arthritis, and Parkinson's disease. For each cohort specific biological samples are collected such as serum, plasma, DNA, feces, urine, cerebrospinal fluid, pancreatic cyst fluid, viable cells, fresh frozen tissues, and/or formalin-fixed paraffin-embedded (FFPE) tissues. Clinical data and sample-specific data are stored in a national disease-specific database. More detailed information about the organization of the Parelsnoer Institute can be found in the article of Manniën et al. [14].

The LUMC participates in the Parelsnoer biobank but also has its own clinical biobanks organized in the Cura Rata biobank (www.lumc.nl/org/cura-rata/) [15]. The Cura Rata biobank is a centralized facility within the LUMC that provides an infrastructure for the collection of biological samples and the storage of associated data for more than 25 disease entities, similar to the Parelsnoer Institute collections. The Cura Rata biobank and the Parelsnoer biobank partially run parallel because samples are stored for both biobanks and sample processing is identical. Furthermore, the Cura Rata and the Parelsnoer biobanks use the same Biobank Information Management System (BIMS), called SampleNavigator®, for the registration of the processing data associated with the sample. In this chapter we will describe the procedures for biobanking within the LUMC [15, 16].

2 Software

2.1 HiX

The healthcare information exchange system (HiX) (ChipSoft, Amsterdam, the Netherlands) is a clinical data repository for the registration of Electronic Health Records (EHRs) that is being used by the LUMC. Unique is the integration of biobanks within HiX at our institution. Registration of informed consents and order management for biosample ordering are integrated in the EHRs which enables physicians and research nurses to include patients into the different biobanks. The process of sample ordering is not part of the diagnostic ordering process but is a separate ordering entity and is solely used for biobank purposes. HiX is used for the registration of relevant clinical data that is associated with the biobank sample. Information is transferred into a database, and the personal data are encrypted in order to provide privacy protection for participants of the biobank. Figure 1 shows an overview of the IT-directed logistics of the biobank.

IT-directed logistics of biobanking in the LUMC

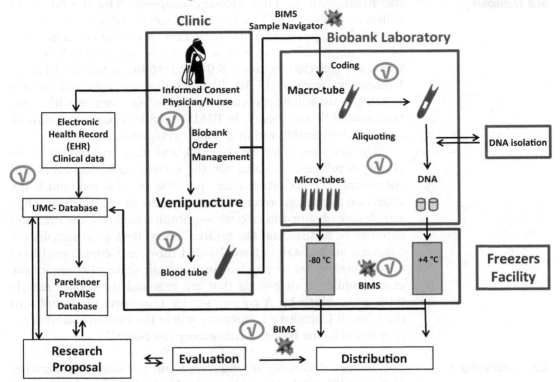

Fig. 1 Schematic presentation of the information technology-directed logistics from the LUMC centralized biobanking facility. *BIMS* Biobank Information Management System, √ electronic registration moment

2.2
SampleNavigator®

SampleNavigator® software (Lelystad, the Netherlands) is employed as the Biobank Information Management System (BIMS) for the LUMC biobanks. SampleNavigator is designed to capture relevant sample information such as track-and-trace, registration of standard operating procedures (SOP) deviations, and user registration (Fig. 1). Track and trace of the biological samples include registration of the time of collection, arrival time at the biobank, processing time, storage location, and aliquot history.

3 Methods (Procedures)

3.1 *Sample Order*

A physician informs a patient about the biobank according to the rules and regulations and agreements of the LUMC. When a patient decides to participate, the patient has to sign an informed consent which is registered in HiX. After registration, a physician orders samples specifically for the biobank through order management in HiX.

3.2 Collection and Transport

The biobank order from order management is directly coupled to the BIMS with an HL7 message coupling. The central blood collection facility (CBCF) receives the biobank order electronically, and labels are printed for the collection tubes and containers just before sample collection. Samples from the outpatient clinics and wards are collected between 8.00 and 16.00, since the Biobank Laboratory closes at 17.30. Samples that are collected outside working hours will be processed the following morning with registration of SOP deviation(s) in BIMS. Sample processing has to be completed preferably within 2 h after venipuncture.

The first time point for the track-and-trace registration of the sample is printing the label for the sample collection tubes and containers. The collection date and time are checked, and if the collection time is incorrect, which can occur in the wards or when samples are obtained from outlying hospital facilities, the time and date will be adjusted on the printed labels. Biological samples are collected in the SOP-defined labeled tubes and container. Blood collection tubes are inverted on an automatic shake platform for the exact number of inversions that are required for proper sample mixing (*see* **Note 1**). A courier service transports the samples to the Clinical Chemistry Laboratory where the biobank samples are transferred to the Biobank Laboratory (*see* **Note 2**).

3.3 Processing

The biological samples are registered and scanned (Handscanner Opticon OPI-2201, Hoofddorp, the Netherlands) at the Biobank Laboratory. This is a second time point in the BIMS track-and-trace registration of the biological samples. The collection date and time are checked and if needed adjusted and described in the BIMS. At that moment patient information is still present on the collection tubes and containers but is then encrypted with a sample code. New labels (Salm and Kipp BY800537 BPT-21-461, Breukelen, The Netherlands) are printed (Zebra printer GX420t, Lincolnshire, Illinois, USA) with a unique sample code. The biological samples are further processed under this specific sample code without any visible personal data.

SOP deviations are registered in the BIMS. The first step is to identify and indicate a SOP deviation by the biobank lab technician and then to specify the deviation in the BIMS. Several predetermined SOP deviations are documented: hemolysis, lipemic, icterus, incorrect tube, inadequate filling of the citrate tube, incorrect temperature before processing, prolonged precentrifugation time, incorrect mixing or homogenizing, incorrect centrifuge setting, storage problems, and others. In "others" the biobank lab technician can describe a SOP deviation that is not listed in the predetermined SOP deviations.

Next, the biological samples are processed at the Biobank Laboratory under the ISO 15189:2012. The biological samples are aliquoted into pre-specified smaller volumes, as indicated in

Table 1. New labels are printed for the cryotubes, which are a third time point in the track-and-trace registration of the sample. SOP deviations for the aliquots are also registered in the BIMS. After registration the samples are processed according to the SOP. Table 1 provides an overview of the processing steps for the various biological samples. In short, the samples are centrifuged at a specific force, time, and temperature. Next, the supernatant is transferred into aliquots with a volume of 100–900 µL per aliquot (*see* **Notes 3** and **4**). Aliquots should be filled starting with the lowest micro-sample number to the highest micro-sample number. The aliquot tubes are closed and stored temporarily at the Biobank Laboratory according to SOP protocol. Short-term storage temperature is dependent on the biological material; DNA is temporarily stored at 4 °C, and the other biological samples are temporarily stored at −20 °C, with the latter of a maximum duration of 4 h (*see* **Note 5**).

EDTA cell pellets from the plasma tubes are used for DNA isolation. DNA isolation is registered in the BIMS, and new labels are printed for the Autopure Qubes (AutoGen, Inc., Holliston, MA, USA). Cell pellet is transferred to the Autopure Qubes, and the EDTA tube is washed with PBS and the suspension added to the Autopure Qubes. The Autopure Qubes must be filled with a minimum of 6 mL and a maximum of 10 mL. The tube is closed and transported to the Department of Clinical Genetics for robotic DNA isolation. After isolation, DNA is transported back to the Biobank Laboratory. At the Biobank Laboratory, new labels are printed for the micro tubes containing DNA, which is registered as a time point for the track-and-trace registration of the sample. At the Department of Clinical Genetics, the quality parameters 260/280 absorbance ratio and DNA concentration are measured which together with the volume of the aliquots is registered in the BIMS. When the DNA stock concentration is >50 ng/µL, DNA is marked as "yes DNA ok" in the BIMS, at 10–50 ng/µL the DNA is marked as "no DNA not ok," and at <10 ng/µL the DNA will not be stored. DNA marked as "not ok" will result in a new venipuncture at the next visit of the patient to again isolate high-quality DNA. DNA is resuspended in Tris-EDTA buffer in the macro tube and divided into two micro tubes. DNA is stored as suspension in the fridge at 4 °C.

3.4 Storage

Aliquots are temporarily stored at −20 °C for a maximum duration of 4 h at the Biobank Laboratory except for DNA which is stored at 4 °C. At noon or at the end of the day, the aliquots are transferred to the long-term storage freezers. The sample boxes are transferred to the long-term freezer facility on dry ice and insulated tempex boxes to ensure limited temperature differences (*see* **Note 6**).

The refrigerator and freezer facility of the biobank are exclusively used for the biobank. The same brand of freezers is used to ensure uniform storage. The samples are predominantly stored at

Table 1
Standard operating procedures at the LUMC biobank for processing of blood and urine

Biological material	Tube type				Max. processing time (h)	Centrifugation			Aliquots		Short-term storage (°C)	Long-term storage (°C)
	Specified	Additives	Color	Volume (mL)		RCF	Time (min)	Temperature (°C)	Number	Volume (µL)		
Blood	Plasma	NaHeparin	Green	10	4	2350	10	20	5	500	−20	−80
		EDTA	Purple	4[a], 10[b]	4	2350	10	20/4	2–3[a], 5[b]	500	−20	−80
		NaCitrate	Blue	9	4	2350	10	20	5	500	−20	−80
	PPP	PPACK/citrate	Blue	3	4	2350	10	20	5	100	−20	−80
	Serum	Serum	Red	2[c], 10[d]	4	2350	10	20	1–2[c], 5[d]	500	−20	−80
		Serum gel	Red	3.5[e], 8.5[f]	4	2350	10	20	2–3[e], 5[f]	500	−20	−80
	DNA	EDTA	Red		4				2	500	4	
	RNA	PAXgene		2.5	4						−20	−80
Urine	Fresh		Yellow	9.5	4	2350	10	4	6	900	−20	−80
	24 h		Yellow	9.5	4	2350	10	4	6	900	−20	−80
	Pot		Yellow	9.5	4	2350	10	4	6	900	−20	−80

Tubes a to f are related to aliquot numbers a to f

−80 °C. BIMS is used as a track-and-trace system for the registration of the drawer, sample box, and position of the samples in the box. Multiple use of an aliquot is also registered in BIMS.

The freezers are cooled with wastewater from the hospital, which saves hundreds of thousands of euros per year on air-conditioning costs. The refrigerators and freezers are connected to the internal alarm and temperature registration system of the biobank facility. In case of an alarm, the infrastructural LUMC emergency and breakdown service will contact the responsible employee(s). In case of an emergency, employees of the centralized biobank facility will carry out the emergency plan (*see* **Note 7**).

3.5 Sample Distribution

Aliquots from the biobank can be requested by researchers after they submit an approved protocol to the biobank. The scientific biobank committee of the specific disease-related biobanks evaluates the protocol and decides if permission is granted.

The data warehouse of the biobank consists of the integration of EHRs and BIMS information (*see* Fig. 1). EHRs data are used for the extraction of relevant clinical information related to the specific biological sample. BIMS can be used for the extraction of data associated with the biological sample such as processing time, type of sample, and previous use of the sample. Researchers can request, for example, samples of a specific gender or age. Researchers can, for example, request samples that have a processing time with a maximum duration of 1 h and never been used before. Biobanks using electronic bioresource data registration enable researchers to request specific samples that are fit for their specific studies and analysis.

4 Notes

1. Serum tubes should be left in an upright position at room temperature for at least 30–60 min before processing the tubes to allow proper clotting.

2. Urine is transferred on ice and should be processed before the ice is melted.

3. Additional aliquots should be filled if more volume of the biological material is left. The extra volume should not be divided over the initial aliquots.

4. If there is less volume than required for an aliquot, one should still fill the aliquot and register a SOP deviation.

5. PAXgene tubes for RNA analyses are directly frozen at −20 °C in an upright position. The patient identification information should be removed for privacy reasons and replaced by a label with a non-identifiable BIMS macro number on the tube.

6. Use cool elements or dry ice in insulated boxes for transport of frozen material.

7. In case of emergency with storage of samples in the freezers, be sure to have an emergency plan to ensure high quality of the samples.

Acknowledgments

We would like to acknowledge Lilian Boonman for her assistance in the management of the biobank organization.

References

1. Vaught J, Kelly A, Hewitt R (2009) A review of international biobanks and networks: success factors and key benchmarks. Biopreserv Biobank 7(3):143–150

2. Vaught J (2016) Biobanking comes of age: the transition to biospecimen science. Annu Rev Pharmacol Toxicol 56:211–228

3. Riegman PH et al (2008) Biobanking for better healthcare. Mol Oncol 2(3):213–222

4. Womack C, Mager SR (2014) Human biological sample biobanking to support tissue biomarkers in pharmaceutical research and development. Methods 70(1):3–11

5. Poste G (2011) Bring on the biomarkers. Nature 469(7329):156–157

6. Button KS et al (2013) Power failure: why small sample size undermines the reliability of neuroscience. Nat Rev Neurosci 14(5):365–376

7. Freedman LP, Cockburn IM, Simcoe TS (2015) The economics of reproducibility in preclinical research. PLoS Biol 13(6):e1002165

8. Simeon-Dubach D, Burt AD, Hall PA (2012) Quality really matters: the need to improve specimen quality in biomedical research. J Pathol 228(4):431–433

9. Bonini P et al (2002) Errors in laboratory medicine. Clin Chem 48(5):691–698

10. Carraro P, Plebani M (2007) Errors in a stat laboratory: types and frequencies 10 years later. Clin Chem 53(7):1338–1342

11. Green SF (2013) The cost of poor blood specimen quality and errors in preanalytical processes. Clin Biochem 46(13):1175–1179

12. Institute N.C. (2016) NCI best practices for biospecimen resources

13. Lori D (2011) Best practices for repositories: collection, storage, retrieval and distribution of biological materials for research. International Society for Biological and Environmental Repositories Vancouver, Canada

14. Manniën J et al (2017) The Parelsnoer Institute: a national network of standardized clinical biobanks in the Netherlands. Open J Bioresour 4:3

15. Available from: https://www.lumc.nl/org/cura-rata/. Cited 19 June 2017

16. Simons J, Verspaget HW (2010) Traceerbaar biobankbeheer in het LUMC. Laboratorium Magazine, 9

Chapter 28

Extracting Knowledge from MS Clinical Metabolomic Data: Processing and Analysis Strategies

Julien Boccard and Serge Rudaz

Abstract

Assessing potential alterations of metabolic pathways using large-scale approaches today plays a central role in clinical research. Because several thousands of mass features can be measured for each sample with separation techniques hyphenated to mass spectrometry (MS) detection, adapted strategies should be implemented to detect altered pathways and help to elucidate the mechanisms of pathologies. These procedures include peak detection, sample alignment, normalization, statistical analysis, and metabolite annotation. Interestingly, considerable advances have been made over the last years in terms of analytics, bioinformatics, and chemometrics to help massive and complex metabolomic data to be more adequately handled with automated processing and data analysis workflows. Recent developments and remaining challenges related to MS signal processing, metabolite annotation, and biomarker discovery based on statistical models are illustrated in this chapter considering their application to clinical research.

Key words Metabolomics, Mass spectrometry, Data processing, Data analysis

1 Introduction

Since its early days, metabolomics has been intensively used to evaluate biochemical signatures and find biological readouts for clinical diagnosis and prognosis [1]. According to the preponderant roles of metabolites in many biological processes, metabolic phenotyping constitutes an essential element for providing mechanistic insights into pathological states, detecting potential adverse effects due to toxic exposure or evaluating beneficial effects of therapeutic treatments [2]. Coupling a separation technique (chromatographic and electrophoretic) with mass spectrometry (MS) today constitutes a reference approach in metabolomics because of the large number of signals that can be recorded at low concentration within a single analytical run [3]. Consequently, raw data present an intrinsic two-dimensional structure with a separation and a m/z dimension. To take advantage from the wealth of biochemical

Martin Giera (ed.), *Clinical Metabolomics: Methods and Protocols*, Methods in Molecular Biology, vol. 1730, https://doi.org/10.1007/978-1-4939-7592-1_28, © Springer Science+Business Media, LLC 2018

information recorded, relevant signals need to be distinguished from artifacts and noisy variables in these massive datasets.

2 Materials

2.1 Availability of the Algorithms

The different tools described in this chapter are available through diverse programing platforms such as R and Matlab or modules integrated in global metabolomic processing workflow, including XCMS, MZmine, apLCMS, xMSanalyzer, mzMatch, MetAlign, MetSign, and OpenMS. The reader is invited to consult the detailed review of metabolomic tools and resources proposed by Misra et al. [4] or specific documentation for more details on the usage of each method.

2.2 Biological Background

The case study of Victor Yushchenko's poisoning with an acute dose of dioxin in 2004, when he was candidate for the presidency of Ukraine, will be used as working example throughout this chapter. Dioxin (2,3,7,8-tetrachlorodibenzo-p-dioxin, TCDD) is a by-product of industrial processes involving chlorine, such as waste incineration and chemical manufacturing. Human populations are mainly exposed to low levels of dioxin through food, and the effects of chronic exposure to dioxin are unknown. However, in case of acute intoxication, hepatitis, neuropathy, and skin damages are just a few examples of pathologies that were observed.

Urine samples were collected at different time points after intoxication and compared with urine of healthy volunteers. All samples were analyzed by ultra high-pressure liquid chromatography (UHPLC) coupled to quadrupole time-of-flight mass spectrometry (Acquity UPLC & Xevo G2 QTOF, Waters). Multivariate analysis was carried out to distinguish the two groups (Comet, Nonlinear Dynamics & SIMCA-P, Umetrics). Detailed methodological aspects of the example (e.g., overall technical modus operandi) can be found elsewhere [5].

3 Methods

3.1 Pre-Acquisition Normalization

Among the multiple steps required for proper sample analysis, ensuring that the total concentration of each sample is in the same range constitutes a prerequisite for any comparison, even in the case of relative quantitation based on peak intensity variations. The sources of this variability include an intrinsic heterogeneity of the samples, problems of stability, or inherent variations arising from prior sample preparation. It is therefore crucial to remove or at least reduce the overall discrepancies by adjusting the sample volume or concentration prior to analysis.

Because these differences are strongly dependent on the nature of the sample, different pre-acquisition normalization strategies should be specifically implemented:

1. Cell cultures and biopsies can be normalized according to the total amount of biological material, evaluated as cell number or total protein weight.

2. Fluctuations associated with biological fluids can be corrected using either global indicators of the total solute concentration (e.g., osmolality or specific gravity), or reference compounds (e.g., creatinine for urine normalization).

3.2 MS Acquisition

The choice of an experimental approach is often determined by the availability of prior knowledge regarding a biological question. An untargeted acquisition can be used to provide a global monitoring of metabolites to derive data-driven hypotheses. Instrumental developments recently improved the acquisition capacity offered by MS. High-resolution MS, such as time-of-flight (TOF) and Fourier transform technologies, today allows the untargeted and simultaneous monitoring of thousands of analytes in full scan mode, with the determination of elemental compositions from accurate mass measurements.

1. Fully nontargeted metabolomic approaches are commonly used for blind biomarker discovery, without any or low prior knowledge about potentially altered metabolic pathways.

2. Alternatively, the abundances of groups of metabolites can be specifically investigated within the full scan MS information using a retrospective examination, namely, post-acquisition data mining or post-targeted analysis (*see* Subheading 3.6).

Both strategies are complementary and can be simultaneously implemented to provide different views of the metabolic information. An example of UHPLC-HRMS data structure generated by the untargeted acquisition of a urine sample from Victor Yushchenko is provided in Fig. 1. Whatever the strategy's choice, signal processing and data mining constitute critical aspects that need to be carefully considered to extract relevant biochemical knowledge in MS-based clinical metabolomics. Accounting for both the characteristics of the analytical setup and biochemical knowledge is mandatory for proper peak detection, alignment, and metabolite annotation.

3.3 MS Signal Processing

Because irrelevant signals, including solvent contaminants and detector noise, may strongly deteriorate downstream data analysis, a number of steps should be sequentially implemented to remove analytical artifacts and emphasize valuable biochemical information related to real analytes. It includes denoising, peak picking, deconvolution, and alignment.

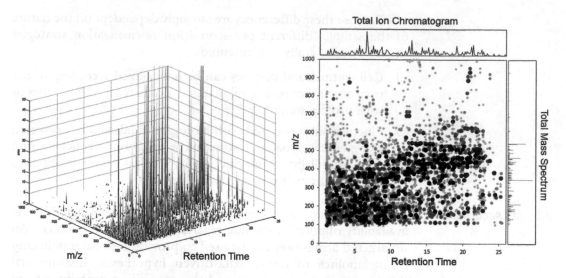

Fig. 1 UHPLC-HRMS data with corresponding total ion chromatogram and total mass spectrum, recorded for a urine sample collected from Victor Yushchenko after intoxication

3.3.1 Peak Detection and Deconvolution

Appropriate feature detection constitutes the starting point of MS data processing, and an additional challenge comes from the presence of multiple signals, such as isotopic peaks, fragment, or adduct ions, produced by analytes in the sample.

1. Signal smoothing and denoising are often used in combination with baseline correction as a first step of processing to increase the data quality. It includes moving average windows, Savitzky-Golay fitting, or wavelet transform to remove random noise, and background subtraction, or Gaussian filters for chemical denoising [6].

2. Peak picking is then achieved to extract relevant signals by generating a series of mass features associated with specific peak areas, m/z, and retention/migration time values. This is done by detecting signals above a specified intensity or signal-to-noise ratio threshold, and mathematical models (e.g., Gaussian) are used to assess peak consistency based on shape constraints and integrate signals between peak boundaries [7].

3. Multiple signals can be generated for each analyte, including isotopic peaks, fragment or adduct ions, n-mers, or degradation products. These derivative ions must be grouped using deconvolution and deisotoping procedures. Peak annotation associates signals sharing similar retention/migration times and peak shapes based on specific m/z differences corresponding to known chemical modifications.

Proper peak detection requires peak shape parameters defining individual peak limits to be carefully adjusted according to the analytical conditions and the type of device (*see* **Note 1**). Other approaches based on empirical peak modeling and isotopic patterns are also available to evaluate the parameters from the data, e.g., from quality control samples.

3.3.2 Alignment

Small differences are usually observed for both the separation and the m/z dimensions for the same signal, when comparing several samples. However, ensuring proper matching between occurrences of the same feature across samples constitutes a prerequisite before further analysis of the data. On the one hand, discrepancies in the m/z dimension are usually small when using calibration ions during MS data acquisition. On the other hand, shifts in the separation dimension can originate from numerous factors including differences of temperature, mobile-phase pH, pump pressure, matrix effects, sample carryover, and column degradation or clogging. Appropriate alignment strategies are therefore used to correct or at least limit variations between runs, while keeping chemical selectivity intact [8].

Two types of strategies can be considered:

1. Warping methods account for the separation dimension as a whole and have usually high computing time and resources requirements. These approaches do not depend on prior peak picking or deconvolution of the signals and lack the ability to account for selectivity changes. Because warping methods maximize the overlap between samples by stretching or shrinking the separation dimension, a reference template needs first to be defined, and a sequential procedure is carried out to align segments of chromatograms/electropherograms defined by a width criterion. Classical warping methods include correlation optimized warping (COW) and dynamic time warping (DTW). Recent developments involve parametric and semi-parametric strategies or piecewise cross-correlation warping approaches [9].

2. Peak matching approaches grouping mass features across samples take advantage of prior deconvolution and peak picking steps to condense the chemical information. Similar mass features falling between the boundaries defined by a tolerance threshold in the separation and the m/z dimension are usually considered as the same analyte (*see* **Note 2**). A specific weight can be associated with errors in each dimension according to the characteristics of the analytical setup [7].

On a practical point of view:

1. Carefully choose the parameters for alignment according to the characteristics of the data.

2. Check the absence of detected peaks in some of the samples, leading to missing values in the aligned data table (*see* **Notes 3** and **4**).

3.3.3 Post-Acquisition Normalization

Post-acquisition normalization strategies can be required to remove remaining unwanted systematic variations due to signal intensity drift over time during the analytical sequence.

Compare sample-based strategies based on the evaluation of a global correction factor including:

1. Total sum normalization (TSN, sum of all features).

2. MS total useful signal (MSTUS, sum of features common to all samples).

3. Probabilistic quotient normalization (PQN, feature-dependent correction).

Quality control samples (QCs) analyzed at regular intervals should be used as landmarks to monitor the stability of the analytical process and guarantee proper data acquisition. By checking the repeatability of the QCs and/or their response to dilution, irrelevant features such as noisy or saturated signals must be removed to ensure data quality and relevance [10].

Investigate the potential drifts according to the processing order of the analytical sequence using QC-based approaches. These methods include an evaluation using local regression models fitted for each feature detected in the QC samples. Use one of the following by evaluating at least five important identified features:

1. Locally weighted scatterplot smoothing (LOESS).

2. Robust spline correction (QC-RSC).

3. Support vector regression correction (QC-SVRC).

3.3.4 Data Scaling

The variability of abundance can strongly differ from one metabolite to another with very different biological consequences, and the most abundant compounds are not necessarily the most relevant. Additionally, low concentrated metabolites are more prone to analytical errors than highly abundant compounds. Hence, at least two scaling procedures need to be evaluated to make the features comparison effective:

1. Individual values should be standardized using a scaling factor, e.g., standard deviation (SD) in the case of unit variance scaling or the square root of the SD for Pareto scaling.

2. If the results are not satisfactory, variables transformations (e.g., log function), are useful to standardize the signals and correct heteroscedasticity [11].

3.4   Multivariate Statistics

Because a large number of signals are recorded for providing a global picture of biological matrices, multivariate methods constitute efficient solutions to highlight metabolic patterns [12]. Two data analysis strategies can be sequentially carried out:

1. Exploratory approaches based on unsupervised learning principles.

2. Predictive models involving supervised algorithms.

3.4.1   Unsupervised Analysis

Unsupervised methods evaluating the dataset without accounting for experimental groups or expected outcome offer reliable tools to extract the major trends in the data (*see* **Note 5**). They provide an efficient way to detect natural groupings among the distribution of samples and should always be carried out as a first step.

For that purpose, principal component analysis (PCA) constitutes the most widely used and convenient method for initial exploratory data analysis. PCA summarizes the sources of variability in the data by building a new subspace of low dimension [13]. Interpretation is usually limited to the first PCs carrying the most salient trends (*see* **Note 6**). Graphical outputs provide a reliable and convenient mean to detect groups of similar samples or signals:

1. The coordinates of the observations (*scores*) allow the distribution of the samples to be easily displayed and evaluated.

2. Relationships between variables can be highlighted by inspecting their contributions (*loadings*) to the model.

These plots allow trends or unexpected patterns in the dataset to be detected:

1. Check for any significant analytical drift by inspecting the QCs projection.

2. Search for potential outliers, i.e., observation(s) outside the Hotelling's T^2 ellipse (95% confidence interval) and consider removing them from further analysis.

3.4.2   Supervised Analysis

In contrast to the previous approaches, supervised methods make use of prior knowledge linked to the experimental setup to drive the analysis. Supervised models aim at predicting a response (dependent variable Y) from the measured metabolomic data (independent variables X).

Partial least squares (PLS or projection to latent structures) regression is one of the largely used methods in metabolomics. The PLS algorithm builds a low-dimension subspace based on linear combinations of the experimental variables. By these means, the PLS model summarizes the X data in a synthetic manner with the ability to predict the Y response [14]. Because the comparison of experimental groups is often the goal of metabolomic studies, PLS discriminant analysis (PLS-DA) is widely used to distinguish

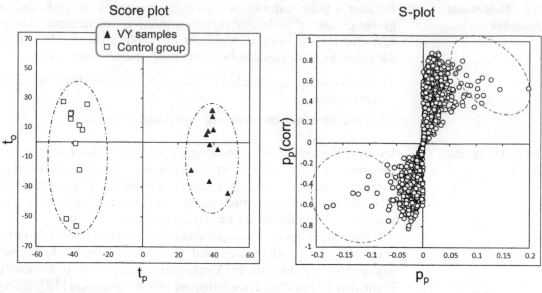

Score plot   **S-plot**

Discriminant features:
Exogenous compounds: propofol, selvofurane, ibuprofen (drug therapy)
Endogenous compounds: mainly steroid metabolites and bile acids

Fig. 2 OPLS-DA model comparing the urine samples collected from Victor Yushchenko with age-matched controls (3682 variables, Pareto scaling). (**a**) Score plot, (**b**) S-plot

known classes of observations. The orthogonal PLS algorithm (O-PLS) [15] facilitating model interpretation constitutes a routinely applied method to analyze metabolomic datasets with efficient diagnostic tools, e.g., the S-plot. Variable contributions to the model can serve as an objective basis for detecting potential biomarkers (*see* **Note 7**). An example of outputs from an OPLS-DA model comparing urine samples from Victor Yushchenko and healthy control volunteers matched for age is provided in Fig. 2.

This multivariate data analysis should be carefully interpreted, particularly when the number of observation is limited (<20 per group). The model validity and its capacity to handle new observations reliably need to be carefully evaluated. To this aim, the data should be divided into a training set of observations, used to build the model capturing the relation between metabolomic data and the response variable(s), and a test set including different samples, used to evaluate the reliability of the prediction.

Common procedures available to assess model validity with respect to perturbations of the training and test sets include:

1. Cross-validation.

2. Permutation tests.

3. Bootstrap.

In any case, the reliable estimation of model generalization ability (validation) requires the prediction of another completely new, independently generated, observations addressing the same biological question (external test set).

3.5 Metabolite Identification

3.5.1 Annotation Levels and Databases

It is largely recognized that metabolite annotation today constitutes the major bottleneck of untargeted metabolomics. Because the biochemical interpretation relies strongly on the quality of metabolite annotation, harmonization efforts including the metabolomic standard initiative (MSI) and the coordination of standards in metabolomics (COSMOS) were made to propose standards for reporting metabolite annotations [16, 17]. Therefore, four annotation levels are considered:

- Level 1 annotation constitutes the highest annotation quality, as it requires the comparison of two or more orthogonal properties with reference material measured in the same experimental conditions.
- Level 2 annotation involves compounds matching a molecular formula and/or the physicochemical/spectral properties of databases entries.
- Level 3 is based on a single molecular property and is thus limited to chemical class annotation.
- Level 4 comprises unknown compounds that cannot be annotated but distinguished from other analytes in the samples based on their specific m/z and retention/migration time values.

Reliable metabolite annotation constitutes the cornerstone of biochemical interpretation, and public databases including endogenous and exogenous metabolites were developed over the last years, e.g., the Human Metabolite Database (HMDB), LipidMaps, METLIN, the Golm Metabolome Database, and KEGG. A detailed review of the different spectral databases was recently proposed by Vinaixa et al. [18].

3.5.2 Annotation Using MS and MS/MS Information

The starting point of the identification process is the definition of an elemental formula, and the development of high-resolution MS devices strongly facilitated this task. Modern instruments today provide high mass accuracy and resolving power, leading to the evaluation of reliable isotopic patterns, thus reducing the number of possible molecular formulae. Additional information is then needed to provide an unambiguous molecular structure:

1. A widely used approach is based on the acquisition of a fragmentation spectrum using MS/MS or MS^n. The obtained data are then compared with entries of experimental MS/MS spectra databases using a similarity score based on matching peaks or differences between spectra. Relative intensities can also be

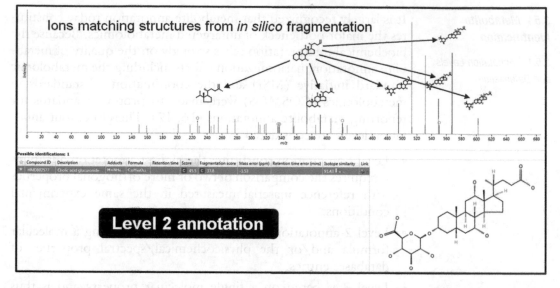

Unknown compound ?

Molecular formula : $C_{30}H_{48}O_{11}$
Neutral mass : 584.3188
Retention time : 13.41 min

Fig. 3 In silico fragmentation leading to level 2 annotation of cholic acid glucuronide from an unknown compound of molecular formula $C_{30}H_{48}O_{11}$, with a *m/z* of 584.3188 and a retention time of 13.31 min

taken into consideration. This approach is straightforward and reliable but remains limited by the number and quality of MS/MS entries in public databases (*see* **Note 8**).

2. By applying fragmentation rules inferred from real reference MS/MS data, a theoretical fragmentation spectrum can be computed in silico for any molecular structure using dedicated algorithms. Candidate compounds are first selected according to their *m/z* and fragmented in silico. A score is then evaluated to rank the candidates according to the similarity between the experimental and predicted MS/MS spectra (*see* **Notes 9** and **10**).

Because they benefit from both strategies, hybrid approaches combining systematic fragmentation and experimental MS/MS spectra comparison constitute also a promising way to help metabolite identification [19]. The in silico fragmentation leading to level 2 annotation of cholic acid glucuronide from an unknown biomarker candidate detected in Victor Yushchenko's urine samples is provided in Fig. 3.

3.5.3 Annotation Using Other Molecular Properties from the Separation

Metabolite annotation tools are based on exact mass, isotopic, and fragmentation patterns, but other physicochemical properties can be used to reduce the number of candidate compounds. When possible, the separation dimension provides retention/migration times that provides complementary orthogonal information to MS. This information can be measured or predicted (*see* **Note 11**):

1. Standard injection under the same analytical conditions allows compounds sharing the same molecular formula and potentially related fragmentation patterns to be distinguished. In vitro synthesis can be carried out from precursors when sources of putative metabolites are scarce or inexistent (e.g., phase II metabolites) [20].

2. Retention/migration time prediction based on quantitative structure-property relationship (QSPR) models offers an alternative to narrow down the number of candidates and improve the reliability of metabolite annotation (*see* **Note 12**).

3. Emerging complementary properties such as the cross-collisional section (CCS) obtained from ion mobility MS constitute also valuable complementary criteria for reliable metabolite annotation [21].

Because unambiguous metabolite identification (level 1) requires the injection of reference material (not always available in exploratory studies), efficient solutions should be found to limit the number of false-positive annotations. Combining experimental information stored in public databases, in silico physicochemical properties predictions, and combinatorial systematic searches provides relatively efficient metabolite annotation.

3.6 Retrospective Analysis

Metabolomic-based approaches focusing on a subset of the metabolome can be carried out by combining nontargeted data acquisition and automatic metabolite annotation. When particular ion features are extracted from full scan MS datasets, specific metabolic information can be obtained in a retrospective manner using post-targeted approaches (*see* **Note 13**) [22]. This reexamination may help to search for novel biomarkers within a chemical class or specific biological pathway of interest. The data can therefore be examined later "on demand" for other compound classes, given additional biological information about possible metabolite perturbations.

4 Notes

1. Peak detection is usually carried out by binning spectral data into intervals that are processed separately, but defining an optimal bin size often constitutes an issue. This parameter is of crucial importance, because small bins will tend to split peaks, while large bins will include unresolved signals.

2. This type of approach remains however somewhat limited, because it does not involve an explicit model for evaluating systematic drifts.

3. Missing values can be due to a signal near the detection limit, a distorted peak shape, or a true difference due to the factors of the experimental setup. In such a situation, it is desirable to use a gap-filling procedure to avoid adding too many zeros in the final data table.

4. Analyzing skewed data including a large number of null values may lead to irrelevant results and wrong conclusions. This is of particular importance when downstream statistical methods assume a normal distribution of the values.

5. Statistical workflows for analyzing metabolomic data should always start with exploratory approaches to ensure data consistency and check for potential outliers or unwanted sources of systematic variability (e.g., analytical drift, batch effect, etc.).

6. Because biological phenomena are not necessarily orthogonal, PCs are sometimes challenging to link with interpretable variations. In that case, other methods such as independent component analysis (ICA) constitute relevant solutions. Alternatively, PCA can serve as a pretreatment step before downstream analysis with clustering and classification methods.

7. The variable importance in the projection (VIP) provides an efficient criterion to assess the predictive merit of each variable in the whole PLS model.

8. The comparison of relative ratios between fragments is not always straightforward when different MS devices are used. In that context, multiple collision energies are usually included in databases to provide a more exhaustive picture of fragmentation patterns.

9. Arbitrary fragmentation rules are not universal, and alternative schemes can be applied for an extensive exploration of potential candidates.

10. Fragmentation cascades can also be considered as a graph that can be optimized for providing biochemically meaningful patterns. These fragmentation trees help to rationalize the way fragments are generated from precursor ions and can be compared to experimental MS/MS spectra of unknown compounds. Because similar fragmentation subtrees can be detected, it may help compound identification by matching a molecule sharing a close molecular structure, even if the true compound is not present in the database.

11. As biochemical reactions give rise to a very large and diverse range of compounds, it is not rare to observe structural, positional, and constitutional isomers. Therefore, retention/

migration times are often considered as a mandatory information in annotation workflows.

12. Predictions remain however limited to fixed analytical conditions, and a specific database is often developed in each laboratory. These differences between platforms and experimental conditions (stationary phase chemistry, mobile phase composition and temperature, running time, etc.) make standardization difficult. Interestingly, this major limitation can be tackled efficiently using knowledge about mechanisms of the separation process. For example, a prediction of retention times related to specific gradient conditions is possible. In the case of reversed-phase liquid chromatography (RPLC) analysis, the linear solvent strength theory model (LSS) can be used.

13. For the sake of clarity, the classification could be summarized according to the acquisition mode: (1) targeted metabolomics (or metabolic profiling), i.e., the detection of a series of specific metabolites, and (2) nontargeted metabolomics (metabolic fingerprinting, global metabolomics), i.e., an analysis based on the nontargeted monitoring of all signals in the samples.

References

1. Ellis DI, Dunn WB, Griffin JL, Allwood JW, Goodacre R (2007) Metabolic fingerprinting as a diagnostic tool. Pharmacogenomics 8 (9):1243–1266. https://doi.org/10.2217/14622416.8.9.1243

2. Clayton TA, Lindon JC, Cloarec O, Antti H, Charuel C, Hanton G, Provost JP, Le Net JL, Baker D, Walley RJ, Everett JR, Nicholson JK (2006) Pharmaco-metabonomic phenotyping and personalized drug treatment. Nature 440 (7087):1073–1077. https://doi.org/10.1038/nature04648

3. Boccard J, Veuthey JL, Rudaz S (2010) Knowledge discovery in metabolomics: an overview of MS data handling. J Sep Sci 33(3):290–304. https://doi.org/10.1002/jssc.200900609

4. Misra BB, van der Hooft JJJ (2016) Updates in metabolomics tools and resources: 2014–2015. Electrophoresis 37(1):86–110. https://doi.org/10.1002/elps.201500417

5. Jeanneret F, Boccard J, Badoud F, Sorg O, Tonoli D, Pelclova D, Vlckova S, Rutledge DN, Samer CF, Hochstrasser D, Saurat JH, Rudaz S (2014) Human urinary biomarkers of dioxin exposure: analysis by metabolomics and biologically driven data dimensionality reduction. Toxicol Lett 230(2):234–243. https://doi.org/10.1016/j.toxlet.2013.10.031

6. Krishnan S, Vogels JTWE, Coulier L, Bas RC, Hendriks MWB, Hankemeier T, Thissen U (2012) Instrument and process independent binning and baseline correction methods for liquid chromatography-high resolution-mass spectrometry deconvolution. Anal Chim Acta 740:12–19. https://doi.org/10.1016/j.aca.2012.06.014

7. Castillo S, Gopalacharyulu P, Yetukuri L, Oresic M (2011) Algorithms and tools for the preprocessing of LC-MS metabolomics data. Chemometr Intell Lab 108(1):23–32. https://doi.org/10.1016/j.chemolab.2011.03.010

8. Lange E, Tautenhahn R, Neumann S, Gropl C (2008) Critical assessment of alignment procedures for LC-MS proteomics and metabolomics measurements. BMC Bioinformatics 9. https://doi.org/10.1186/1471-2105-9-375

9. Tomasi G, Savorani F, Engelsen SB (2011) Icoshift: an effective tool for the alignment of chromatographic data. J Chromatogr A 1218 (43):7832–7840. https://doi.org/10.1016/j.chroma.2011.08.086

10. Dunn WB, Broadhurst D, Begley P, Zelena E, Francis-McIntyre S, Anderson N, Brown M, Knowles JD, Halsall A, Haselden JN, Nicholls AW, Wilson ID, Kell DB, Goodacre R, C HSMH (2011) Procedures for large-scale metabolic profiling of serum and plasma using gas chromatography and liquid chromatography coupled to mass spectrometry. Nat Protoc 6

(7):1060–1083. https://doi.org/10.1038/nprot.2011.335

11. van den Berg RA, Hoefsloot HCJ, Westerhuis JA, Smilde AK, van der Werf MJ (2006) Centering, scaling, and transformations: improving the biological information content of metabolomics data. BMC Genomics 7. https://doi.org/10.1186/1471-2164-7-142

12. Boccard J, Rudaz S (2014) Harnessing the complexity of metabolomic data with chemometrics. J Chemometr 28(1):1–9. https://doi.org/10.1002/Cem.2567

13. Hotelling H (1933) Analysis of a complex of statistical variables into principal components. J Educ Psychol 24:417–441. https://doi.org/10.1037/h0071325

14. Wold S, Sjostrom M, Eriksson L (2001) PLS-regression: a basic tool of chemometrics. Chemometr Intell Lab 58(2):109–130. https://doi.org/10.1016/S0169-7439(01)00155-1

15. Trygg J, Wold S (2002) Orthogonal projections to latent structures (O-PLS). J Chemometr 16(3):119–128. https://doi.org/10.1002/Cem.695

16. Sumner LW, Amberg A, Barrett D, Beale MH, Beger R, Daykin CA, TWM F, Fiehn O, Goodacre R, Griffin JL, Hankemeier T, Hardy N, Harnly J, Higashi R, Kopka J, Lane AN, Lindon JC, Marriott P, Nicholls AW, Reily MD, Thaden JJ, Viant MR (2007) Proposed minimum reporting standards for chemical analysis. Metabolomics 3(3):211–221. https://doi.org/10.1007/s11306-007-0082-2

17. Salek RM, Neumann S, Schober D, Hummel J, Billiau K, Kopka J, Correa E, Reijmers T, Rosato A, Tenori L, Turano P, Marin S, Deborde C, Jacob D, Rolin D, Dartigues B, Conesa P, Haug K, Rocca-Serra P, O'Hagan S, Hao J, van Vliet M, Sysi-Aho M, Ludwig C, Bouwman J, Cascante M, Ebbels T, Griffin JL, Moing A, Nikolski M, Oresic M, Sansone SA, Viant MR, Goodacre R, Gunther UL,

Hankemeier T, Luchinat C, Walther D, Steinbeck C (2015) COordination of Standards in MetabOlomicS (COSMOS): facilitating integrated metabolomics data access. Metabolomics 11(6):1587–1597. https://doi.org/10.1007/s11306-015-0810-y

18. Vinaixa M, Schymanski EL, Neumann S, Navarro M, Salek RM, Yanes O (2016) Mass spectral databases for LC/MS- and GC/MS-based metabolomics: state of the field and future prospects. Trends Anal Chem 78:23–35. https://doi.org/10.1016/j.trac.2015.09.005

19. Lynn KS, Cheng ML, Chen YR, Hsu C, Chen A, Lih TM, Chang HY, Huang CJ, Shiao MS, Pan WH, Sung TY, Hsu WL (2015) Metabolite identification for mass spectrometry-based metabolomics using multiple types of correlated ion information. Anal Chem 87(4):2143–2151. https://doi.org/10.1021/ac503325c

20. Jeanneret F, Tonoli D, Hochstrasser D, Saurat JH, Sorg O, Boccard J, Rudaz S (2016) Evaluation and identification of dioxin exposure biomarkers in human urine by high-resolution metabolomics, multivariate analysis and in vitro synthesis. Toxicol Lett 240(1):22–31. https://doi.org/10.1016/j.toxlet.2015.10.004

21. Paglia G, Williams JP, Menikarachchi L, Thompson JW, Tyldesley-Worster R, Halldorsson S, Rolfsson O, Moseley A, Grant D, Langridge J, Palsson BO, Astarita G (2014) Ion mobility derived collision cross sections to support metabolomics applications. Anal Chem 86(8):3985–3993. https://doi.org/10.1021/ac500405x

22. Jeanneret F, Tonoli D, Rossier MF, Saugy M, Boccard J, Rudaz S (2016) Evaluation of steroidomics by liquid chromatography hyphenated to mass spectrometry as a powerful analytical strategy for measuring human steroid perturbations. J Chromatogr A 1430:97–112. https://doi.org/10.1016/j.chroma.2015.07.008

Perspectives

The field of clinical metabolomics can be defined as the use of metabolomics-driven techniques in a clinical context, studying disease-related changes of the metabolome, which ultimately reflect possible alterations in genome, transcriptome, and proteome. With the metabolome seen as "the closest link" to the phenotype, it is not surprising that many researchers attribute a key role to metabolomics in the context of personalized medicine. Nevertheless, metabolomics-driven clinical research has long struggled fulfilling the high expectations and hopes which had been put into it. In the beginning, mainly biomarker discovery-driven case-control studies dominated the field of clinical metabolomics. While such research was mainly successful on the discovery of gene mutation-related metabolic markers, giving rise to strong metabolic correlations [1], it did frequently fail in the context of less pronounced metabolic phenotypes. However, a recent paradigm shift [2], an advanced analytical technology, and a much more quantitative understanding of human metabolism undoubtedly allow to paint a bright future for the field.

With respect to the analytical technologies and protocols applied in clinical metabolomics, mass spectrometry and nuclear magnetic resonance spectroscopy stand central. Furthermore, it has become recognized that a detailed molecular analysis is important for a sound understanding of (patho-)biochemical mechanisms. Such is reflected in advanced analytical methods, providing a more detailed mapping of chemical isomers, not solely limited to diastereomers but also including geometrical isomers as well as enantiomers and appreciating differential biological effects [3]. Additionally there has been a shift from exploratory metabolic profiling toward large panel (quantitative) targeted metabolomics approaches. The latter allowing for a more quantitative view of clinical data sets in correlation with possible disease mechanisms. Analytically speaking, advanced mass spectrometric techniques, increased quantitative capabilities and robustness, and the integration of different analytical techniques play an important role. In this respect, ion mobility mass spectrometry has recently had a large impact particularly on clinical lipidomics, allowing for lipid class and isomer separation [4, 5] and further more resulting in the first "plug-and-play" commercially available lipidomics platform, designed for nonexperts in the field, quantifying up to 1000 lipid species. It will be interesting to see if such approaches might enter the clinical routine in the future. However, applicability, robustness, and ease of data interpretation are key features for driving clinical metabolomics forward. In addition it will be of importance to facilitate data integration of several omics layers as well as pathway analysis [6] as only in combination we will gain a better understanding of disease mechanisms and in turn hopefully become capable of designing advanced treatments.

Leiden, The Netherlands   *Martin Giera*

Martin Giera (ed.), *Clinical Metabolomics: Methods and Protocols*, Methods in Molecular Biology, vol. 1730,
https://doi.org/10.1007/978-1-4939-7592-1, © Springer Science+Business Media, LLC 2018

References

1. van Karnebeek CDM, Bonafe L, Wen X-Y et al (2016) NANS-mediated synthesis of sialic acid is required for brain and skeletal development. Nat Genet 48(7):777–784. https://doi.org/10.1038/ng.3578. http://www.nature.com/ng/journal/v48/n7/abs/ng.3578.html#supplementary-information
2. Johnson CH, Ivanisevic J, Siuzdak G (2016) Metabolomics: beyond biomarkers and towards mechanisms. Nat Rev Mol Cell Biol 17(7):451–459. https://doi.org/10.1038/nrm.2016.25. http://www.nature.com/nrm/journal/v17/n7/abs/nrm.2016.25.html#supplementary-information
3. Serhan CN, Petasis NA (2011) Resolvins and protectins in inflammation resolution. Chem Rev 111 (10):5922–5943. https://doi.org/10.1021/cr100396c
4. Jónasdóttir HS, Papan C, Fabritz S et al (2015) Differential mobility separation of leukotrienes and protectins. Anal Chem 87(10):5036–5040. https://doi.org/10.1021/acs.analchem.5b00786
5. Lintonen TPI, Baker PRS, Suoniemi M et al (2014) Differential mobility spectrometry-driven shotgun lipidomics. Anal Chem 86(19):9662–9669. https://doi.org/10.1021/ac5021744
6. Ritchie MD, Holzinger ER, Li R a (2015) Methods of integrating data to uncover genotype-phenotype interactions. Nat Rev Genet 16(2):85–97. https://doi.org/10.1038/nrg3868

INDEX

Martin Giera (ed.), *Clinical Metabolomics: Methods and Protocols*, Methods in Molecular Biology, vol. 1730,
https://doi.org/10.1007/978-1-4939-7592-1, © Springer Science+Business Media, LLC 2018